Secrets of Women

Secrets of Women

Gender, Generation, and the Origins of Human Dissection

Katharine Park

ZONE BOOKS · NEW YORK

2006

Zone Books
1226 Prospect Avenue
Brooklyn, NY 11218

Printed in the United States of America.

Distributed by The MIT Press,
Cambridge, Massachusetts, and London, England

Library of Congress Cataloging-in-Publication Data

Park, Katharine, 1950–
    The secrets of women : gender, generation, and the ori-
gins of human dissection / Katharine Park.
        p. cm.
    Includes bibliographical references and index.
    ISBN 1-890951-67-6 – ISBN 1-890951-68-4 (pbk.)
    1. Human dissection – Italy – History. 2. Science,
Renaissance – Italy. 3. Women – Miscellanea. 4. Human
anatomy – Italy – History. 5. Women – Health and hygiene
– Italy – History. I. Title.

QMM33.4.P37 2006
611–dc22
                                                2006044710

Contents

To Caroline Walker Bynum, Joan Cadden, and Nancy G. Siraisi

Acknowledgments

I began the research that would eventually result in this book at the Mary Ingraham Bunting Institute at Radcliffe College in 1990, when I was supposed to be working on quite another project; I finished writing the book at the Radcliffe Institute for Advanced Study in 2004. In the interim the topic changed as much as Radcliffe did. I had originally planned to study the roots of human dissection in medieval Christian devotional and funerary practice, a project that resulted in two articles on the subject, "The Criminal and the Saintly Body" and "The Life of the Corpse," neither of which foregrounded the question of gender.[1] It took me the four years these articles were in press to understand that the ubiquity of women and women's bodies in my sources was more than an accident of survival. With the benefit of hindsight, I now find this delay incredible; I can explain it only as a result of my own inability to step outside the current historiography of early anatomy, which focused single-mindedly on dissection as an institutional practice performed by male masters employed by universities and medical colleges, for the benefit of male medical students and practitioners, on the mostly male bodies of condemned criminals.[2] So powerful was this definition of the topic that it took me years to see what was literally before my eyes: that the opening

9

and inspection of cadavers was a much more diverse cultural phenomenon in which women played a much larger part than I ever could have imagined.

In the process, I had to redefine my topic and ultimately the broader subject to which this book aimed to contribute. Dissection, as I explain in the Introduction, is an anachronistic category, although I have kept the term in my title as the clearest way to communicate my project to modern readers. My view of what the history of medicine is the history of also changed. While I began by embracing the description of women (wives, mothers, sisters, daughters, domestics, and female empirics) as "marginal" to the work of healing in late medieval and early modern Europe, like other informally trained or unlicensed practitioners, I can now only see women as standing at its heart, as the first line of defense against illness and as those primarily entrusted, in the normative context of the family, with the ongoing management of health, healing, sickness, infertility, complications of childbirth (many and frequent), and death.

I could never have rethought my topic in this way without the work of the three historians to whom I have dedicated this book. In the course of thirteen years of higher education, I had only two female teachers, Caroline Walker Bynum and Joan Cadden, and their influence on my topic and my methods will be obvious to anyone familiar with their writings. Although I never studied formally with Nancy G. Siraisi, her work on Italian medicine in the late Middle Ages and the Renaissance has served as the foundation of my research. Like Bynum and Cadden, she, too, has been a personal and intellectual model and a treasured friend.

My debts extend far beyond these three historians, however. I am enormously grateful to Radcliffe College and the Radcliffe Institute for Advanced Study for providing financial support and an intellectual community in which this book could take shape. I also owe a great deal to the community of Villa I Tatti, the Harvard University Center for Italian Renaissance Studies, where I spent a number of very happy months as a visiting professor in spring 2001 and January 2003, and to Susana, Lady Walton, for her gracious hospitality at the Giardini La Mortella, where this book took final form in summer 2005. I am grateful to the staff at the various Florentine libraries where I did most of my Italian research — the Archivio di Stato di Firenze, the Biblioteca Medicea Laurenziana, the Biblioteca Riccardiana, and the Biblioteca Nazionale Centrale di Firenze — as well as to the staffs of the Houghton, Widener, and Countway libraries at Harvard University, the Biblioteca Comunale dell'Archiginnasio di Bologna, and the National Library of Medicine in Bethesda, Maryland.

I have also been fortunate to find helpful listeners and readers. I am very grateful to the many departments, seminars, and colloquiums that gave me opportunities to float ideas and prose and helped me to refine my work. In addition, this book has benefited enormously from the input of scholars who generously read and commented on part or all of its many drafts: Monica Azzolini, Montserrat Cabré, Karen Encarnción, Mary Fissell, Monica H. Green, Jeffrey Hamburger, Cynthia Klestinec, Michael McVaugh, Vivian Nutton, Lesley Pattinson, Gianna Pomata, Lisa Pon, Charles Rosenberg, Jessica Rosenberg (my sterling Radcliffe Research Partner), Nancy G. Siraisi, and Joseph Ziegler. I owe them more than I can say. The editors and staff of Zone Books, Ramona Naddaff, Meighan Gale, Gus Kiley, Amy Griffin, and Sierra Van Borst have been ideal partners, and I have greatly appreciated their competence and enthusiasm for the project and the freedom they gave me to let it develop as it would.

Finally, I thank my husband, Martin Brody, who took time out from his own work as a composer and scholar to help me think through the various stages of this project and to travel with me to visit many shrines and churches and to contemplate many mummified corpses. Without his patience, support, and critical intelligence, this would have been a very different book.

Introduction

A nun with visions of Christ's Passion. A blind, crippled, homeless holy woman. Four patrician wives and mothers. Two prophetesses, one of them a married, lactating virgin. An executed criminal. These very different women had one thing in common: their bodies were opened and their viscera examined after their deaths. This book uses their stories to write a history of human dissection in late medieval and Renaissance Italy. It begins in the late thirteenth century, when human dissection emerged for the first time in Western Europe as an established, though relatively infrequent, practice with roots in a variety of secular and ecclesiastical institutions. It ends in the mid-sixteenth century, when anatomical knowledge based on human dissection was generally accepted as one of the foundations of learned medicine and natural philosophy and would soon be adopted by lay writers as an important way to understand the body and the self. Surprisingly, given the relative absence of documentation concerning women in other social and medical contexts, I could not have written this history in this way using male subjects, since the materials for detailed case studies are lacking for men. This is not because fewer men's bodies were opened than women's during the period in question, although that may in fact be the case. Rather, it reflects the special emphasis given to the opening of female bodies in late medieval Italian culture and to dissection as one of the best techniques for knowing what was most important about those bodies, as well as the way this became a model for knowing human bodies in general, regardless of their sex.

The story of how and when western Europeans first began to open and inspect the insides of human bodies has been written many times.[1] Although my interpretation builds on the work of earlier historians, I have tried to redefine the topic by including a wider range of practices, contexts, and people. To date, whether they have used visual or textual sources, historians have tended to focus on one type of procedure: the opening and inspection of human bodies in university medical faculties and other corporate institutions (notably colleges of physicians or surgeons) in order to teach anatomy to medical practitioners or to further anatomical research. They have concentrated on the motivations and actions of one type of knower, the learned physician or surgeon, and on one type of human cadaver, the executed criminal, to whom this sort of dissection was limited by statute and custom. These cadavers were for the most part male, not only because so few women were executed for capital crimes, but also because anatomy was about knowing the generic human body, which was understood as male. But it was not just female bodies that were in short supply; until the years around 1500, when anatomists began to turn increasingly to local hospitals for anatomical material, dissection in the service of medical teaching and study was rare, whatever university and college regulations might have said. Relatively few criminals were executed in this period, and fewer still were eligible for dissection, which in most cities was confined to the bodies of foreigners of low standing.[2] Until at least the early sixteenth century, broadening the pool of cadavers was not a priority because anatomy was not deemed an important component of medical training. The requirement of annual dissections that appears on the books of many late fourteenth- and fifteenth-century Italian medical faculties was more often ignored than observed.

Outside of colleges and universities, however, human dissection proceeded apace. Beginning around 1300, it developed quickly and spontaneously out of a set of *ad hoc* cultural practices that had nothing to do with medical instruction: funerary ritual (most notably embalming by evisceration), the cult of relics of the Christian saints, autopsies in the service of criminal justice and public health, and a birth practice that eventually became known as Caesarean section. (This last, which involved extracting a living fetus from the body of a woman who had just died in childbirth in order to allow it to be baptized, is better called by its contemporary name, *sectio in mortua* — cutting open a dead woman.)[3] While all these practices required opening human bodies and were often (though not always) performed by surgeons and physicians, they had little else in common with academic dissection. Except for the occasional public display and dismemberment of a saint's body in order to multiply its relics, these other dissectionlike practices — embalming, autopsy, fetal excision — generally took place in secluded or domestic settings. None of them involved the profound dishonor associated with the public academic dissection, in which an unnamed and naked corpse was not only exposed in front of a group of unrelated viewers but also largely dismantled; this violated both its personhood and its social identity by rendering it unrecognizable and unsuiting it for a conventional funeral, in which the clothed body was displayed on its bier.[4] The other procedures, which involved opening only the abdomen, left the corpse largely intact.[5] Because they did not assault the honor of the person in question or of his or her family — and indeed were often performed at the initiative of family members or other close associates — they inspired little or no resistance. On the contrary, in fact: embalming, which seems to have been both the earliest practice and the precursor of other forms of evisceration, was reserved for revered, even sacred, dead.

By paying at least as much attention to these more private, less invasive procedures as to the formal dissection of criminals by university lecturers, I aim to restore the latter to the social and religious context of which it was a part. Even the words contemporaries used to describe the opening of human bodies reveal the strong continuities between the world of teachers and students and the worlds of childbirth, murder trials, chronic illness, state funerals, and Christian cult. Medical writers used the Latin noun *anatomia*, together with its variants — *nothomia, anathomia* — and its vernacular cognates, to refer indifferently to the practices known now as dissection (the opening of a corpse to learn about human bodies in general) and autopsy (the opening of a corpse to make a determination about the state of an individual body, usually the cause of death). Both practices fell under the often-cited definition of anatomy given by the seventh-century Byzantine medical writer John of Alexandria in his influential commentary on Galen's *On the Sects* (*De sectis*): "Anatomy is the artifical cutting and elucidation of things that are concealed in the hidden body."[6] But the term also occasionally appears in texts referring to embalming. The relevant verbs allowed even fewer distinctions among these various practices. Latin writers tended to use nontechnical words — *incidere* ("cut"), *aperire* ("open"), and even *exenterare* or *eviscerare* ("eviscerate") — to describe not only dissections and autopsies but also embalming and the opening of women who had died in childbirth. These four practices were even more closely associated in Italian texts written by people who were not medical professionals, who almost always used the verb *sparare* (there is no corresponding noun), which referred more commonly to preparing animals for cooking, as in gutting fish or pigs.[7]

The various procedures involved in opening the human body were also closely linked in practice. Consider, for example, the *Commentaries on the Anatomy of Mondino*, a massive anatomy textbook published by Jacopo Berengario of Carpi in 1521. Although Berengario referred intermittently to the formal public dissections he conducted as professor of anatomy and surgery at the University of Bologna, much of his information came from casual observations made in the course of his thriving surgical practice, which included both formal autopsies and incidental ones made when operations went wrong. (He also dissected miscarried or stillborn fetuses obtained from midwives.) One of his most detailed accounts of such an autopsy involved the sudden death of a pregnant woman. Berengario was called to open her corpse in hopes of finding "two fetuses [*faetus*] if not completely alive, then at least alive enough to be baptized." Instead, he discovered one, outside the woman's uterus, lodged in her intestines. "This greatly astonished me," he wrote, "and the fetus was baptized by the women of the house."[8] Proceeding then to open the uterus, he found a large swelling (*apostema*), which had ruptured it and ejected the fetus. In this case, a planned *sectio in mortua* turned into an impromptu autopsy as well.

Embalming and autopsy were particularly linked. The two early fourteenth-century holy women whose stories form the basis of Chapter One were initially eviscerated so that their corpses could be preserved. Only several days later were their viscera inspected and holy objects found inside. Beginning in the late fifteenth century, the easy association between embalming and autopsy — having extracted the abdominal organs in order to preserve the corpse, why not examine them to determine cause of death? — is also demonstrated in the cases of secular notables such

17

as Lorenzo di Piero de' Medici, "the Magnificent," (d. 1492) and Isabella of Aragon (d. 1533).[9] Surgeons called to embalm the body of an illustrious person might take advantage of the opportunity to make casual observations for their own use; for instance, Berengario noted in his *Commentaries* that fat people tended to have quantities of fat adhering to their hearts, citing the example of Giovanni Francesco della Rovere, archbishop of Turin, whose corpse he had eviscerated and prepared.[10]

By treating all these practices together rather than looking at academic dissection in isolation, I aim to restore their cultural coherence. This is a fundamental point: assuming anachronistically that opening the human body is in the first instance a medical procedure, historians have ignored the broader phenomenon of which it was a part — or reduced these other, related procedures to the status of "background" or "cultural context." In contrast, I consider the opening of the human body as a whole. Its variants (dissection proper in the modern sense, embalming, autopsy, fetal excision, the "recognition" or inspection of the corpses of holy women and men) are like a set of angled mirrors: each illuminates and reflects the others. No one is primary, least of all dissection, which was by any measure the most arcane. In order to emphasize their commonalities and the degree to which they were associated in the minds of contemporaries, I have for the most part used the words "dissection" and (preferably) "anatomy" to refer to all of them, except when clarity has demanded a more precise term.

At the same time, however, I underscore the specificity of particular practices within this bundle of related activities: the different concerns that motivated them, the different spaces in which they took place, and the different moral economies of personhood that they implied. For example, as I have already mentioned, dissections for the exclusive purposes of medical research and teaching were not only relatively rare but also uniquely dishonoring, which explains their restriction to the corpses of executed foreign criminals, hospital patients, and animals. Most of the other practices, in contrast, were associated with social and cultural elites; embalming was generally reserved for prospective saints, princes, popes, and other ecclesiastical and civic leaders, while those who were autopsied at their own or their families' request had, by definition, access to high-level professional medical care. The actual extent of fetal excision is unknown, although there is no reason to assume it was limited to the well-to-do; when the Dominican preacher Giordano of Pisa described calling (and paying for) doctors and midwives to perform the operation on a woman who had just died in childbirth in a house attached to the Dominican convent in Pisa, he was almost certainly referring to an act of charity.[11]

Even the notorious theatricality of formal, public anatomies shows the importance of placing practices involving the opening of bodies in specific contexts and attending to their specific meanings. Displaying an unclothed corpse to a large group of unrelated people had various effects. On the one hand, he embalmed bodies of holy men and women, not to mention popes, were often stripped by their admirers while they lay in state; this trope of hagiography served to emphasize the intensity of popular devotion inspired by a potential saint's body and the magical power of objects that had been in contact with it.[12] The public exposure of the corpse of an executed and dissected criminal, on the other hand, was an occasion of dishonor and shame for the individual and his or her family.[13] Yet even this ritual had strongly positive associations; as an opportunity for university towns such as Bologna and Padua to flaunt their intellectual resources, it had by the late sixteenth century become a focus of civic pride.[14]

The specificity of cultural contexts and meanings have led me to restrict this study to northern Italy, whereas most histories of anatomy have tended to adopt a universalizing, chronological approach.[15] Ignoring linguistic and cultural boundaries obscures not only the important part played by specific lines of influence and local traditions in anatomical practices, texts, and illustrations. It also overlooks the fact that during the first two hundred years of its continuous history, from the late thirteenth to the late fifteenth century, human dissection was confined to Italy and (to a lesser extent) southern France.[16] The early interest in opening human bodies sprang partially from conditions specific to Italian medicine: the long tradition of animal dissection associated with the southern Italian city of Salerno (an important center of medical teaching as early as the eleventh century), the sophistication of thirteenth-century surgical practice in the Po valley, and the intensified attention devoted to the anatomically informed works of Galen on the part of medical masters at the University of Bologna in the years around 1300.[17] But it also reflected specifically Italian funerary practices and attitudes toward human corpses. Italians began to eviscerate their revered dead for embalming relatively early, in the second half of the thirteenth century, in connection with papal funerals and the cults of "new saints." (The latter were contemporary holy men and women, as opposed to long-dead martyrs, whose bodies were valuable sources of civic prestige and prosperity, as well as sites of healing power.)[18] At the same time, however, Italians were much less inclined than inhabitants of areas with a Germanic cultural heritage to ask for their corpses to be dismembered for repatriation if they had died abroad, or to be divided for burial in multiple locations.[19] All these circumstances contribute to explaining why most of the practices that involved opening the human body and examining its contents (rather than merely dividing it) began in Italy and flourished there considerably earlier than in most other parts of Europe.

As my emphasis on saints' cults and funerary practices suggests, I argue that social and, especially, religious practices were far more central to the early history of dissection than many other histories of anatomy would lead one to believe. A long historiographic tradition, dating back to at least the middle of the nineteenth century, presents religion and science as diametrically opposed cultural enterprises and the Church as deeply hostile to dissection.[20] This misconception is still widespread. Generations of Italian tour guides, not to mention playwrights, journalists, and historical novelists, have waxed eloquent over the supposed moral and intellectual courage of such late fifteenth- and sixteenth-century heroes as Leonardo da Vinci, Michelangelo, and Andreas Vesalius, author of *On the Fabric of the Human Body*, published in 1543, whose famous title page celebrated the study of anatomy based on dissection rather than on ancient texts (figure I.1). These men, the story goes, defied religious superstition and braved persecution and censure in the service of art or science, pursuing their intellectual passion in dark cellars and back rooms with trapdoors in the floor for the quick disposal of corpses when the police (or the Inquisition, or whoever) arrived.

Like the familiar story associated with Christopher Columbus, whose courageous voyage of 1492 purportedly proved to a doubting public that the earth was round, this story has been debunked repeatedly by medievalists to no avail.[21] The power of such fictions to weather frequent and detailed disproof testifies to the important cultural work they perform by supplying foundation stories that confirm deep-seated Western intuitions about the scientific origins of modernity — intuitions that continue to inform the writing of even specialists in the field.[22] Equally deep-seated is the unwarranted assumption that, just because twentieth- and twenty-first-century Western understandings of the body are dominated by medical models and medical discourses, this was

21

Figure I.1. Vesalius dissecting the body of a female criminal. Andreas Vesalius, *De humani corporis fabrica* (Basel: Joannes Oporinus, 1543), title page.

also true in earlier periods; in this view, the history of the body has at its core a history of anatomy and physiology, to which a variety of "cultural meanings" (regarding, for example, gender, shame, and sexuality) are appended.[23] My research suggests, however, that the men and women whose lives and work I describe in this book, the inhabitants of northern Italian cities from the mid-thirteenth to the mid-sixteenth century, understood their bodies primarily in terms of family and kinship, on the one hand, and religion, on the other. Medical models — even in this world of highly developed medical institutions and practices — came in a distant third. Family and religious concerns underpinned procedures such as embalming, autopsy, and "Caesarean section," which were generally performed at the initiative of laypeople (in the sense of nonprofessionals). Because these procedures were so closely associated with the practice of dissection in the service of medical research and teaching, and because they played such an important part in its history, the concerns that informed them shaped dissecting practices as well.

In terms of religious culture, several important and distinctive elements are at play, among which the most important is probably the absence in medieval Christian culture of anything resembling a belief in corpse pollution.[24] Fears concerning the impurity of the corpse worked powerfully to shape the cultures of ancient Greece and Rome and to limit human dissection in those cultures, with the exception of a single generation of Greek medical writers in early third-century BCE Alexandria. (Alexandria lay in Egypt, where longstanding traditions of embalming reflected a very different set of attitudes toward dead bodies.)[25] From very early on, however, Christian culture defined itself in opposition to Mediterranean paganism in this regard. This change was due in large part to the consolidation of the cult of the saints. Saints were understood as present in their mortal remains, even after death,

23

and their corpses, far from being sources of pollution, were reservoirs of protection and magical power. As a result, their tombs were privileged places; anchored by these precincts, as Peter Brown has put it, the "familiar map of the relations between the human and the divine, the dead and the living, had been subtly redrawn."[26] Nascent ideas of bodily resurrection further strengthened the ties between the beloved dead and their mortal remains.[27]

This new relationship to dead bodies challenged many third- and fourth-century Christians raised in societies that saw human corpses as horrifying and impure. Over time, however, the new attitudes became second nature, and by the thirteenth and fourteenth centuries, when dissection was first emerging as a western European practice, few traces of the older fears remained.[28] Like Christians all over Europe, Italians enthusiastically embraced the cult of relics — in the form of fragments or even entire corpses — as well as funerary practices that involved disembowelling, dismembering, and mutilating dead bodies, although those practices differed significantly according to region and class. Attitudes toward human corpses continued to be complicated; handling recently dead bodies was not pleasant, to be sure, and manual labor of any sort was considered shaming to those of elevated birth. It took several centuries before university-trained physicians regularly opened human bodies themselves, rather than leaving the job to lower-status barbers and surgeons, but this had nothing to do with defilement by dead bodies or attempts to legitimate a "polluting" activity.

Rather, late medieval Christianity saw the human body as one of the principal elements connecting the natural and the supernatural worlds. Like the body of Christ, who died like a criminal, mutilated on the cross, or like the scattered bones of long-dead martyrs, the body was a conduit for divine grace. This was also true of living bodies, such as those of the murderers whose public humiliation and physical suffering at the hands of the executioner might exempt them entirely from the pains of Purgatory, or the visionaries and ecstatics — mostly but not entirely women — who bore on their bodies the marks of Christ's wounds.[29] For thirteenth- and fourteenth-century Christians, sanctified bodies were often mutilated in life and available for evisceration and dismember- ment after death — processes that served to multiply and diffuse their power. (God in his omnipotence would reunite the scattered fragments of their bodies at the Last Judgment.)[30] At the same time, the belief that possession by the holy spirit, like possession by the devil, could leave its marks on and even within the body provided a motivation to examine and explore the corpses of extra- ordinary people, inside as well as out.[31]

The logic that connected dissection to ideas concerning fam- ily and kinship is even more direct than to religion, at least to modern eyes. Late medieval Italian society, especially urban elite society, was profoundly patriarchal; even more than northern Europeans, Italians — at least Italian nobles and patricians — under- stood family membership primarily in terms of blood relation- ships defined by biological descent through the male line.[32] This emphasis on paternity collided with the realities of conception, gestation, and childbirth, all of which foregrounded the mother's contribution to generation and the physical tie between mother and child. Equally unnerving, men could never know for certain if their children were in fact their own; paternity, constructed this way, was fragile, dependent on the sexual fidelity of women, whose untrustworthiness was the stuff of a thousand fables, jokes, and songs. The precarious nature of fatherhood, and thus of the family itself, centered on the uterus, the dark, inaccessible place where the child's tie with its father was created, its sex determined, and its body shaped. Generations of Italian medical writers, not to mention legal theorists, theologians, and natural philosophers, struggled to describe this mysterious process in terms that could ground kinship securely in patrilineal descent.[33] Ordinary men and women, in the meantime, wrestled with equally pressing but more practical questions: Is she a virgin? Why can't we (she) get pregnant? Is it a girl or a boy? How do I know the child is mine? The answers to all these questions lay inside the female body and, more specifically, inside the womb.

Male writers often referred to matters of this sort as "the secrets of women," a phrase that had multiple and sometimes conflicting connotations.[34] On the one hand, it implied that women had access to knowledge concerning sexuality and generation that men did not, and that they hoarded this knowledge for their own, often unsavory purposes. On the other hand, it simply described a topological situation: precise information on matters equally important for men and women was inaccessible to both. Indeed, in the minds of physicians — as well as many of their female patients — women stood to benefit as much as or more than men from the advances of medical knowledge in this area, since it was vital to their health and to their very survival in an age when astounding numbers of adult women died from complications of childbirth.[35] Understanding the secrets of women became one of the principal goals of fifteenth- and sixteenth-century medical writers, both because the topic was important in its own right and because it was thought that anyone who could probe the complicated and mysterious workings of the uterus would have little trouble understanding the rest of the comparatively simple human frame. This is why the womb appears as a — arguably *the* — privileged object of dissection in medical images and texts.[36]

By this, I do not mean that the number of folios or illustrations devoted to the female anatomy outnumbered those devoted to the male. This was patently not the case; the male body was the generic body, as I have already noted, and women's bodies served

26

to demonstrate only the female reproductive system. Rather, the uterus acquired a special, symbolic weight as the organ that only dissection could truly reveal, and as a result, it came to stand for the body's hidden interior. This is vividly illustrated in an influential group of anonymous texts and images of German origin that circulated widely in manuscript during the fifteenth century, providing information useful for medical practitioners in graphic form. These were published in Venice in 1491 as an attractive, large-format Latin book called the *Medical Compilation* (*Fasciculus medicine*), attributed to one Johannes de Ketham.[37] Like its manuscript models, this book used male figures to illustrate a variety of general topics, including points for bloodletting (figure I.2), the association of the signs of the zodiac with different members, and ways to treat various types of wounds (figure I.3). Although the last such figure is even titled "On Anatomy," it shows only the outside of the body, together with a list of the diseases to which it is heir (figure I.4). The female figure ("On Woman"), in contrast, has a decidedly inward cast. While it, too, shows the locations of a variety of diseases, carefully labeled on the affected body parts, it also demonstrates the pregnant womb, located among a variety of ill-defined structures meant to represent the other visceral organs (figure I.5). Thus the female figure has come to illustrate internal anatomy in general, apparently by association with the uterus: where the male bodies are mostly surfaces, the woman is identified with a visualizable inside.

Emboldened by the evident success of the 1491 *Fasciculus*, its publishers issued an Italian translation in 1494, the *Fasiculo de medicina*. This included several additional texts and images, including a completely reworked version of the female figure, now titled "Figure of the uterus from nature" (figure I.6). This appears to be the first image in a printed book ever to show an internal organ on the basis of direct inspection of a dissected body, testifying to the

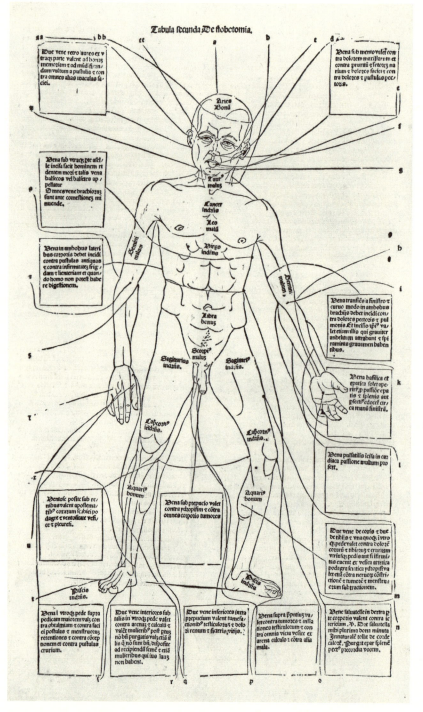

Figure I.2. "On Phlebotomy." *Fasciculus medicine*, attributed to Johannes de Ketham, ed. Giorgio Ferrari da Monferrato (Venice: Giovanni and Gregorio de' Gregori, 1491), sig. aii v.

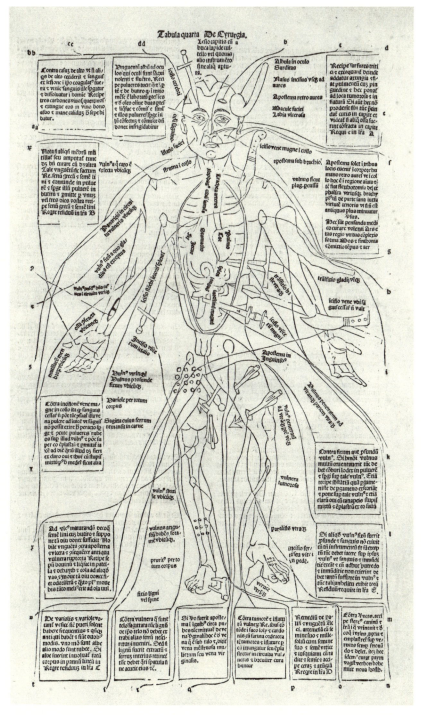

Figure I.3. "On Surgery." *Fasciculus medicine*, attributed to Johannes de Ketham, sig. bi r.

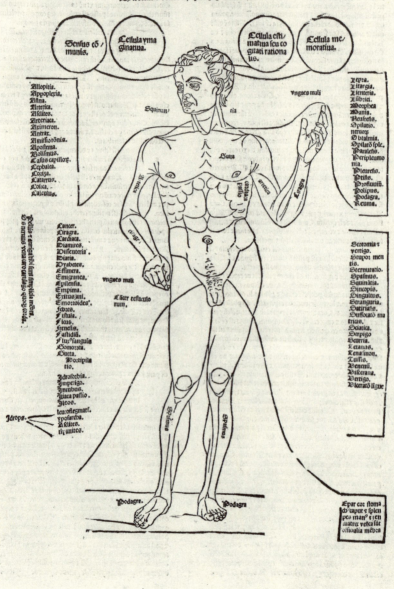

Figure I.4. "On Anatomy." *Fasciculus medicine*, attributed to Johannes de Ketham, sig. bii r.

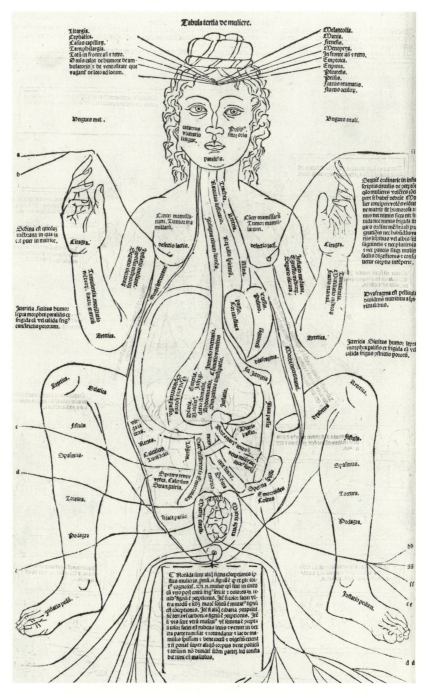

Figure I.5. "On woman." *Fasciculus medicine*, attributed to Johannes de Ketham, sig. av v.

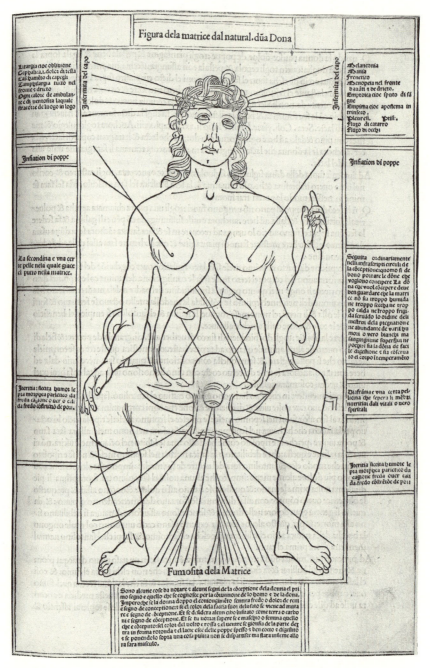

Figure I.6. "Figure of the uterus from nature." *Fasiculo de medicina*, attributed to Johannes de Ketham, ed. and trans. Sebastiano Manilio (Venice: Giovanni and Gregorio de' Gregori, 1494), sig. d1r.

implicit link between the female body (the body defined by its interior) and dissection (the technique by which that interior might be revealed).[38] I don't mean to suggest that men were not thought to have internal organs; although this was not typical of earlier manuscript versions of this image, the "wound man" in the 1491 and 1494 editions has clearly defined viscera, themselves vulnerable to sword, dagger, and lance. But the male body was not reduced to and identified with its interior, like the female body.

The female figure in the *Fasiculo* served as both model and foil for Leonardo da Vinci. Although it is doubtful that he ever personally dissected a woman, his single most elaborate and finished image shows a female torso. This was evidently intended to correct and supplement the *Fasiculo*'s image and to show what the inside of the human body was really like (figure I.7).[39] Some of Leonardo's notes concerning his projected book on anatomy also foreground the female body by placing it at the very beginning. "This work must begin with the conception of man, and describe the nature of the womb," he wrote around 1489, "and how the child [*pucto*] lives in it, up to what stage it resides in it, and in what way it quickens into life and feeds. Also its growth and what interval there is between one stage of growth and another. What it is that forces it out from the body of the mother, and for what reasons it sometimes comes out of the mother's belly before the due time."[40] Vesalius adopted a similar strategy in his famous *Fabrica*, which also begins with the female womb; although the sections on internal anatomy include only a few images of the female body, the title page showcases the open cadaver of a woman, with her uterus exposed (figure I.1).[41]

Therefore, although women's anatomy was reduced functionally to their organs of generation, these had an emblematic status as the exemplary object of dissection: representations of the female body came to stand both for the interior of the human body

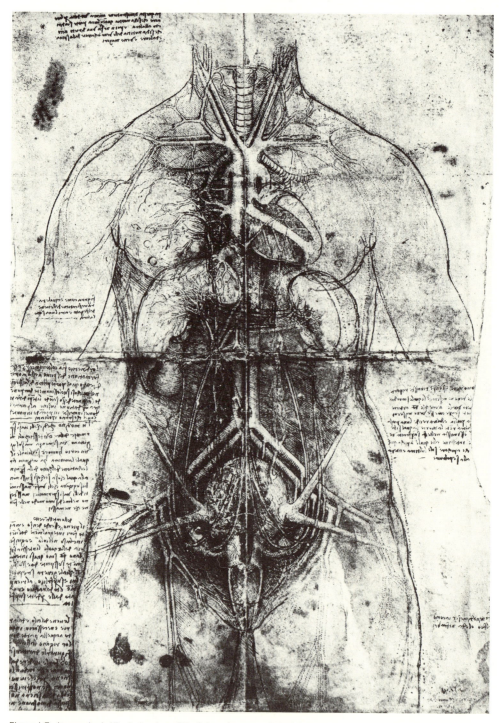

Figure I.7. Leonardo da Vinci, drawing of the internal organs of the female body, Windsor Castle, Royal Library, ms. W.12281r, ca.1508.

and for the powers of dissection-based anatomy to reveal its hidden truths. At the same time, however, the uterus retained its specific identity as the enigmatic space where both life and knowledge began and within which the male seed was mysteriously transmuted into a human child.

Although these matters seem to lie far from the religious issues concerning saints and relics I discussed above, the two topics are more closely related than they might appear, for the bodies of mothers and those of holy women were understood in analogous terms. Male writers described the latter as holy vessels, uniquely suited — in ways that men were not — to receive the gift of divine grace, which often took the form of intimate conversations and visions.[42] There was a strong corporeal dimension to these peculiar abilities. Not only did many women manifest their experiences of the divine in strikingly corporeal ways (including extreme fasting, stigmata, levitation, and long periods of rigidity and insensibility), but both they and their male supporters also imagined their divine possession in physiological terms, as the reception of the holy spirit into their hearts.[43] Like the father's seed in the mother's uterus, Christ's presence in the heart created new life; this might manifest itself materially in the form of objects impressed with his likeness, which only dissection could reveal. Given this emphasis on both the corporeality and the inwardness of women's religious inspiration, it is no coincidence that the bodies of holy women were opened and inspected beginning in the early fourteenth century, while the first known autopsy of a holy man (Ignatius Loyola) took place two hundred fifty years later, in 1556.

In order to explore these linked themes of generation, holiness, and female corporeality in connection with the early history of human dissection, I have arranged this book as a series of case studies. Each begins with the opening of a woman's body and moves outward from that event to consider its surrounding circumstances — where she was opened, by whom, and for what reasons — and its relationship not only to the history of anatomy but also to the social and religious practices that framed and informed the act of dissection. In order to reconstruct these circumstances, I have drawn on many different kinds of sources: devotional texts and images, chronicles, records of canonization procedures, works of fiction and mythology, diaries and letters, and treatises by medical writers. My first case study (Chapter One) focuses on the story of an Umbrian abbess, Chiara of Montefalco, whose corpse was opened in 1308 by her fellow nuns. After an interlude (Chapter Two) in which I analyze changes in the idea of "women's secrets" over the course of the fourteenth and fifteenth centuries, I begin Chapter Three with the dissection of Fiametta Adimari, the wife of one of the richest and most powerful men in Florence, who died in 1477. Chapter Four centers on the 1520 anatomy of Elena Duglioli, a Bolognese visionary whose corpse was repeatedly opened and inspected by, among other people, Jacopo Berengario of Carpi, the professor of surgery and anatomy whose experience with extrauterine pregnancy I described above. Finally, in Chapter Five, I use the body of the unnamed female criminal on the title page of Vesalius's *Fabrica*, who was probably executed in 1542, to show how foregrounding the gender of the cadaver forces us to broaden and rethink the cultural meanings of dissection in what is often presented as an icon of modern science. The modernity of this image is debatable, I argue, but there is no question that it encoded many contemporary commonplaces regarding gender, generation, knowledge, and holiness with dramatic force.

By ending rather than beginning my story with Vesalius, I aim to bridge the artificial divide between the late Middle Ages and the early modern period that shapes so much of the historiography of early science and medicine. This divide is largely the product of the kind of selective reading that created, among other things, the myth of medieval resistance to human dissection. By emphasizing the opening of human bodies outside the context of university medical study, I aim to demonstrate the continuity between the work of academically trained physicians and surgeons and the actions and decisions of the many men and women who employed them to eviscerate and inspect the corpses of their masters, mistresses, siblings, parents, spouses, and children. Finally, in focusing on women's bodies, I aim to show the very specific ways gender shaped one important area of natural inquiry. It is a commonplace of feminist science studies that those who study the natural world understand both it and their own enterprises in gendered terms; it is often claimed that the sixteenth and seventeenth centuries marked a kind of watershed in this respect, fashioning a new view of nature as the feminized and objectified focus of male inquiry.[44] Yet no one has studied how this process of gendering might have worked in particular branches of early natural inquiry: how and to what extent it functioned in terms of both context and content, whether it was stronger in some disciplines than others, and whether it was incidental or fundamental to their development. Such questions cannot be answered by looking at early modern texts and images in isolation, without considering their medieval antecedents as well.

This book attempts to address these problems for anatomy in the period between the late thirteenth and the mid-sixteenth centuries. It argues that women's bodies, real and imagined, played a central role in the history of anatomy during that time. Urgent questions about where babies came from and how they were conceived spurred physicians and surgeons to open human corpses and to write about them. In the process, what male writers knew as the "secrets of women" came to symbolize the most difficult intellectual challenges posed by human bodies: challenges that dissection promised to overcome. This story does not figure in any of the histories of late medieval and Renaissance anatomy, which focus on it as a university subject rather than as one of the many tools used by men and women to make sense of their experiences and to advance their interests in the world. It remains to be seen whether anatomy is idiosyncratic in this respect — since it is the science of bodies, the centrality of gender to its history is overdetermined — or whether similar histories, for other disciplines, are hiding in plain sight.

Whatever else it may be, this book is less a history of women than I would have liked, for although I have tried hard to track down materials that would reveal women's experiences and understandings of their own bodies, I have found articulate sources difficult to obtain. (I have speculated on the topic where I thought I could.) It is to a greater degree a history of the health care available to women, or at least to the mostly elite women whose anatomies I have been able to discover and explore. Above all, however, it is a study of the rich hoard of texts and images in which learned men — mainly physicians, surgeons, and clerics — reflected on women's bodies and were alternately alarmed, inspired, attracted, repelled, and fascinated by them. This book, then, is about women's bodies and men's attempts to know them, and through them to know their own.

CHAPTER ONE

Holy Anatomies

In August 1308, Chiara of Montefalco died in the monastery of which she was the abbess and in which she had lived for many years. Present at her deathbed were her fellow nuns and two friars from the local Franciscan convent: her chaplain and her brother. A renowned ascetic and visionary, she had been ill for some time. Several hours before her death, however, she regained her color, appetite, and energy, and she passed away peacefully in a seated position, as shown in a fresco painted twenty-five years later for the monastery's chapel (figure 1.1).[1] Shortly afterward, the nuns decided to embalm her corpse, which they already considered a precious relic. In the words of Sister Francesca of Montefalco, who testified a decade later at Chiara's unsuccessful canonization procedure, they agreed "that [her] body should be preserved on account of her holiness and because God took such pleasure in her body and her heart."[2] The nuns appear to have been familiar with the practice of embalming, for they knew they had to eviscerate the corpse and fill its interior with herbs and spices.[3] Accordingly, they ordered "balsam and myrrh and other preservatives" from the town's apothecary, Tommaso di Bartolomeo of Montefalco, as he recounted in his own testimony.[4]

Sister Francesca described what happened next: "And after leaving the others, Sister Francesca of Foligno, who is now dead,

39

Figure 1.1. Chiara of Montefalco on her deathbed. Montefalco, Church of Santa Chiara, Chapel of the Holy Cross, 1333.

and Illuminata and Marina and Elena, who is now dead, went to cut open [*ad scindendum*] the body, and the said Francesca cut it open from the back with her own hand, as they had decided. And they took out the viscera and put the heart away in a box, and they buried the viscera in the oratory that evening."[5] The next day, the nuns continued their exploration of Chiara's body. As Sister Francesca recalled, "after vespers or thereabouts, the said Francesca, Margherita, Lucia, and Caterina went to get the heart, which was in the box, as they later told the other nuns. And the said Francesca of Foligno cut open the heart with her own hand, and opening it they found in the heart a cross, or the image of the crucified Christ," as well as something that looked like the scourge with which he had been beaten during the Passion.[6] Later examination of the heart revealed other Passion symbols, including the crown of thorns, nails, and lance.[7]

Intrigued by these discoveries, Sister Francesca and her companions disinterred and explored the rest of Chiara's entrails, discovering, still in the words of Francesca, that

> inside Chiara's gallbladder there were three things that seemed to be round, so that they could not relax or rest until they knew what they could be. So they consulted with Master Simone of Spello [the monastery's physician], in order to ask him if these objects could have been caused by some illness. And they placed the gallbladder in his hand so that he could open it up. He did not want to do this because, as he said, he did not feel himself worthy. So Francesca cut open the gallbladder, and found in it three small stones.

The physician assured the nuns that there was no natural explanation for these structures, which they understood as referring to the Holy Trinity.[8] They were confirmed in their belief in the miraculous nature of these structures when Chiara's heart began almost immediately to perform wonders, curing a man with a diseased leg and causing another, who had disparaged its powers, to wound his arm with a brick.[9] These miracles, along with the holy objects the nuns found inside her body, augmented Chiara's already significant reputation for holiness, which eventually resulted in the apostolic canonization proceedings of 1318 and 1319, at which Sister Francesca and more than four hundred other witnesses testified.

In the course of her testimony, Sister Francesca gave two separate reasons for opening Chiara's body: the nuns wanted to preserve her corpse by embalming it, because God "delighted" in it; and they "believed they would find something wonderful" inside her heart.[10] Although the first explanation may seem paradoxical, given the nuns' later testimony concerning the corpse's persistent incorruption, it did not appear contradictory to contemporaries. Embalming was understood as a short-term measure, regularly performed on the bodies of notables such as the pope, who were put on display after their deaths; it was not expected to stabilize the corpse for more than a matter of days.[11] Therefore, while incorruption was neither a necessary nor a sufficient sign of sainthood, any holy person whose body failed to rot even after the effects of preservation were thought to have worn off might be revealed as having died in the odor of sanctity. One such case was Margherita of nearby Cortona, who died in 1297 and whose body was both embalmed and incorrupt.[12]

Embalming by evisceration seems to have come into regular use on the Italian peninsula over the course of the thirteenth century in connection with the proliferation of the cults of "new saints" such as Chiara of Montefalco — local holy men and women, venerated during their lifetimes — in the northern and central Italian city-states in which they lived and died. These cults were promoted by members of new religious orders, particularly the Dominicans and (as in Chiara's case) the Franciscans, but they also expressed the aspirations of local citizens and officials, who used them to promote civic loyalty and municipal prestige.[13] These cults centered on miracle-working corporeal relics in the novel and striking form of complete corpses, as opposed to the dispersed bodily fragments of ancient saints and martyrs, which could serve as the focus of pilgrimage and devotion. Thus the embalmed corpse of Margherita of Cortona, which I have already mentioned, very quickly became the object of an thriving local cult.[14] One of the scenes from the early fourteenth-century fresco cycle that decorated her chapel — as documented in an eighteenth-century watercolor — shows the importance of her preserved body in mobilizing popular piety and setting the stage for her eventual canonization (figure 1.2). In the foreground on the right, the sick and lame pray to Margherita for healing; in the center, those whose prayers have been answered testify to the bishop while swearing on the Bible in the presence of churchmen, local notables, and townspeople. On the far left, a notary records each miracle, creating a document that could serve as the basis of a canonization procedure. Chiara's fellow nuns evidently had similar hopes for their abbess. Whether because or in spite of their ministrations, Chiara's mummified corpse, like Margherita's, is still extant and in good condition; it, her heart, and her heart relics can still be seen over one of the side altars of the church of the monastery of Santa Chiara in Montefalco (figures 1.3 and 1.4).[15]

Regarding Sister Francesca's second explanation for the nuns' opening of their abbess — their eagerness to examine the contents of her heart — other witnesses were more explicit. They reported a vision described by Chiara some years earlier, in which Christ planted his cross in her heart, another scene commemorated by a fresco in the monastery chapel (figure 1.5).[16] In the words of Chiara's official biography, or *vita*,

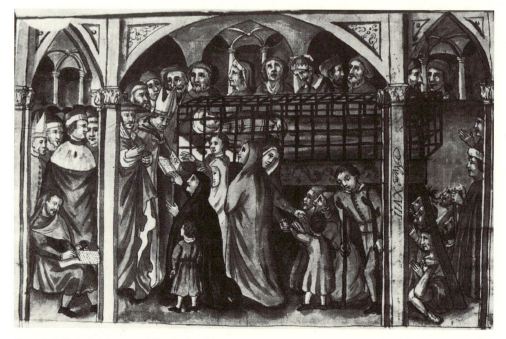

Figure 1.2. Embalmed corpse of Margherita of Cortona, with petitioners seeking healing (*lower right*) while people who have already received miracles present sworn testimony to the bishop in the presence of a notary (*lower left*). Adriano Zabarelli, watercolor reproducing one of the frescoes that decorated Margherita's chapel in the fourteenth century, attributed to Pietro and Antonio Lorenzetti, Cortona, Biblioteca Communale e dell'Accademia Etrusca, cod. 429, no. xvii, ca. 1629.

Figure 1.3. Reliquary cross containing the crucifix, scourge, and three gallstones found in Chiara's corpse in 1308. Montefalco, Church of Santa Chiara.

Figure 1.4. Detail of figure 1.3, showing the crucifix from Chiara's heart.

Figure 1.5. Chiara's vision in which Jesus planted his cross in her heart. Montefalco, Church of Santa Chiara, Chapel of the Holy Cross, 1333.

the Lord Jesus Christ, in the form of a beautiful young man, clothed in white garments and carrying on his shoulder a cross the same in shape and size as the cross on which he was crucified, appeared to Chiara in prayer. And he said to her: "I am looking for a sturdy spot on which to set the cross, and I find here a suitable place on which to set it." Then he added: "If you want to be my daughter, you must die on the cross."[17]

From this point on, Chiara referred repeatedly to having Christ in her heart.

The nuns' public deposition, taken by a city notary only five days after Chiara's death, when their memories were fresh and their testimony less reworked than in the formal canonization procedure ten years later, is particularly clear on this latter point. Before a host of witnesses, including local clerics and municipal officials and citizens, they swore that Chiara had rebuked a nun who had tried to make the sign of the cross over her on her death-bed, saying, "'This is unnecessary, since I have the cross of our lord Jesus Christ fixed in my heart.' And on account of this and other things they heard from her," the deposition continued, "they were moved to cleave and see [*ad rimandum et videndum*] her heart and viscera, saying among themselves, 'The dead blessed Chiara says these things to us, nor have we ever heard anything from her that was untruthful or vain; therefore, let us cleave and see the things that she has told us.'"[18]

Among other things, these events clearly demonstrate the absence in this period of either an effective ecclesiastical prohibition or a cultural taboo regarding the opening of human bodies, even in the immediate wake of Pope Boniface VIII's famous bull of 1299, *Detestande feritatis*, which is often invoked in this connection (even though it forbade only boiling flesh off bones).[19] From the very beginning, the nuns were under suspicion from religious and civic authorities, as the notarized deposition required of them indicates, but the authorities were concerned only that they had fabricated the objects found inside Chiara's heart. At no point, either immediately after her death or during any phase of the extended canonization proceedings, which took place under papal auspices and involved countless theologians, canon lawyers, and prelates, as well as a number of secular officials, was there any hint that it was inappropriate — let alone blasphemous or illegal — to open Chiara's corpse and examine her viscera. Nor did her family, her neighbors, or the notables of the region raise any objection to the procedure. Rather, there is every indication that embalming was considered an expression of special reverence and respect.

Chiara's anatomy should also be seen in its medical context. Her death, the opening of her body, and the later evaluation of her case by secular and ecclesiastical authorities took place in an environment in close touch with contemporary medical learning and practice. (It may even be relevant that Sister Francesca of Montefalco, who cut open Chiara's body, was the daughter of a physician.)[20] Chiara had been attended in her various illnesses by at least two doctors: Master Gualtieri of Montefalco (Sister Francesca's father) and the monastery's regular physician, Master Simone of Spello, who also served as Montefalco's publicly salaried doctor and attended Chiara on her deathbed. Master Simone was the first person consulted by her nuns about the objects they had found inside their abbess and was present when they cut open her gallbladder to extract its stones.[21] He was the only person other than the nuns to swear to the collective deposition of August 22 describing the opening of Chiara's body, and, like them, he testified at the canonization proceedings of 1318–1319.[22] The episcopal vicar, Béranger of Saint-Affrique, the impresario of the early fourteenth-century canonization attempt and the author of Chiara's vita (composed between 1315 and 1317), repeatedly emphasized Master Simone's presence, describing the nuns' opening of the gallbladder as having been done "on medical advice [*medico consilio*]."[23] Of the stones themselves, he wrote, "after lengthy examination of them, physicians and experts in the study of nature [*medicorum et naturalium*] issued the opinion that they could not have been naturally formed, but only by divine power."[24]

In this related set of contexts, devotional, funerary, and medical, the opening of Chiara of Montefalco appears much less surprising than it did at first sight. Although unprecedented — it was the first documented case of a holy person whose internal organs were inspected and found to reveal corporeal signs of sanctity — it was hardly unprepared. Inspired by Francis of nearby Assisi (d. 1226), who had received the stigmata of the Passion in the course of a religious vision, people had begun to look for external marks of sanctity on the bodies of holy men and women.[25] The new practice of embalming by evisceration provided both a rationale and an opportunity to look for marks on internal organs as well. Chiara's was the first of several such cases. Something quite similar occurred in the nearby town of Città di Castello, where the body of another local holy woman, Margherita, was opened immediately after her death in 1320, only a year after Chiara's elaborate canonization proceeding came to an inconclusive end. Like Chiara, Margherita was an ecstatic and, in her own way, a visionary — her visions all the more remarkable because she had been blind from birth — and the story of her corpse was in many ways a patent imitation of the events surrounding Chiara's. According to her *vitae*, she was initially eviscerated in order to preserve her body; only several days later did it occur to the Dominicans who had custody of her corpse to pay special attention to her heart. When they disinterred it, in order to put it in a reliquary, they discovered three small stones impressed with images of Mary, the infant Jesus,

49

Joseph, the Holy Spirit, and a kneeling penitent who appeared to be Margherita herself (figures 1.6 and 1.7).[26]

The practice of holy anatomy was by no means common in this early period, which is understandable, given the failure of the attempts to canonize both Chiara and Margherita, which suggests that ecclesiastical authorities found the evidence of internal organs unconvincing. (Chiara was canonized only much later, in 1881, and Margherita remains a *beata* to this day.) I know of only two other prospective saints who had their bodies opened and their viscera inspected during the two centuries following Margherita's death, as I will describe in Chapter Four: the prophetess Colomba of Rieti (d. 1501) and the lactating virgin of Bologna, Elena Duglioli (d. 1520). One of the most striking things about these four procedures is that they all involved women: I have not found a single case of the internal inspection of the body of a holy man before the middle of the sixteenth century, at which point, in the new religious and political context of the Counter-Reformation, the practice began also to be applied to the corpses of male religious — famous examples include Ignatius Loyola (d. 1556), Carlo Borromeo (d. 1584), and Filippo Neri (d. 1595) — although in a less systematic way.[27]

Given that holy anatomies, though rare, were uncontroversial, how can we account for their early restriction to female bodies? In the rest of this chapter, I will argue that it is no coincidence that the first potential saints to be opened in this way were women; in the context of the early fourteenth century, certain elements of Franciscan devotion, along with contemporary assumptions about the processes of vision and generation, gave a special meaning and plausibility to the production of holy objects inside the female body, while female sanctity raised particular questions and issues that made the inspection of the body's interior especially urgent. The cases in question were complicated, involving the collaboration of female religious, secular men and women, friars and eccle-

Figure 1.6. Five fourteenth-century Tuscan and Umbrian holy women associated with the Third Order of St. Dominic: (*left to right*) Giovanna of Florence, Vanna of Orvieto, Catherine of Siena, Margherita of Città di Castello, and Daniella of Orvieto. Andrea di Bartolo, *Corpus Domini Altarpiece*, Murano (Venice), Museo Civico Vetrario, ca. 1400.

Figure 1.7. Detail of figure 1.6, showing Margherita of Città di Castello holding her heart with its three stones before an altar with a crucifix.

siastical officials, and medical men, and it is clear that the various parties involved had different motivations, different understandings of what they were doing, and different interpretations of the meaning of the objects they found. Nonetheless, all seem to have shared a set of assumptions about the power of vision to induce physical alterations in the body of the seer, and the attendant importance of reading bodily signs. These assumptions in turn reflect a model of vision and visuality that differs dramatically from modern views, which stress the impassivity of the viewer and establish a hierarchical relationship between subject and object; the stories about opening up Chiara and Margherita, in contrast, repeatedly emphasize the transformative power of the object of vision and the intimate and reciprocal relationship between viewer and viewed.

Authenticating Sanctity

Although they began as embalmings, the anatomies of Chiara of Montefalco and Margherita of Città di Castello also recall contemporary legal practice, for the early fourteenth century saw the first appearance of what we might call judicial anatomy: the examination of an opened corpse by medical experts at the request of a judge in order to determine the cause of death. This practice had developed gradually in Italy, beginning in Rome and Bologna.[28] The first reports, from the early thirteenth century, indicate that postmortem inspections by doctors were initially confined to the exterior of the body. Pope Innocent III seems to have been the first to have repeatedly ordered such inspections; his decretals include two cases where he called on the "judgment of experienced doctors" to assign responsibility for the deaths of two men who had suffered head wounds.[29] This practice of judicial medical examination was institutionalized in a secular context. In the mid-thirteenth century, the government of Bologna began to appoint a pool of respected local doctors who could be called on to testify as to the cause of death in trials involving assault and suspected murder. Although these men initially based their judgments on external inspection, by the early fourteenth century they were beginning to open victims' bodies in order to look for hidden signs and internal causes of death.[30] In 1302 — in the first known case of this sort — the judge ordered a commission of two surgeons and two physicians to examine the entrails of one Azzolino degli Onesti to determine whether he had died from natural causes or from poison. Five years later, in 1307 (one year before the opening of Chiara of Montefalco), the physician Bartolomeo of Varignana, who had also participated in the earlier commission, reported that he had opened the body of a woman named Ghisetta at the request of another judge and found that she had died from a wound that had continued to bleed internally, although it appeared superficially to have healed.[31]

These anatomies should be seen against the background of the emergent practice of human dissection in the medical faculty at the University of Bologna, where Bartolomeo of Varignana was a professor, which I will discuss in the next chapter. They also provide a context for the use of evidence derived from the nuns' anatomy of Chiara in her canonization procedure, where the parallels with contemporary judicial practice are unmistakable. Indeed, formal canonization proceedings as conducted in this period had the legal and bureaucratic structure of a trial. This forensic tenor is evident as early as the first half of the thirteenth century, when popes began to instruct their agents to interrogate local witnesses regarding the lives, miracles, and local reputations of prospective saints.[32] As the process evolved, it split into a preliminary inquiry and a large-scale proceeding formally authorized by a papal bull. These inquiries generated mountains of written documentation, including local records, duly authenticated; the

53

notarized testimony of witnesses, taken both in person and by deposition, in response to an officially established series of questions (the "interrogatory articles"); a formal account of the saint's life, or *vita*; and a summary assessment of the case by three cardinals. In Chiara's case, the large-scale, or "apostolic," phase of the proceedings took place over the course of eleven months, from September 1318 to July 1319, and involved the testimony of 486 witnesses on 315 interrogatory articles.[33]

Given the many parallels between contemporary judicial practice and canonization procedures, it is not surprising to find that the testimony of doctors was called upon in both. But who exactly was on trial in Chiara's case? Why, for the first time ever, did physical evidence — in the form of signs found *inside* a potential saint's body — occupy such a decisive position in authenticating her sanctity?[34] The answer to these questions lies, at least in part, in the specific form her piety took and the unprecedented nature of the religious movement with which she and her fellow nuns were associated.

The thirteenth-century Italian founders and early leaders of the mendicant orders had called for a new spiritual ideal, organized around a life of penance, poverty, and urban religious activism, in place of the traditional model of monastic enclosure. To their surprise — and to the puzzlement of later ecclesiastical leaders — this ideal took deepest root not among men, as they had clearly expected, but among laywomen. The second half of the thirteenth century saw an exuberant flowering of female religious life, especially in Umbria and southern Tuscany, largely as a result of the local influence of Francis of Assisi. Women embraced this ideal in significant numbers — Francis's follower Chiara of Assisi was one of many — committing themselves to lives of strenuous poverty and asceticism, and coming together in spontaneous lay communities, often clustered around Franciscan or Dominican convents.[35]

This feminization of the penitential ideal in the late thirteenth century posed a challenge for ecclesiastical authorities, who had doubts about the wisdom and appropriateness of large numbers of laywomen leaving their families to live autonomous and sometimes highly visible penitential lives. These doubts were magnified by the strong visionary and ecstatic element in this movement, which began to produce charismatic women known not only for their extreme asceticism but also for their trances and prophetic revelations.[36] Over the course of the late thirteenth and early fourteenth centuries, this movement grew and diversified, as the women in question elaborated new ways of life in more open communities, known as Beguinages, and in enclosed houses, such as that of Chiara of Montefalco. Church authorities had considerable success in controlling and institutionalizing this unexpected phenomenon, and they did so in collaboration with the women themselves, many of whom sought more formal ecclesiastical recognition. A number of communities, including Chiara's, were recast as monasteries; when Chiara became abbess, her community had only recently received the Rule of Saint Augustine. In other cases, women joined the newly created third orders associated with the Dominican and Franciscan convents that already served as the focus of their devotional lives.[37] Overall, it is clear that male clerics and ecclesiastical authorities continued to be simultaneously inspired and concerned by this female religious movement and somewhat uncertain about its nature and meaning.[38]

It was in this context that the events following Chiara's death acquired such significance. Margherita of Cortona (d. 1298), Angela of Foligno (d. 1309), and Chiara were the most prominent representatives of this new female piety in Umbria and southern Tuscany, and the popular pressure for Chiara's canonization, channeled and reinforced by her champion Béranger of Saint-Affrique, posed a number of novel problems. First and foremost, like that of a number of her holy female contemporaries, Chiara's reputed sanctity had manifested itself not in the highly visible acts of mercy and moral, intellectual, and religious leadership that were typical of male saints like Saint Francis — preaching, begging, bearing public witness — but in the private elevations and ecstasies recounted in Béranger's *vita*.[39] Hearing voices and having visions left no tangible evidence and could not be witnessed directly by others. Furthermore, because Chiara observed enclosure, the indirect witnesses to these acts were her fellow nuns, whose testimony, if not actually suspect, was seen (not implausibly) as biased and requiring strict verification. In addition, charismatic women whose reputation and influence were primarily local fit poorly with the style of sanctity promoted by the contemporary papacy, which privileged laymen and clerics of high ecclesiastical, social, and political status.[40] Finally, the charismatic nature of Chiara's piety itself aroused suspicion. As Nancy Caciola has shown, holy women like Chiara were widely understood by themselves and others as inhabited or possessed by supernatural forces.[41] Under such circumstances, these women were doubly suspect; not only might they be faking the visions and physical behavior that lent them such authority, but, far more dangerous and disturbing, they also ran the risk of mistaking inhabitation by demons for inhabitation by God.

All these circumstances — Chiara's poor "fit" with dominant, masculine models of sanctity, the lack of a public record and male witnesses, and the elusive and risky nature of her spiritual life — made her a legitimate object of skepticism and suspicion, and her canonization procedure was the first in the Church's history to attempt to authenticate systematically the visions and revelations of a holy person.[42] Even before her death, there had been attempts to verify her sanctity medically. According to Béranger of Saint-Affrique, one of her many visitors was the doctor Filippo of

Spoleto, who had "heard a great deal during this period about Clare's holiness but . . . did not believe it, because the things he heard seemed naturally impossible." Once in Montefalco, he was told that Chiara had subsisted for as long as two months on four ounces of bread and that she had lost her sense of taste, so that she could not even distinguish among water, vinegar, and wine — claims that Filippo proceeded "cautiously to test [*caute . . . experiri*]" in ways that Béranger did not specify.[43] (Although unconvinced by his findings, he changed his mind on the way home after encountering a furious storm, which he interpreted as the saint's justified retribution for his lack of faith.) Even Béranger, who soon became Chiara's most ardent supporter, claimed to have had initial doubts. Immediately after Chiara's death, on receiving the first reports from Montefalco about the miracles associated with her body, he had hastened to Montefalco in his official capacity as vicar of the bishop of Spoleto; "afire with indignation," as he put it, he believed that the nuns had fabricated the objects they had found in her heart. Only after he had examined the heart personally, together with those he described as "many grave persons, and especially Franciscan friars and judges and lawyers and rectors of the [local] churches," did he conclude that the objects were miraculous rather than artificial.[44]

Not all the skeptics were so easily convinced, however. Ten years later, one of the witnesses called during the apostolic canonization proceedings, Brother Tommaso di Buono of Foligno, contested Chiara's sanctity on every possible front. A Franciscan friar in Montefalco, he said that he had been told by the nuns' chaplain and several other friars that the holy objects found in her heart had been faked. "From this," he testified, "I have a violent suspicion that, as people say, the signs in her heart were artificially made by a sister from Foligno [presumably the late Sister Francesca of Foligno], who was a nun in the monastery and who was skillful with her hands." In addition, he said that he had heard that Chiara often fell to the ground as if possessed — "whether by a good or a bad spirit I do not know." Finally, he claimed that the nuns had lied about Chiara's ascetic habits, and he strongly suggested that her mystical states resulted from uterine suffocation, the falling sickness, or some other physical disorder.[45]

Brother Tommaso's testimony illustrates the various problems posed by female visionaries like Chiara of Montefalco and the pressing need to exclude other explanations for the bodily phenomena associated with her holiness — natural illness, human fraud, and demonic possession — before her sanctity could be accepted as confirmed. Such comments were in accordance with the conventions of contemporary hagiography, where they served to frame and validate the ultimate triumph of the saint. The level of distrust in Chiara's case was unusual, however, as was the extended inquiry into every aspect of her life and miracles and into the nuns' actions after her death. The appearance of a new form of religiosity, different from and in many respects clearly opposed to the model of official sanctity dominant at the papal court — male, elite, and learned — transformed the canonization inquiry into a kind of trial. In this context, the nuns' opening of Chiara's body, heart, and gallbladder were recast as a kind of anatomy to reflect the judicial tone of the proceedings, on the model of the forensic anatomies used in Bologna to determine the cause of death, and thus culpability, in cases of assault and poisoning. In Chiara's case, too, the objects found in her heart served as physical evidence with the capacity to exonerate all the nuns of credulity, misrepresentation, or fraud.[46]

The church authorities' need for physical proof in the case of Chiara of Montefalco forms part of a broader phenomenon already noted by historians of late medieval female spirituality: the predominance of women among holy visionaries and male

58

hagiographers' persistent tendency to translate the internal mystical experiences of their female subjects into perceptible corporeal signs, such as stigmata, bleeding wounds, lactation, and other paramystical phenomena. Caroline Walker Bynum has related this phenomenon to contemporary attitudes on the part of both men and women that associated women with the body and men with the mind or soul, although it is worth noting that men endowed with mystical gifts, such as Saint Francis, might also exhibit corporeal signs of sanctity.[47] This gave special importance to the bodies of holy women, which were seen as vessels of their supernatural powers, and it may be why contemporaries appear to have embalmed the bodies of prospective female saints more often than those of male saints — a fact that does much to explain the restriction of holy anatomies in this period to female corpses.[48] At the same time, however, this focus on the dead bodies of holy women seems also to have reflected a special need for proof on the part of male clerics when female holiness was involved. As Amy Hollywood has argued, whereas late medieval female mystics stressed their visions and their interior struggles and experiences, the authors of their *vitae* insisted on visible marks; this reflected not only the men's general sense that "the identification of women with the body demand[ed] that their sanctification occur in and through that body" but also their particular need for "externally sensible signs of visionary and mystical experience in order to verify the claims to sanctity of the woman saint."[49]

Caciola goes one step further and explains why such signs might also appear *inside* women's bodies, as in the cases of Chiara of Montefalco and Margherita of Città di Castello. Theologians understood possession, whether by God or by demons, in physiological terms, as a kind of corporeal indwelling to which women's open and porous bodies were particularly prone. One important way to distinguish divine from demonic inspiration involved identifying the affected organ: demons inhabited the bowels, while God in-habited the heart.[50] This is why the holy objects in Chiara's and Margherita's hearts acquired such importance as evidence not only that their visions were neither fraudulent nor natural but also that they were divine.

Generating Relics

In addition to these concerns about evidence, authenticity, and female corporeality, a second factor helps explain why anatomies were performed principally or exclusively on holy women: the perceived similarities between the production of internal relics and the female physiology of conception. Women, after all, generated other bodies inside their own. God's presence in the heart might be imagined as becoming pregnant with Christ.[51] This association gained special force over the course of the thirteenth century, with the flowering of a powerful strand of devotional writing and preaching associated particularly strongly with the tradition of Francis of Assisi, a stone's throw from Montefalco, that described spiritual experience in terms of pregnancy and birth. "We are [Christ's] mother, when we carry him in our heart and body through pure love and sincere conscience," reads one of Francis's letters, "and we give birth to him through a holy process [operationem], which must shine as an example to others."[52]

Francis did not invent this metaphor. The birth of Christ in the soul had been a common trope in Christian writing since late Antiquity, often expressed in arrestingly concrete form. Writing in the fourth century, for example, Ambrose of Milan referred to Jesus as "the boy who is birthed by the one who accepts the spirit of salvation in the uterus of his mind [utero suae mentis]"[53] and described the failure of this process in terms of miscarriage and inability to bring a fetus to term: "Not all give birth, not all are perfect.... For there are some who miscarry the Word [abortivum *excludant Verbum*] before they give birth, and there are some who have Christ in their uterus but never give him form."[54] Following Augustine, it became common to identify this metaphorical uterus with the heart, a rhetorical figure that was greatly elaborated by Bernard of Clairvaux.[55] Thirteenth- and fourteenth-century writers developed the analogy of conversion with birthing, holding up Mary as a model for all Christians, and fused it with devotion to the Passion. Although Mary bore Christ without pain, according to this tradition she later felt that pain at the foot of the cross, in the form of a second childbirth, where it became part of her "compassion" — her "co-suffering" — with her son.[56] The fresco of the Crucifixion that dominates the Chapel of the Holy Cross in the church of Chiara's monastery contains a clear reference to this tradition; commissioned in 1333, it shows Mary to the left of the cross, supported by one of her companions in a position strongly identified with parturient and midwife (figure 1.8).

The long tradition of assimilating the conversion of the Christian to the conception and birthing of Christ in the heart, together with the tradition of identifying his or her travails with Mary's compassion for Christ as manifested in the pains of childbirth, were developed by male writers, who clearly intended these associations to apply to both men and women — an example of medieval men using female as well as male corporeality to think about Christian spirituality, as Caroline Walker Bynum has described.[57] But this trope must have seemed particularly well suited to model and describe the experiences of women. Certainly Béranger used it this way in his *vita* of Chiara of Montefalco, in which he narrated a vision that appeared to Giovanna of Montefalco, Chiara's successor as abbess, sometime after Chiara's death:

> During the night of Christmas, while Giovanna ... was praying after
> Matins, she suddenly saw Santa Chiara, holding Christ in the form of

Figure 1.8. Crucifixion scene showing Mary suffering the pains of childbirth (lower left). Montefalco, Church of Santa Chiara, Chapel of the Holy Cross, 1333.

a very beautiful baby in her arms, and she was greatly amazed at the sight. Her first thought was that the baby was Chiara's child. She began to think how this could be if Chiara had been a virgin. And Chiara immediately responded to her thought by saying, with great fervor of spirit: "Love has made me conceive him, love has made me give birth to him, and love has made me possess him forever."[58]

In this, as in other aspects of the story of Chiara of Montefalco, it is difficult to differentiate the voice of the male recorder from the voice of the female witness and to determine whether the identification of Chiara with Mary originated with Béranger or with Sister Giovanna and the other nuns in Chiara's monastery, or whether it evolved in the course of what was obviously a long collaboration between the two parties in the service of Chiara's canonization. There is nothing specific to suggest that Chiara saw herself in this way, despite the currency of the idea of the pious heart as uterus, and there are other cases in which male interpreters of female religious experience read female role models such as Mary into women's more Christ-centered accounts.[59] We have no direct access to the thoughts and experiences of Sister Francesca and her companions during those hot days in August when they repeatedly opened their abbess's corpse and internal organs; their testimony was taken a decade after the events in question, in the course of what was obviously a highly orchestrated set of hearings, and their Italian responses were translated into scribal Latin, yielding a highly mediated account. Nevertheless, if they did in fact entertain the idea that Chiara had conceived Christ in her heart — whether this occurred to them immediately after her death or at some later point in their manipulations of her body — it suggests yet another way of understanding the opening of her corpse: not merely as an anatomy, in which her internal organs were examined, but also as related to a contemporary birthing practice in which women who had died during childbirth might be cut open to extract the baby for the purposes of baptism.

This operation, which had been occasionally practiced before 1300, began to be performed with some frequency in southern France and northern Italy in the late thirteenth and early fourteenth centuries.[60] (The first known description in a medieval medical textbook dates from 1305.)[61] The procedure figured prominently in the vernacular histories of the Roman Empire that were circulating in France and Italy in this period, including the extraordinarily influential *Deeds of the Romans* (ca. 1213–1214), which begins with the extraction of Julius Caesar from his mother's body — the eponymous Caesarean section (figure 1.9).[62] More immediately relevant to the case of Chiara of Montefalco, this procedure was strongly promoted by the Church in the late thirteenth and fourteenth centuries, in connection with the salvation of infant souls.[63] Word of it was certainly circulating in lay and ecclesiastical circles in early fourteenth-century Tuscany, as is clear from a public sermon delivered in the Florentine church of Santa Maria Novella by the Dominican friar Giordano of Pisa in April 1305, only three years before Chiara's death. Speaking on the theme of baptism, Brother Giordano discussed the conditions under which laypeople, rather than priests, might christen a child. According to the scribe reporting on the sermon,

> he then spoke of women who die in childbirth, with their fetus [*criatura*] alive in their belly. He sternly criticized people who bury them thus, and he said that it was a great sin. He told of a woman who was in Pisa, in their place [presumably a house attached to the Dominican convent], who died in childbirth and had a living fetus in her belly, so that he had her opened [*isparare*]. And he said: I sent for four doctors [*medici*] and midwives [*balie*], and I paid them very well. And in this way we opened her and drew the boy from her belly, and

64

Figure 1.9. Birth of Caesar. *Li Fet des Romains*, Venice, Biblioteca Nazionale Marciana, cod. marc. fr. 3, fol. 2r, Italian, early fourteenth century.

he was alive, and we baptized him, and his soul was saved. Was this not a great mercy? Many [souls] are lost in this way and are in Limbo, through your fault. [The women] should be opened up, and it is a great mercy. Thanks be to God. Amen.[64]

This sermon resonates with the testimony of Sister Giovanna of Montefalco, the monastery's next abbess, who described finding the crucifix inside Chiara's body in words that suggest a similar scene: "Asked which sisters extracted it from the body of the said holy Chiara, she said that it was Sisters Francesca, Marina, Illuminata, Caterina, and Elena, who is now dead; Francesca opened and cut the body with a knife like a razor, while the other aforenamed sisters stood by, looking on and helping Francesca."[65] Even the language used by Sister Giovanna to describe the crucifix found in Chiara's heart suggests a fetus inside the womb (compare figure 1.4):

the cross was like a little human body [*corpusculum humanum*], though not completely well formed, since that little body appeared to have a little head made of flesh, of about the size of a small bean. But that little head did not have [fully] formed and distinct little members, like eyes or nose, as far as the witness could recognize, and it had arms but no perforated hands with fingers, ... and it had thighs and legs, but not feet formed with toes.[66]

Further evidence that the process by which relics were generated inside a female saint's body might be imagined as analogous to the physiology of human reproduction appears in the case of Margherita of Città di Castello, whose heart yielded what the author of her first *vita* called "three wonderful stones impressed with various images [*diversas ymagines impressas habentes*]" of the holy family, the holy spirit, and Margherita herself.[67] As the use of the word "impressed" suggests, these stones evoked the process of stamping or sealing, embodying a verse from the Biblical Song of Songs: "Put me as a seal upon thy heart."[68] These stones would also have resonated with contemporary theories of generation, which was described in terms of the stamping of form (from the father's semen) on matter (the blood in the mother's uterus), as a seal makes an impression on soft wax. This process of stamping, sealing, or impressing was used to explain many other physiological phenomena, including sensation and perception.[69] It also functioned in the outside world; for example, the thirteenth-century Dominican natural philosopher Albertus Magnus described fossils and cameos, both of which bore clear analogies to the stones in Margherita's heart, as resulting from the natural impression of images from the constellations on malleable mineral matter.[70]

In particular, the figures on the stones generated in Margherita's heart could be seen as products of a particularly widely acknowledged kind of imprinting in which a pregnant woman was believed to mark her fetus with images of things she had seen or imagined during conception and pregnancy.[71] Late medieval scholars and theologians explained this process in terms of contemporary theories of vision, as the impression on blood, brain, and spirit of forms emitted by external objects, which penetrated the viewer's body through the eyes.[72] These forms then traveled through the blood vessels to the uterus — or, in Margherita's case, the heart — where they stamped themselves on the soft fetal body to produce marks in the shape of strawberries or other objects strongly present in the mother's imagination. Drawing on Augustine's influential discussion of the three levels of vision (corporeal, spiritual or imaginative, and intellectual), the Dominican preacher Jacobus de Voragine had used this mechanism to explain the physical appearance of the stigmata on Saint Francis's body. [73] Its relevance was even more obvious in the case of marked objects generated inside the entrails of a holy woman. Béranger emphasized that Chiara's vision, like Francis's, belonged to this second level; she saw, as he put it, "in the spirit," meaning that her imagination was engaged.[74] In Margherita's case, this process had a special, supernatural quality. Although she had been blind from birth, the marvelous capacities attributed to her — in addition to miraculous healing and levitation — included the ability to see Christ incarnate during Mass at the elevation of the host. "Nor is it wonderful if [Christ] wished to show himself to her pure sight, although he had deprived her of the vision of all earthly things," wrote the anonymous author of her first *vita*, "so that divine mercy might shine in an earthen and abject vessel ... for she had always appeared to have in her mind and mouth the birth of the glorious Virgin, the nativity of Christ, and the help of Joseph, of whom she often spoke."[75] These were precisely the images impressed on the stones in Margherita's heart, and they confirmed her sanctity by authenticating her capacity for supernatural sight.[76]

In the end, perhaps the most striking thing about the stories of Chiara and Margherita is the way they made literal and material the metaphor of the birth of Christ in the heart. Although this metaphor was a commonplace of late medieval devotional writing, I know of no case before Chiara's in which it occurred to anyone to explore it anatomically, although there was a model for this in the story of the martyrdom of Saint Ignatius of Antioch in Jacobus de Voragine's mid-thirteenth-century *Golden Legend*, an influential collection of saints' lives. (During his tortures at the hands of the Romans, Ignatius repeatedly told his persecutors that he had the name of Jesus written on his heart; after his death, his body was cut open to reveal the inscription in letters of gold.)[77] The decision by the nuns of Montefalco to interpret this metaphor physically reflects the power of Franciscan elements in the spirituality of late medieval Umbria in general and Chiara's monastery in particular. Francis had arguably been the first holy person to incarnate a scriptural metaphor — in his case, Galatians 6:17: "I carry the stigmata of the Lord Jesus Christ on my body." Thus when Sister Francesca and her fellows interpreted Chiara's claim to have Christ in her heart literally, they were simply extending and further feminizing the model already embodied by Francis himself.

Vision and Visibility

Chiara of Montefalco and Margherita of Città di Castello were visionaries — seers of divinely inspired sights whose elements (the Nativity, the crucified Christ, the instruments of the Passion) were impressed on the matter of their hearts, just as things desired by pregnant women might be impressed on the matter of their fetuses. In this respect, they illustrate the highly visual nature of thirteenth- and early fourteenth-century Italian spirituality, which Hans Belting has described as focused on the "need to see." This manifested itself in a florid visual culture centered on the display of images, reliquaries, and relics (including holy corpses), which became the focus of religious contemplation; Christians were encouraged to identify with and assimilate themselves to holy models — saints and, above all, the suffering Christ — of whom these objects were the visual sign.[78] In addition to being seers themselves, however, Chiara and Margherita became the objects of others' vision after their deaths. The resulting exposure of their bodies raises a broad set of issues regarding female modesty and the contemporary understanding of sight as a primary mover of the passions, especially sexual desire.

The tension between the demands of feminine modesty and the highly visual nature of late medieval piety figured most spectacularly in the story of Margherita, whose naked body, unlike Chiara's, was exposed to public view.[79] The authors of Margherita's two *vitae* concur on the basic facts. Abandoned by her parents after a pilgrimage to Città di Castello failed to cure her blindness and lameness, she became a Dominican tertiary at the age of fourteen and died twenty years later, in 1320, in the home of two pious citizens who had taken her in and raised her. Immediately afterward, her corpse was brought to the convent of San Domenico, accompanied by a large and enthusiastic crowd. "When the friars wished to bury her body in their cloister," according to the author of her first *vita*, "a huge cry went up from the people, as if divinely inspired: 'Not in the cloister — rather, we want her to be buried in the church, because she is a saint!'"[80] Moved by their ardor, the friars placed her corpse in a casket and brought it into the church, where it immediately healed a mute and lame girl. Impressed by this miracle and by Margherita's fame and popular support, the city fathers volunteered to subsidize the embalming of her body by two local surgeons, Master Vitale of Castello and Master Manni of Gubbio. The initial evisceration took place in the church itself,

69

in front of the high altar. Margherita's first *vita* describes the dramatic scene: "A huge crowd of friars, clerics, and laypeople had come, and when the said friars and surgeons had extended her arms, wishing to open her body, the blessed Margherita placed her arms in the form of a cross on her body, covering her genitals [*sexum fragilitatis humane*], in the sight of everybody there. Then, after a short interval, when they had extracted her viscera, there was an enormous earthquake."[81] This does not seem to have disrupted the proceedings, however, and the embalming went on as planned. Afterward, the friars at first buried Margherita's internal organs in the cloister, in an earthen vessel, but they later exhumed them and placed them beside her preserved body in the church.

As the miracles associated with Margherita's remains multiplied, the friars decided to move her viscera to a gold reliquary so that they might be displayed in the sacristy. Those present at the transfer included two named friars, Brother Niccolò di Giovanni Santi of Città di Castello and Brother Gregorio of Borgo Cresci; three medical men, Masters Giovanni and Jacopo of Borgo and Ugolino Verde; plus many others, both clerics and laymen. Having removed Margherita's heart from the vessel, Brother Niccolò cut into the "tube [*cannam*]" from which it hung, in order to separate it from the other organs, at which point

three wonderful little stones fell from the said tube, impressed with different images. On one was seen to be sculpted the face of a very beautiful woman with a golden crown, which certain people interpreted to be the effigy of the blessed glorious Virgin Mary, to whom Margherita was enormously devoted. On another appeared a small child in a cradle, surrounded by cattle, which certain people said signified Christ or the nativity of Christ. On the third was sculpted the image of a bald man with a grey beard and a golden cloak on his shoulders and with him a kneeling woman dressed in a Dominican habit, and they said that this represented the blessed Joseph and the blessed Margherita. On the side of this stone was a pure white dove, which they said represented the Holy Spirit, through which Mary conceived the Son.[82]

The opening of Margherita in 1320 was no doubt inspired by that of Chiara, whose canonization process had ended in failure less than a year earlier; it bespeaks the rivalries not only among Umbrian cities but also between the Dominican and Franciscan orders.[83] Where the crucifix in Chiara's heart reflected the Franciscan emphasis on the Passion, the imagery of Mary and the Nativity on Margherita's heart relics echoes the special devotion of Dominicans to the Virgin and Child.[84] The remarkably public nature of the events surrounding the opening of Margherita's body and the discovery of the miraculous stones in her heart — both witnessed by a crowd of men with impeccable professional, municipal, and ecclesiastical credentials — may well have been intended to build a case for Margherita's sainthood that avoided the weaknesses in Chiara's, which relied on the testimony of a handful of nuns. But the fact of so many male witnesses is nonetheless striking in a culture where respectable women were expected scrupulously to avoid being seen by men, especially in circumstances that might awaken sexual desire. Chiara in particular was known for her extraordinary modesty. When it was her turn to go out begging, before she embraced strict enclosure, "she wrapped herself in her cloak in such a way as to make her face and body unrecognizable," according to Béranger's *vita*, and "as she went, she shielded herself from the sight of men so as not to see any of them and not to be seen by any."[85] She continued this behavior after enclosure, keeping her face and body hidden behind a wall or cloth even when a male relative came to the window to speak to her.[86] The description of the opening of Margherita's naked body calls special attention to this issue; when uncovered, the corpse miraculously moved its arms to cover its genitals.

Although Margherita's gesture registers the cultural discomfort involved in exposing the naked body of a holy virgin to public view, this event should not be understood simply as the display of a female body to an erotic and objectifying male gaze. (This theme does not emerge as an explicit part of the verbal and visual discourse of anatomy until the sixteenth century, as I will argue later in this book.)[87] In late medieval Italy, nakedness was used as a symbol of humility and poverty as much as of sexual availability. Thus Margherita's lack of clothing, like her outspread arms, assimilated her more directly to the naked male body of Christ on the cross or the penitent Mary Magdalene than to, say, Eve, and it evoked the model of viewing as identification to which I have already referred.[88] Nor should this scene of public evisceration, which strikes modern readers as at best macabre, be read as sadistic or even unambiguously violent. Margherita was, after all, dead when her body was opened, and her embalming (like that of Jesus) was an act of reverence intended to acknowledge her holiness and enhance her power.[89] More than anything else, it recalls the ritual exposure of the naked bodies of dead popes, which was part of papal funerary ritual, and the spoliation and even mutilation of the corpses of holy men and women by a devoted populace in search of relics, which was a topos in many saints' lives.[90] Furthermore, medieval understandings of the relationship between seer and seen differed from the ideas of subject and object that structure modern ideas of vision. Not only is the equation of the feminine with passive objectification and the masculine with active subjectivity unconvincing in light of the active visionary powers of many late medieval holy women. It is also not obvious that medieval approaches to vision — in a culture where images spoke, bled, and palpably altered the bodies of their viewers — should be reduced to modern categories of object and subject at all.

As a number of historians have emphasized, late medieval understandings of vision placed the seer and the seen in a relationship not of opposition or of domination and subordination but of reciprocity and mimesis. Sight was simultaneously active and passive, and the eye was both an instrument of penetration and a point of vulnerability. In the words of Suzannah Biernoff, "vision, in the medieval world, did not leave the viewer untouched or unchanged. The consequences of looking when one *shouldn't* were well rehearsed; but one could also expect to be positively changed by judicious looking.... To see was to become similar to one's object."[91] Looking was understood as a physical encounter, in which the subject reached out and touched the object, while the object — as I have already described — impressed its form on the viewer's body like a seal on soft wax. This account of vision was shaped by contemporary understandings of gender. Although the general model applied to both men and women, the female viewer, like the female object of vision, expressed both the powers and the vulnerabilities of sight in their starkest form. Possessed of porous, soft, moist bodies, they were particularly impressionable as viewers, as when they received the visual forms of external objects and impressed them onto even more malleable fetal flesh.[92]

This impressionability was understood as moral and spiritual as well as physical, and it found expression in many texts and images that celebrated late medieval holy women. One late fourteenth-century altarpiece by the Sienese painter Andrea di Bartolo shows Margherita, together with four other Tuscan and Umbrian female visionaries associated with the Dominican order (figure 1.6); each woman appears twice, in a larger magisterial image and in a smaller scene underneath, which depicts her in a characteristic attitude of devotion.[93] The three middle scenes show the physical power exerted on the viewer by the object of vision, in each case a crucifix. Vanna of Orvieto (second from left, d. 1306) stands transfixed in the pose into which she fell involuntarily every time she saw an image of the crucified Christ — a distressing and often inconvenient behavior that caused her companions to hide such images — while Catherine of Siena (center, d. 1380) is shown receiving her (invisible) stigmata from the crucifix that appeared to her in Pisa.[94] Next to her kneels Margherita, holding up her heart with its figured stones (figure 1.7), which had taken the impression of her visions of the Incarnation. The action of the object of vision of the viewer is even more vividly illustrated by the fresco in the chapel of Chiara's monastery that shows Christ setting his cross in Chiara's heart (figure 1.5); as Chiara gazes on the figure of Jesus, he literally pierces her body, impressing his image on her heart in a corporeal reconstruction of the interpenetration of viewer and viewed.[95]

In addition to being particularly impressionable viewers, women exemplified the active, potentially perilous power of sight. They could wound others merely by looking at them; examples included the evil eye and the poisonous gaze of a menstruating woman, which was thought to spot mirrors.[96] In the literature of courtly love, the eye of the (usually female) beloved wounds the (usually male) lover with a dart that enters him through his own eye.[97] As objects of vision, too, women posed particular dangers for men, moving them to lust and other forms of bodily temptation not only by inflaming their senses but also through the power of sight to assimilate viewer and viewed: in looking at women, men were drawn to resemble them in their moral weakness and their vulnerability to sin.[98]

Understanding these gendered notions of vision illuminates the spectacular public opening of Margherita of Città di Castello.

74

Since Margherita was blind, she ran no risk of visual corruption, at least by the sights of this world, but she could nonetheless serve as a source of sin for others. Thus when her corpse moved to cover its genitals, it acted to defend not only its own chastity but also that of its male viewers, many of whom were Dominican friars. Even in death, her naked body kept a strong and active influence over its beholders; the pudic gesture expressed both her modesty and her power. This combination of vulnerability and force was emphasized by the earthquake that accompanied the opening of her body, which in turn acknowledged the analogies between the woman naked on the table and Jesus naked on the cross.

The descriptions of Margherita's anatomy by the authors of her *vitae* reflect the omnipresent play of similarity and difference in the relationship between male and female bodies as they are portrayed in male-authored texts and images of the late Middle Ages. Viewed in light of the metaphors that shaped the imaginations of late medieval Christians, the spectacle of Margherita's open and naked body calls attention to the ways in which female corporeality served as a spiritual model for both men and women.[99] The malleability and impressionability of women's bodies figured the converted self's ability to conform itself to God, which was described in terms of the birth of Christ in the heart, just as Mary's labor pains at the Crucifixion modeled the Christian experience of compassion. In this sense, the friars, municipal officials, and doctors who participated in Margherita's anatomy could be expected to identify with its object. At the same time, however, Margherita's naked corpse was palpably different from the bodies of the hundreds of male spectators who observed it in the church of San Domenico; it was female while they were male, supine while they were upright, lame while they were able, blind while they strained to see. These differences created a certain distance between viewer and viewed that was absent when Chiara of Montefalco was opened by her fellow nuns. The interplay of resemblance and dissimilarity in the story of Margherita will continue to characterize the perceived relationships between male and female bodies — and between anatomists and their objects — throughout the period covered by this book.

Although the bodies of Chiara and Margherita were dissected because they were women, as I have argued in this chapter, their anatomies also set them apart from the rest of their sex. Their opened bodies and hearts figured women with nothing to hide: no concealed vices, no dirty secrets. Their external words and behaviors conformed with their innermost selves. In this respect, they differed profoundly from most women, who were defined, as I will show in the next chapter, by secretive bodies and secretive minds.

CHAPTER TWO

Secrets of Women

The year 1286 began inauspiciously in the cities that marked the
southern edge of the Po valley. February brought with it bitter
cold, frost, and snow, followed by a serious epidemic. According
to Brother Salimbene de Adam, who was spending his twilight
years in a Franciscan convent not far from his natal city of Parma,
the disease manifested itself in "swellings" (*apostemata*). What
struck Salimbene most was that it affected chickens as well as
people. "In the city of Cremona, one woman had forty-eight
chickens die in a short period of time," he wrote. "And a certain
physician had some of the chickens opened and found a swelling
on their hearts. There was a kind of abscess [*vescicula*] on the
tip of the heart of each chicken. Likewise, he had a certain dead
man opened and found the same thing on his heart." Although this
was the only anatomy recorded in connection with the epidemic,
medical experts outside Cremona were equally concerned with
addressing the growing threat to public health. In May 1286,
Salimbene noted, "the physician Master Giovannino, who was
living in Venice, where he was on the municipal payroll, sent a
letter to his fellow citizens back in Reggio telling them that they
should eat neither vegetables nor eggs nor chickens for the entire
month of May" — advice that led to a precipitous decline in the

77

price of poultry. The women who kept the chickens responded to this threat to their livelihood in their own way. According to Salimbene, "some wise women fed horehound to their chickens, either ground or crushed and mixed with water and bran or flour. And by the power of this antidote, the chickens were cured and escaped death."[1]

Salimbene's reference to the operation performed in 1286 by the unnamed physician of Cremona, twenty-two years before the opening of Chiara of Montefalco, looms large in the historiography of the origins of human dissection, where it is widely cited as the first credible and specific mention of the anatomy of a human body in western Europe.[2] Although nowhere near as detailed as the eyewitness testimony of the nuns of Montefalco, Salimbene's account is plausible, even if there is no particular reason to believe that it was the first procedure of this sort. The event he described took place in a city with which he had longstanding personal and institutional connections, and another local chronicle confirms the presence in 1286 of an epidemic that affected small animals as well as humans.[3] Salimbene's narrative is entirely consistent with evidence regarding the practice of dissection in the nearby city of Bologna, already an important center of medical teaching and practice: external (and possibly internal) postmortem examinations were being performed to determine the cause of death in criminal cases, as I described in the preceding chapter, and anatomies were taking place in connection with medical study.[4] Writing around 1275, the preeminent master in the city's nascent medical faculty, Taddeo Alderotti, noted that he could not describe the structure of the placenta in the case of multiple births, since the texts at his disposal did not address the matter satisfactorily; nor, he noted, "have I seen an anatomy of a pregnant woman" — a formulation that suggests that he may well have witnessed dissections of nonpregnant women, men, or pregnant animals.[5]

Salimbene's matter-of-fact account of the opening of a human body, as in the case of Chiara of Montefalco two decades later, confirms the absence of cultural or religious taboos governing these matters.[6] More informative than his meager description of the anatomy, however, is the broader context in which he placed it: the written prescription promulgated by Master Giovannino of Reggio, and the measures taken by the women of Cremona to care for their chickens. These framing details are intrinsically no more implausible than Salimbene's reference to the anatomy, and it is these very details, rather than the simple fact that a man's body was opened, that illuminate the meanings of the event as Salimbene understood it.

In this chapter, I will explore these meanings, which center on the problem of who knew what about the internal workings of the body, how they knew it, and to whom and in what ways they communicated that information. Salimbene's account does not focus on the anatomy *per se*, which occupies a subsidiary position in his narrative. Rather, it foregrounds these more general issues, even as it proposes that knowledge of the body's internal structures and processes can help people understand how diseases can be prevented or cured. The passage suggests two different models for gaining and disseminating knowledge of this sort. The first is personified by two male medical experts, the anonymous physician in Cremona and Master Giovannino of Reggio. The former performed the anatomy and most likely communicated its results to civic and medical authorities. (There are a number of records of municipally mandated and funded anatomies during the first epidemics of plague in the mid- to late fourteenth century, and it is quite possible that the procedure described by Salimbene was an early, less formalized version of this public-health practice.)[7] Salimbene implies that the physician's findings made their way to Master Giovannino, who then wrote a letter of preventive advice

— contemporaries would have called it a *consilium* — to his fellow citizens back home in Reggio.[8]

Salimbene's second model of gaining and communicating knowledge about the internal workings of the body and the treatment of disease is represented by those he called the "wise women [*mulieres sagaces*]" of Cremona. Lacking the formal medical training of Master Giovannino and the anonymous physician in Cremona but experienced in raising chickens, these women had apparently already figured out that they could treat the illness, at least in poultry, by administering horehound in chicken feed. In contrast to Salimbene's description of the physicians' actions, the friar does not specify how the women gained this knowledge or whether they made any effort to disseminate it. However, his account strongly suggests that their measures were not anatomically based and were the result of experience, rather than of study and theoretical reflection.

Taken as a whole, then, Salimbene's narrative concerning the epidemic of 1286 not only records one of the first anatomies for which we have plausible testimony but also demands to be understood in the context of contemporary medical culture. As in the case of the opening of Chiara of Montefalco, this context was highly localized and strikingly complex, revealing the coexistence and interaction of learned and lay forms of healing, oral and written modes of communication, and varieties of medical knowledge and practice, all of which had gendered associations. Taking the Cremona autopsy as its point of departure, this chapter explores the gendered lens through which contemporaries understood the knowledge of the inside of the human body, which dissection was developed to explore. I will argue, first, that knowledge of the body's interior based on anatomy and dissection was represented in late thirteenth- and early fourteenth-century Italian learned discourse as male and public, in opposition to characteristically female and secret forms of knowing, and, second, that the female body emerged during the course of this period as the ideal type of body, with a hidden and secret interior, the paradigmatic object of dissection.

The idea of secrecy was central to the way learned men wrote and thought about such issues. Although individuals could guard secrets in their bosoms, to be discerned by holy men and women like Chiara of Montefalco, secrets were more commonly understood as collective — the property of particular groups. The metaphor of secrecy, as Helmut Puff has claimed about the related metaphor of unspeakability, was a "trope of communication" that signaled the presence of a boundary between (at least) two communities of knowers.[9] Boundaries delineated by the invocation of secrecy could be established within a single linguistic community, as in Latin magical and alchemical writing, where they were maintained by the purposeful use of opaque allegories and verbal codes, or in the vernacular world of urban crafts, where proprietary techniques in fields such as glassmaking and textile production were protected by guilds and civic authorities.[10] In other cases, as with the "women's secrets" of this chapter, such boundaries marked the interface between learned expertise and lay knowledge, which corresponded to the interface between Latin and vernacular or written and oral culture — between the world of textual knowledge and the world of experience-based practice. In cases of this sort, in which secrecy was attributed by one group to another rather than claimed by the group in question, the idea often had sinister connotations, calling forth associations of power, on the one hand, and withholding or deception, on the other. In the hands of untrustworthy people — Jews, Muslims, and unlettered women, according to Christian men of learning — secret knowledge could be monopolized and kept for unsavory ends.[11]

As Puff has emphasized, secrecy was often invoked by learned writers not only to identify the knowledge of particular communities but also to facilitate or to control its dissemination. The problem of control gained particular urgency when the knowledge in question dealt with sexuality, a highly charged topic in the context of medieval Christian sexual morality and social practice.[12] When this chapter begins, in the second half of the thirteenth century, the secrets known by women were understood quite broadly as including a wide, if ultimately limited, repertory of therapeutic remedies and operations related to bodily health. Over the next two hundred years, however, these secrets began to be identified closely with sexuality and generation. In this context, women increasingly became the objects of knowledge rather than knowers themselves.

Secret Knowledge

Although Salimbene did not use the word "secrets" to describe the orally transmitted, experience-based, concrete, and bodily oriented therapeutic knowledge possessed by the wise women of Cremona, his learned contemporaries would have found the term appropriate. One of the best-known and most influential examples of the use of the term is in the Latin treatise *De secretis mulierum*.[13] (I will refer to it by its Latin title to differentiate it from the other treatises on the same topic that I discuss below.) Attributed to the Dominican theologian and natural philosopher Albertus Magnus, but probably composed in the late thirteenth century by a German follower, this work circulated well into the early modern period, mostly in Germany, in multiple, highly variable versions. The preface of the Latin text lays out the work's premise: "Since you asked me to bring to light certain hidden and secret [*occulta et secreta*] things about the nature and condition of women," the author wrote, "I have set myself to the task of composing this short and compendious treatise."[14] The text deals with the

"secrets of women" in two different but related senses: as knowledge about the processes of generation and the functioning of
the female reproductive system (these were literally "secret,"
since they were simultaneously hidden inside the body and unfamiliar to men, especially celibate male clerics); and as specialized knowledge possessed by at least some women, which they
might choose to withhold from men.[15] I will defer the problems
raised by the former meaning of women's secrets — as knowledge
of female bodies, generation, and sexuality — to the next section.
Here, I will focus on women's secrets in the second sense: what
and how women were thought to know, and how this knowledge
was thought to differ from that of learned men, especially male
physicians, which is the issue framed by Salimbene's account of
the epidemic and autopsy of 1286.

The author of *De secretis mulierum* was obviously a learned man —
indeed, he was a professed religious — writing for his peers. He
addressed his audience as companions (*socii*) and brothers (*fratres*)
and composed his work not merely in Latin but in the technical
language of natural philosophy and theoretical medicine, with
copious references to textual authorities such as the Arabic medical writer Avicenna and, especially, Aristotle. At the same time,
however, he presented himself as versed in women's knowledge —
"learned in various experiments [*experimentis*] from women," in
the words of his earliest known commentator.[16] In the context of
late medieval medical and natural philosophical writing, *experimenta* had a very specific meaning. Medieval "experiments" bore
little relation to the controlled tests of theoretical propositions
fundamental to modern scientific practice; rather, they were
"singular discoveries." In the words of Jole Agrimi and Chiara
Crisciani, they were born solely of experience: recipes, remedies,
and procedures found, often by trial and error, to accomplish a
particular result.[17] Commonly referred to as "secrets" (*secreta*) by

Latin writers, they were strongly associated with popular medical and artisanal practice as well as with the magical tradition. The horehound-based preparation used by the women of Cremona to cure their chickens during the 1286 epidemic, as described by Salimbene, was a classic *experimentum* in the medieval sense.

In the text attributed to Albertus Magnus, however, women's *experimenta* were not therapeutic. Rather, they were formulas and procedures known to certain women — the author referred specifically to "prostitutes and women ... learned in the art [of abortion]" — and exploited by them to the detriment of men.[18] For example, the author claimed that these women knew that vigorous activity, such as traveling, dancing, and having frequent sex, provoked miscarriage, while a piece of iron placed in the vagina at the time of the new moon caused serious injury to the penis.[19] He identified the latter fact as an *experimentum* — one commentator called it a *secretum*[20] — and strongly implied that women were loath to communicate this kind of knowledge except to other women, since it represented one of the most important ways they had to gratify their inordinate sexual appetites, exercise sexual autonomy, and exert power over the men in their lives.

As Agrimi and Crisciani have shown, the two types of knowledge about the physical operation of natural substances, knowledge based on causes or first principles and knowledge of contingent effects derived from observation, had gendered associations. The former was normally the province of the learned male scholar, versed in the theoretical structures of natural philosophy and medical theory, which qualified in scholastic epistemology as *scientia*, or certain knowledge; this was contained above all in works by Greek and Arabic writers (especially Aristotle, Galen, and Avicenna) that had been translated into Latin beginning in the twelfth century.[21] The latter type of knowledge, in contrast, did not require familiarity with the texts of ancient authorities,

Aristotelian categories of explanation, or academic techniques of hermeneutics and disputation — or indeed with any texts at all. Based on the accurate but untheorized perception of phenomena accessible to the senses — "I feed chickens horehound and they recover" — it was the province of the "rustic" (*rusticus*), the "simpleton" (*simplex*), and the "unlettered" (*ydiota*), which were all contemporary terms used to designate the illiterate layperson. This entire category of illiterate knowers was frequently assimilated to women, often further denigrated polemically as "old women" (*vetule*) or "little women" (*muliercule*). Just as Aristotle was *the* philosopher, so were women *the* mistresses of this sort of useful, if fallible, experience-based knowledge.[22]

Italian medical writers contemporary with Salimbene and the author of *De secretis mulierum* developed these ideas with greater specificity, though without the sexual paranoia that characterizes the latter work. They acknowledged the possibility of learning many useful remedies and therapeutic practices from women. In his *Surgery*, for example, Lanfranco of Milan announced his debt not only to "reverend medical doctors" but even to "women...., who are all without doubt experienced in the things that befall them."[23] At the same time, however, learned medical authors emphasized the inevitable shortcomings of this kind of knowledge. Another surgical scholar, Bruno of Longobucco, identified (Latin) literacy as a minimum requirement for practice and lamented that "vile and presumptuous women have usurped and abused this art [of healing], since, although they treat [patients], as Almansor notes, they possess neither art nor wit."[24] This complex of themes was particularly charged for teachers and practitioners of learned medicine in the region around Bologna in the decades after 1260. In this precise period medicine, originally an artisanal system of teaching and practice, transmitted privately, orally, and more or less informally from master to apprentice, began to develop an increasingly academic structure, in the form of medical faculties and a body of Latin instructional texts. This shift was first institutionalized at the University of Bologna around the figure of Taddeo Alderotti of Florence, who was the first Italian medical writer to claim implicitly to have witnessed the anatomy of a human being, as I mentioned above.[25]

One of the most eloquent spokesmen for the aspirations of learned medicine in this period was Guglielmo of Saliceto. A close contemporary and near neighbor of Salimbene, Guglielmo lived and practiced in Piacenza and Cremona, as well as elsewhere in the Po valley, although he probably died before the epidemic of 1286.[26] Educated in medicine and surgery in Bologna before the establishment of a formal medical faculty, he stood on the cusp of the transition from oral to written transmission of medical learning and of the emergence of medicine as a university discipline in Italy. His teachers, Ugo Borgognoni of Lucca and Bono del Garbo of Florence, worked in a craft tradition and left no writings, while their students, including Guglielmo, were the authors of important medical texts. In Guglielmo's case, these included a *Surgery*, composed in two phases, around 1268 and 1275, and a *Summa on the Preservation [of Health] and Healing*, a physic textbook, probably written between the two versions of the former work.[27]

In these two books, Guglielmo exhibited a characteristic suspicion of women's experiential knowledge of medical and obstetric practice. Like Lanfranco, he acknowledged that they could be a useful source of information; for example, he noted that some things concerning generation and birth — specifically, the structure of the placenta and umbilical cord — could be known only through the reports of midwives. But he also argued that such knowledge, based exclusively on experience and sense perception, required verification and interpretation by learned doctors.[28] Equally important, Guglielmo emphasized, was the role of literate physicians in committing to Latin script the otherwise transitory discoveries of women and other artisanal practitioners; as he declared in his preface, he had decided to record the fruits of his studies so that after his death "that which has been shut away in my soul and acquired over a long period of time will not disappear from common use."[29] In fact, one of the principal things for which he reproached his teachers was their practice of transmitting their learning orally, which compromised its permanence, its availability, and ultimately, therefore, the public good. He saw the new orientation toward the publicity of writing as the proud accomplishment of his generation of medical scholars. His own teacher Ugo Borgognoni had practiced an elliptical and evanescent form of oral teaching, as we know from Ugo's son Teodorico, who castigated his father for his pedagogical methods in his own work on surgery.[30] In contrast, Teodorico, like Guglielmo, emphasized his commitment to openness and publicity, so that the "secrets of the art of surgical medicine" might no longer be "hidden and implicit [*occulta et implicita*]."[31]

In the eyes of the new learned medical writers and practitioners, then, women stood rhetorically for the bad old ways. The fundamental issue involved not gender *per se* but generational and, increasingly, class difference. The "usual practice, pursued by old women," in the words of Taddeo Alderotti, stood for an earlier way of doing things, according to which healing was a craft tradition, organized around knowledge discovered casually and transmitted orally from master or mistress to apprentice or from parent to child.[32] The professional ideology of men like Guglielmo of Saliceto did not in principle reject the efficacy of this kind of practice but argued that it did not serve the interests of patients or practitioners. Within this general intellectual context, the logic of *De secretis mulierum*, which identifies secrecy with women's knowledge, is less peculiar than it might at first appear. Like

Guglielmo of Saliceto and his learned contemporaries, the author of the treatise saw women as the bearers of certain valid but limited forms of knowledge, derived from experience and unmodulated by rational criticism and analysis; these included, especially, the anatomy and physiology of generation and the practice of childbirth. For him, as for Guglielmo, it was precisely the fact that women communicated their knowledge only to their close associates, orally and in the vernacular, that made it secret; the same information committed to Latin script, which was in fact comprehensible only to a minuscule, overwhelmingly male fraction of the inhabitants of Europe, acquired, paradoxically, a public and "manifest" character.

This context illuminates in turn Salimbene's description of the Cremona anatomy. Like judicial and university dissections, this fit neatly into the epistemological regime developed by Guglielmo and his contemporaries, where both functioned as prime expressions of learned, public science in the sense I have described. Throughout the later thirteenth and fourteenth centuries — except in the extraordinary case of the nuns of Montefalco, enclosed in their monastery and working on a model derived more from obstetrics and embalming than from anatomy — the practice of dissection was the exclusive province of physicians, medical professors, and surgeons. Anatomies performed in the interest of public health, such as the Cremona anatomy of 1286, and forensic anatomies, like the ones documented in early fourteenth-century Bologna and described in the preceding chapter, were civic events; recorded in municipal account books, local chronicles, notarized depositions, and judicial decisions, they served the common good. For the same reason, such anatomies were typically mandated and paid for by municipal authorities and entrusted to established physicians and surgeons — often, indeed, to physicians and surgeons already on the public payroll (*medici condotti*), charged with the care of the general population or the poor.[33] This was the case with the opening of Margherita of Città di Castello in 1320, described in Chapter One, while the notarized oath sworn by Sister Francesca and her colleagues several days after the opening of Chiara of Montefalco in 1308 was arguably a belated attempt to bring a private event into the public realm. Municipal physicians and surgeons were invariably male — although a small number of women matriculated in local medical guilds, I know of none who was ever given a public commission — and as far removed on the spectrum of healers from "old women" and other private and irregular practitioners as it was possible to be.[34]

Much the same can be said of those who performed anatomies in the context of high-level medical instruction, such as (presumably) Taddeo Alderotti at the University of Bologna, and his pupil, Mondino de' Liuzzi, author of the first anatomy textbook based on human dissection, which he composed in the second decade of the fourteenth century.[35] Because the university was in part a municipal institution — some faculty members received salaries from the city — the anatomies that took place under its auspices had a public character as well. Mondino implied that his cadavers were typically executed criminals, "dead from beheading or hanging," which were no doubt supplied through official channels, although it is unlikely that this was his only source.[36] Conversely, the only people known to have been prosecuted in connection with an anatomy in this period were four students at the University of Bologna, who attempted to appropriate a corpse for private study by secretly disinterring the body of a recently executed criminal in 1319.[37] In general, however, late thirteenth- and early fourteenth-century autopsies and dissections epitomized the system of learned medicine championed by Guglielmo of Saliceto, oriented toward publicity and the common good.[38] Trained in logic and natural philosophy and versed in the works of Aristotle,

Galen, and Avicenna, the men who performed them subscribed to an epistemology that rejected the "secrecy" of private, lay knowledge about healing and located certainty in the abstract and intellectual understanding of causes — the hallmark of the academically educated physician — rather than the more concrete and empirical knowledge of effects and *experimenta* that were the province of craftspeople and women.[39]

The study of anatomy played an important role in this epistemology, since, as Galen had repeatedly argued, knowledge concerning the bodily interior was essential to understanding both the workings of the healthy body and the causes of particular illnesses, which were in turn essential to knowing how to treat them. Galen had stressed the need for anatomical knowledge in various works that circulated under his name in late medieval Latin versions, of which the most influential were *On the Uses of the Members* (*De juvamentis membrorum*) and *On Interior Things* (*De interioribus*).[40] The former explained why particular organs had the forms they did, while the latter addressed the internal causes of diseases as revealed through external signs and symptoms. One of the two extant commentaries on *On Interior Things* was probably the work of Bartolomeo of Varignana, a student of Taddeo and contemporary of Mondino, who not only studied and taught medicine at the University of Bologna but also participated in the first two documented forensic anatomies involving internal inspection of the body, performed in the city in 1302 and 1307.[41] In his commentary, Bartolomeo stressed the importance of knowing the internal parts of the body where illnesses were located, since these were the key to understanding both the material and the final causes of those illnesses.[42]

In this way, anatomy became emblematic of a particular kind of knowledge, different from and opposed to the "secrets" of women. Learned medical writers acknowledged that both wise women and anatomically trained physicians aimed to determine the proper treatment for particular illnesses. But whereas women's knowledge relied on chance and superficial observations and was, as a result, contingent and uncertain, the anatomical knowledge of university-trained physicians allowed them access to causes hidden inside what Bartolomeo of Varignana called the "deep places" (*locis profundis*) of the body.[43] During the epidemic of 1286, the wise women of Cremona just happened on a remedy for their sick chickens. In contrast, the physician who performed the autopsy used a technique that, despite its novelty, had a clear theoretical and epistemological warrant. It gave him access to the interior of the body, not only allowing him to reveal the true nature of the disease — a cardiac abscess that afflicted its human and animal victims — but also laying the foundation for a publicly disseminated set of recommendations to prevent its spread.

Secret Bodies

So far, I have treated the secrets of women as an epistemological concept, focused on different ways of gaining knowledge about the body and communicating, or not communicating, that knowledge. But the phrase "women's secrets" had a second, equally important set of associations in this period, when it was used primarily to refer to what the author of *De secretis mulierum* called certain "hidden, secret things about the nature of women" — specifically, information relating to generation and the female genitals. In this connection, women appeared primarily not as possessors of a particular form of knowledge about the body but as the objects of knowledge itself.

Initially, this latter meaning of "women's secrets" tended to be subsumed within the former; not only did the processes of conception and gestation — indeed, heterosexual intercourse itself — take place inside women's bodies, but women were also considered the principal experts on those matters, "experienced in the things that befall them," in Lanfranco's words.[44] They knew when they or others were pregnant or not pregnant, virgins or "corrupted," or carrying a male or a female fetus, and women managed childbirth as well. Beginning in the second half of the thirteenth century, however, the cities of the Italian peninsula saw the appearance and consolidation of a well-established body of formally trained (and overwhelmingly male) medical practitioners who served the burgeoning market for medical expertise, as well as the development of medicine as an academic discipline. In the competitive and well-paid world of urban medicine, male physicians had strong incentives to develop practices that included the treatment of what were known as "women's illnesses" (diseases of the female genitals and reproductive system) as well as the management of infertility and eventually even childbirth, as I will show in Chapter Three.[45] In this way, the interior of women's bodies became a matter of interest to medical scholars and practitioners as well as to literate laymen, and by the middle of the fifteenth century male medical writers had developed enough confidence in their mastery of the subject that they began to present themselves as equal or even superior to female experts, notably midwives.

These changes took place against the background of a broad shift in learned writing about female bodies, which has been described by Monica H. Green.[46] As Green has shown, the language of secrecy was increasingly used to refer to women's genitals and women's sexual and reproductive functions, so that by the middle of the thirteenth century the phrase "women's secrets" became widely understood as referring specifically to such matters. The new terminology reflected a growing audience for this kind of information, consisting of male physicians, theologians, and natural philosophers to whom women's bodies really were secret; "the language of concealment thus comes into play at exactly the moment when women's bodies need to be revealed to men."[47] At the same time, Green argues, there was a shift in the contours of the literature that treated these topics. Whereas writing on women's illnesses within a medical framework had previously been quite distinct from writing on sexuality and generation, which was typically the province of theology and natural philosophy, the late thirteenth and fourteenth centuries saw the drifting together of the two traditions. Thus texts on generation began increasingly to incorporate medical ideas — the *De secretis mulierum* attributed to Albertus Magnus is a good example — while works on women's diseases focused on problems connected with menstruation and reproduction, rather than on the broad range of women's illnesses.

The permeation of literature on women's medicine by an intense preoccupation with generation had significant ramifications. Not only were women and their medical needs increasingly defined primarily or exclusively in relation to their reproductive organs and functions — which had not been the case in earlier writing on women's illnesses — but the topics of sexuality and generation were also couched in terms shaped by Aristotelian natural philosophy and Christian theology, both of which had strong antifeminist elements. The ultimate effect, as Green has argued, was to provoke a new and "unsavory, misogynist taint to the whole enterprise of writing publicly about women's bodies."[48] Like the mistrust of women's knowledge about the body, this mistrust of women's bodies themselves became encoded in the language of secrecy.[49]

This process is exemplified in two different Italian treatises that circulated under the title *Secrets of Women* during the fourteenth and fifteenth centuries, most notably in Florence. In some respects, these two works were quite distinct in their approach to women and their bodies. Like the Latin *De secretis mulierum* attributed to Albertus Magnus, the text known as *I segreti delle femine* belongs generally to the realm of philosophical writing on sexuality and generation and is in fact a very literal late fourteenth- or early fifteenth-century translation into the Florentine vernacular of *Les secrés des dames*, a French abridgement of that work.[50] Following this, *I segreti delle femine* presents an intimidating and repellent picture of women. Focusing on menstruation, which it consistently calls an "illness" (*malactia*), it describes the menses as a corrosive and poisonous fluid, emitted through the eyes as well as the vagina.[51] At the same time, it describes women as intensely lustful, craving sex at the expense of their male partners' health, and able to cause grave injury to men through the practices of anal sex or sex by the light of the moon.[52] The second vernacular text, which circulated under the title *Le segrete cose delle donne*, is very different in tone.[53] A fourteenth-century Florentine reworking and condensation of Latin texts associated with the name Trotula, it belongs to the medical branch of the "women's secrets" tradition that focused on therapeutics.[54] Whereas *I segreti delle femine* portrays women's bodies as poisonous and dangerous to others, *Le segrete cose delle donne* speaks of women respectfully, representing them as merely frail and fundamentally imperfect — a danger to themselves. "Because women are naturally weaker [than men], and because they are often ill, we must know that they abound with diseases of the members devoted to generation, which they do not dare to reveal," the author writes. "They hide the weakness of their condition out of shame."[55]

The two vernacular texts are similar, however, in denying women the expertise regarding their own bodies that one finds in earlier works on the topic, including the Latin *De secretis mulierum*, whose author attributed considerable knowledge — even if it was used malevolently — to "some women," specifically prostitutes, abortionists, and procuresses. *I secreti delle femine*, on the

other hand, following its French source, transforms the premise of the Latin work to de-emphasize women's knowledge. In his prologue, for example, the author implies that women need to read his work at least as much as men:

> A lady asked me for courtesy's sake to write something profitable. And although I say I have little knowledge of this topic and am not used to write or to dictate, nonetheless, I will undertake it for her love, because her love has divided my heart and my mind, so that I wish only to do things that please her.... And similarly, I hope to have better treatment from her, or at least some comfort. And because she is a perfect lady, who in all good things concerning honor and beauty is without peer in the world, I will tell you [first] what I have found in a book called *The Secrets of Women*. I pray this lovely lady not to be angry with me when she reads this book, nor to love me less because I tell here their secrets as briefly as I can.[56]

Drawing (perhaps satirically) on the formulas of courtly love, the author reverses the logic of the original Latin *De secretis mulierum*. Rather than revealing to men what women already know about their internal processes, he suggests that women — specifically, his beloved — need instruction and would "profit" from familiarity with this male-authored work. His implicit acknowledgment of women's ignorance about the facts of generation thus transforms the meaning of his reference to women's "secrets" in the last sentence. Such matters are indeed secret because information about them is hidden inside women's bodies, inaccessible to both men and women, not because such knowledge is withheld by women from men.

Furthermore, although the author of the French text on which *I segreti delle femine* is based retains the claims in the Latin text that "some women" injure men during intercourse and that "evil women" provoke miscarriages by intensive partying, the Italian translator has removed all references to these techniques as part of a larger body of women's knowledge.[57] Where the Latin text refers to women "learned [*docte*] in these evil deeds and other similar practices" and to "prostitutes and women learned [*docte*] in this art,"[58] the Italian version (like its French original) eliminates all references to learning (*doctrina*), implying that the women in question may well be acting out of ignorance or self-indulgence.[59] Even the section on childbirth, the only place in the Latin text in which women's knowledge was invoked in entirely positive terms, has been qualified; where the Latin refers approvingly to "competent midwives [*obstitrices discrete*],"[60] *I segreti delle femine* notes only that "there are very few wise midwives who know [their art] perfectly ... and on account of this many children are lost."[61] All in all, this work presents women as incompetent, malicious, lustful, physically disgusting, and prone to use their privileged knowledge of the female body to deceive and manipulate their unwitting husbands and partners.

Despite its milder tone regarding women's nature, the other vernacular treatise, *Le segrete cose delle donne*, rejects even more decisively the idea that women possess a body of secret knowledge about sexuality and generation. There are no evil and lascivious women here, bent on provoking abortions. Rather, women appear almost exclusively as patients; their bodies are the objects of learned male solicitude, and it is assumed throughout that their concerns focus on infertility and the difficulties of childbirth, rather than on unwanted pregnancy. Even the chapter on childbirth puts male physicians clearly in charge.[62] Not only does the author of *Le segrete cose delle donne* imply that women have no special knowledge regarding their sexual and reproductive powers, omitting even the role of the midwife in childbirth, but, unlike the lovestruck author of *I segreti delle femine*, he gives no indication that he even considers them potential readers; the book presents itself as a specialized work by a male medical writer for an audience of his peers, and the manuscripts themselves confirm that the treatise was used mainly by surgeons and, especially, physicians.[63]

Although the particular group of manuscripts in which *Le segrete cose delle donne* appears reflects a local Florentine tradition, it is evidence of the broader urban phenomenon I mentioned earlier: the increasing involvement of medical men in the specialized treatment of women's illnesses, especially in the areas of sexuality and generation. There were strong incentives for urban doctors to pursue these interests. From at least the early fourteenth century, they had been consulted by secular and ecclesiastical authorities charged with adjudicating suits involving petitions for divorce or contested inheritance, which often hinged on determinations of virginity, pregnancy, and legitimacy. In such cases, male medical experts increasingly supplemented or even replaced the all-female juries formerly charged with matters of this sort. Thus, whereas in 1241 the archbishop of Pisa had ordered a number of "good and honest women of good repute" to examine one Ricca for signs of virginity — she was suing to divorce her husband on grounds of impotence — by the early fourteenth century, at least in cities with well-developed medical institutions such as Bologna, physicians might be called on to make determinations regarding pregnancy in collaboration with midwives. In 1302, the year in which Bartolomeo of Varignana participated in a judicial anatomy, he was also asked to consult on a case of this sort.[64]

Even more compelling were the reasons for male doctors to develop expertise in treating women for illnesses relating to generation, in the context of the growing market for such services. Men had long been involved in the medical care of women, including problems with their genitals and reproductive systems.[65]

But the scale and intensity of their involvement seems to have increased dramatically over the late fourteenth and fifteenth centuries. This reflected in part the growing confidence of the urban population, particularly the well-to-do, in professional medical services, and in part concerns about family continuity in the decades after 1348, when the advent of plague sent mortality levels soaring, inspiring a surge in natalist sentiment on the part of both municipalities and individuals.[66] The pressure on elite married couples to reproduce was huge, and those who failed to bear a child — specifically, a male child — might go to great lengths to secure their lineage, consulting practitioners of all sorts. One well-documented case involved Francesco di Marco Datini, a wealthy silk merchant from Prato, and his wife, Margherita. When Margherita had not managed to conceive after several years of marriage, their family and friends suggested a variety of remedies, ranging from a magical belt and a poultice made by a female empiric to the services of a well-known local physician, Master Naddino di Aldobrandino of Prato.[67]

Among the wealthy, there is ample evidence for the increasing involvement of male physicians in treating infertility and in supervising pregnancy and the care of newborn infants, fields that had long been the special, if not the exclusive, province of midwives and other experienced women.[68] One of the fifteenth-century Florentine manuscripts containing *Le secrete cose delle donne*, for example, reflects a specialized interest in these matters by its physician-owner; the last four folios are devoted to a large number of inserted Latin recipes, *experimenta*, *secreta*, and *consilia* relating exclusively to conception and fertility, written in what was evidently the owner's hand.[69] By the second half of the fifteenth century, as I will show in the next chapter, pregnant women fearful that they were about to miscarry consulted their physicians rather than their midwives, while husbands recorded

98

in their account books payments made to doctors for the care of their pregnant, parturient, and postparturient wives. In sum, we see an increasing inclination on the part of well-to-do urban inhabitants of late fourteenth- and fifteenth-century Italy to entrust their reproductive lives to the supervision of male physicians. While husbands and wives invariably retained midwives and female assistants to manage the birth itself, these women seem increasingly to collaborate with — and perhaps to some degree under the supervision of — medical men.[70] While I have found no evidence of male midwives, in the sense of men whose work was largely or entirely confined to the delivery of babies and who were identified as such by both themselves and others, it is worth noting that Benedetto de' Reguardati, one of the principal physicians at the Sforza court in Milan, had a reputation for specialized expertise in the management of infertility and childbirth. In a letter from 1469, on the occasion of the birth of the future duke, Galeazzo Maria Sforza, he went so far as to describe himself, somewhat facetiously, as "no less a midwife than a doctor [*non meno balyo che medico*]."[71]

In addition to encouraging enterprising physicians to develop practical expertise in the areas of fertility, sexuality, pregnancy, and birth, this lucrative market for medical services led to a surge in the number of original writings on these topics by fourteenth- and fifteenth-century Italian physicians. These included short items such as recipes and *consilia* produced in the context of high-level practice, as well as sections on generation that formed part of longer medical works and, beginning in the fifteenth century, specialized treatises on gynecology and obstetrics, such as those of Antonio Guaineri and Michele Savonarola, which I will discuss below.[72] Such works served both to educate students and other practitioners in the growing field of reproductive medicine and to enhance their authors' professional reputations for this kind of

99

expertise. Male authors had long been active in this area; as early as the second half of the thirteenth century, Guglielmo of Saliceto's *Summa* included a section on gynecology, which dealt with fertility and pregnancy (though only marginally with childbirth), along with diseases of the female genitals.[73] However, the extensive circulation of these free-standing works, freshly composed and specifically focused on reproductive issues, is new, and their existence dramatizes the increasing involvement of male doctors, especially physicians, in the theory and practice of reproductive medicine. Writing in the second half of the thirteenth century, Guglielmo of Saliceto presented himself as knowledgeable in such matters, but he seems not to have involved himself in the process of birth, or indeed in any aspect of treatment that involved inspecting or manipulating women's genitals, which he left to midwives and other female practitioners. Furthermore, he was well aware that these restrictions limited his knowledge in important ways. "These details about membranes [that envelop the fetus], the umbilical cord, the generation of the heart and liver and brain . . . , the excretions retained together with the fetus until the hour of its birth, and other such things pertain to physic," he wrote; "They are verified by anatomy [probably a reference to animal dissection], by the reports of midwives, and by women with children and those who have experienced many births and miscarriages. And without miscarriages, it seems to me that they cannot be verified in any other way by the physician."[74]

By the first half of the fifteenth century, however, physicians had more direct ways of obtaining information on such topics. According to Antonio Guaineri (d. 1440), in his *Treatise on Uteruses* (*Tractatus de matricibus*), he not only examined the bodies of his female patients, sometimes even touching and inspecting their genitals, but also personally attended births.[75] He collaborated closely with midwives, directing them as they manipulated pro-

lapsed uteruses and treated women afflicted with uterine suffocation using vaginal suffumigation and masturbation.[76] He had
considerable respect for the knowledge of midwives and other
female practitioners, whom he — like the author of the Latin
treatise *De secretis mulierum* — frequently described as "learned"
(*docte*) in the matters that concerned them, although he noted
that the errors of midwives might create serious problems, such as
uterine abscesses and prolapses. Michele Savonarola's vernacular
treatise on childbirth and the care of infants, written in the 1450s,
goes further still in this direction. Whereas Guaineri treated
midwives as expert, albeit fallible, and indicated that physicians
had much to learn from them, Savonarola consistently asserted his
superior knowledge and authority. He assumed that normal pregnancies and births would be managed by midwives but that anything out of the ordinary — failure to conceive (including failure
to conceive sons), illness during pregnancy, complications in
childbirth — was the province of doctors. On preventing miscarriages, he wrote, "I wish to leave it to the practicing doctor
[*medico pratico*], because it is not a job for the layperson [*vulgare*]"
— a category in which he clearly included midwives.[77] In lively
prose, he modeled what he believed to be proper hierarchies of
knowledge regarding reproduction: the husband asks the physician,
"Teach me, master, to make sons," while the midwife inquires,
"Master, how should I take care of [the woman who has just given
birth]?"[78] (Although the treatise claims to be written for "the
women of Ferrara," the voices of Savonarola's female patients are
never heard.) In connection with this medicalizing of reproductive matters, he redefined pregnancy and childbirth as "natural"
but "acute illnesses" — the first time I have seen them patholo-
gized in a way clearly intended to bring them into the purview of
the physician.[79]

Like other professional medical writers of the fourteenth and fifteenth centuries, Savonarola made no reference to women's secrets. In part, this was certainly because, once defined as referring to women's sexuality and reproductive functions, especially in vernacular treatises such as *I segreti delle femine*, the phrase had acquired a salacious, nonscholarly tinge that worked against the dignity of the physician's learning. In part, too, physicians rejected the terminology of women's secrets because it was in their interest to break down traditional conventions of female modesty, at least in this particular area, in order to attract a female clientele. In part, finally, it reflected a thoroughgoing rejection of women, whether they were prostitutes, midwives, or mothers, as experts in the areas of sexuality and reproduction. Savonarola went much further than precursors such as Guaineri or the compiler and translator of *Le segrete cose delle donne* in this respect, presenting himself as someone who instructed midwives rather than someone who had been instructed by them.[80] "With respect to the midwife (*comare over ostetrice*)," he wrote, "I say that here I will attempt to lay out clear teaching (*doctrina*) to all midwives and their assistants, so that they may become learned in the rules that must be observed in birthing the fetus, because it is certain that many babies and mothers either die or suffer harm on account of their ignorance."[81]

Savonarola's claim to independent expertise in these matters was no doubt exaggerated, for it is clear that the detailed instructions he gave for the management of normal childbirth can only have come directly from observation of and conversation with midwives.[82] Nonetheless, his treatise marks a new stage in the attempts by male medical authorities to penetrate the realm more vulgarly known as women's secrets, which was no longer identified as women's knowledge regarding sexuality and reproduction but now meant the physiological functions associated with the uterus. Given the growing visibility of women's genitals to their physicians, the secrecy in question was less sociological than ana-

102

tomical; women's bodies required intensive scrutiny and study, not because women were modest or malicious but because their bodies were intrinsically opaque, since, to their own disadvantage, the reproductive processes that were increasingly thought to define them took place deep inside.[83]

The Uterus Revealed

Even for physicians such as Guaineri and Savonarola, however, the uterus was frustratingly inaccessible.[84] The male genitals presented themselves for easy inspection and consideration, and their reproductive functions were, by comparison, fairly simple: they produced (or didn't produce) fertile seed apt to generate sons. Furthermore, despite the social and political importance of producing sons to continue the lineage, the male genitals were not considered as fundamental to men's health and physical identity as the uterus was to women's. If a man's overall complexion — the balance of hot, cold, wet, and dry qualities that shaped his corporeal and psychological being — depended on the complexion of his heart, a woman's complexion depended on that of her womb; in addition, the uterus, as Mondino had already noted in his early fourteenth-century *Anatomy*, was physically connected to almost all of women's higher organs: the brain, heart, and liver, not to mention the breasts, diaphragm, bladder, and colon.[85] This also made women more difficult to understand than men, as Guaineri emphasized, since the complexion of the uterus varied over the course of the menstrual cycle.[86] Finally, from a cultural point of view, the female genitals simply had more important things to say than the testicles and penis. It didn't really matter if men were virgins (though it did matter if they were fertile); they rarely menstruated and they never got pregnant, let alone got pregnant with boys or girls.[87]

All these things mattered desperately in the case of women, and it was therefore deeply frustrating that the uterus was concealed inside the body, so that its state and contents could be inferred only indirectly, through the woman's testimony or through external signs. Given the suspicion of the former that permeated the natural philosophical and medical literature, those signs became the focus of intense scrutiny in the texts on sexuality and generation I described above. Whether written in Latin or in the vernacular, produced for lay readers or for learned practitioners, couched in terms of "women's secrets" or in the sober language of professional therapeutics, these works devote a great deal of space to the external indicators of virginity, pregnancy, and fertility: for example, the color of a woman's face, the shape of her body, whether her feet feel cold after sex. In this sense, the entire project of writing about women's health in the fourteenth and fifteenth centuries was haunted by what Claude Thomasset has called the "enigma of the interiority of the female body."[88]

As Guglielmo of Saliceto had already pointed out, however, there were ways to obtain information about the uterus and the development of the fetus without relying on women's notoriously unreliable word, notably through the examination of miscarried fetuses and through "anatomy."[89] While Guglielmo was almost certainly referring to animal dissection — animals, especially pigs, had been used to demonstrate the internal organs to medical students since at least the mid-twelfth century — Taddeo Alderotti, his slightly later contemporary at the University of Bologna, seems also to have worked with human bodies. The fact that Taddeo regretted not having had the opportunity to see the anatomy of a pregnant woman, as I have already mentioned, suggests not only that human dissection was an emergent practice in late thirteenth-century Bologna but also that he shared Guglielmo's opinion about the general importance of direct inspection for exploring the mysteries of the pregnant uterus.

One of the striking things about late medieval learned medical writing on the pregnant uterus is the conviction that this was an area in which the work of Galen and Arabic medical authorities was lacking. Neither *On the Uses of the Members* nor *On Interior Things*, the two principal works on human anatomy and physiology that circulated under Galen's name in this period, dealt with the topic at all. (In fact, the former, which omitted the last four books of Galen's *On the Use of Parts*, did not deal with any aspect of the genitals, male or female.) The pseudo-Galenic *On Seed* said only that the uterus had seven symmetrical subdivisions, three on the right for male fetuses, three on the left for female fetuses, and the middle one for a hermaphrodite.[90] As a result, late thirteenth- and fourteenth-century medical writers in search of more detailed information on fetal development and the changes of the uterus in pregnancy had to rely on the relevant chapter of Avicenna's *Canon* (3.21.1.2); this became the object of commentaries by not only Mondino de' Liuzzi, whose anatomy textbook I have already mentioned, but also Mondino's colleague Dino del Garbo, Dino's son Tommaso, and Tommaso's contemporary Jacopo of Forlì — an indication of continuing interest in the subject.[91] These treatises are among the very few to contain references to the use of human dissection not merely as an instrument of medical instruction but also as a tool of medical research. Mondino himself invoked it to determine how fetuses breathed inside the womb, arguing that "if the fetus breathed through its mouth, then the air would make its lung white; but experience shows its lung is red, when an unborn [excised?] fetus is anatomized."[92] (The term *non natus*, typically used for children delivered by opening dead mothers rather than miscarried, implies that Mondino was referring to humans rather than animals.) A generation later, Tommaso del Garbo also recommended the study of miscarried or stillborn fetuses in his own commentary.[93]

The unavailability of pregnant female corpses for anatomy that so frustrated Taddeo was long to vex medical scholars, given the practice of delaying executions until the condemned had given birth. This forced them, like Mondino and Tommaso del Garbo, to rely on miscarried or excised fetuses and pregnant pigs. The remarkable thing about all these texts, however, is the intensity of their concern with matters relating to the uterus. In this connection, it is notable that the first two specific references to dissections in an Italian academic context — Taddeo's mention of the anatomy he had never seen and Mondino's mention of two dissections performed in January and March 1316 — both involved not only women's bodies but the uterus in particular.[94] This is further evidence that the uterus was the exemplary object of anatomy, the ideal type of the organ whose truth could be plumbed only by opening up the body.

The history of anatomical illustration reinforces this point, for the pregnant uterus was the first human internal organ to be represented in an Italian medical work on the basis of personal observation (figure 2.1).[95] The image in question, a woodcut of a seated woman whose abdomen has been opened for inspection, appears in the *Medical Compilation* (*Fasiculo de medicina*), an influential vernacular collection of medical texts and figures attributed erroneously to Johannes de Ketham and published by an important Venetian press, that of the de' Gregori brothers, in 1494.[96] The *Fasiculo* was an expansion and translation of the original Latin edition of the compilation, published by the same press three years earlier, and this figure is the only one of the original woodcuts to be completely redesigned for the 1494 edition. (Three woodcuts were added, including a depiction of an academic dissection, which I will discuss below.)[97] The image's title calls attention to the element of direct observation, describing the image as the "Figure of the uterus of a woman *from nature* [*Figura dela matrice dal natural d'una dona*]."

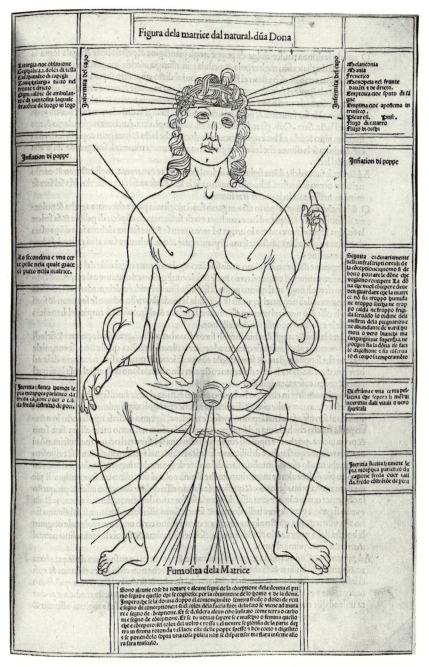

Figure I.6. "Figure of the uterus from nature." *Fasiculo de medicina*, attributed to Johannes de Ketham, ed. and trans. Sebastiano Manilio (Venice: Giovanni and Gregorio de' Gregori, 1494), sig. d1r.

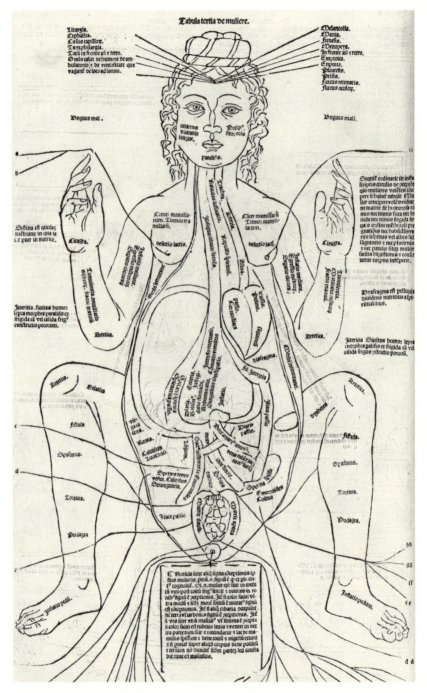

Figure 2.2. "On woman." *Fasciculus medicine*, attributed to Johannes de Ketham, ed. Giorgio Ferrari da Monferrato (Venice: Giovanni and Gregorio de' Gregori, 1491), sig. av v.

Nothing specific is known regarding the draftsman who created the original sketch on which this woodcut was based (perhaps through one or more intermediary drawings), or who the
woman was and where she was dissected. If she was pregnant, as
is often claimed, she is unlikely to have been executed as a criminal; she may have died and been anatomized in one of the Venetian hospitals.[98] The novelty of the woodcut depicting her opened
body is apparent when compared with the corresponding woodcut from the 1491 Latin edition of the *Fasciculus* (figure 2.2).
Based on a German manuscript tradition dating to the late fourteenth century, this woodcut makes no visual or verbal allusion to
dissection.[99] It is presented as a diagram — there is no sign of an
incision, as in the later image — and its generic title, "Third picture, on woman [*Tabula tertia de muliere*]," emphasizes neither the
uterus nor the element of direct experience alluded to in the
phrase "from nature [*dal natural*]" attached to the 1494 woodcut.[100] The vaunted naturalism of the latter was further height-
ened by the removal of many of the identifying captions (references
to parts of the body and the diseases to which they are subject)
that adorn the figure in the 1491 diagram.

The 1494 illustration of the "uterus from nature" is particularly striking in the context of the tradition of anatomical illustration in Italy, which was less developed than its northern European
counterpart.[101] The images from the rich fifteenth-century manuscript tradition on which the *Fasciculus* was based, including its
female figure, do not appear to have circulated south of the Alps;
rather, almost all of the few anatomical illustrations in Italian
manuscripts come from much earlier — the thirteenth or fourteenth century. Like the *Fasciculus* woodcuts, they are diagram-
matic rather than descriptive and make no reference to the process
of dissection: they are illustrations of texts rather than records of
observations. Of these, the most elaborate are the three known

Italian manuscripts of the *History of Cutting*, or *Historia incisionis*, which is associated with a set of nine images of internal organs once called the "five-figure series"; none of these manuscripts includes the eighth figure, a highly geometric schema of the womb, although this is most likely an accident of survival.[102] The other principal type of older uterine diagram appears in the series of images of fetal presentations associated with the *On Matters Pertaining to Women* (*De genecia*) attributed to Muscio, which do not focus on uterine anatomy *per se* (figure 2.3).[103]

The only image with Italian associations to refer to the dissected uterus before the 1494 *Fasiculo* woodcut is something of an anomaly. It appears in a sumptuous manuscript, the *Book of Notable Things*, which was commissioned in 1345 for presentation to Philip VI of France by Guido of Vigevano, an Italian physician at Philip's court. The manuscript includes a number of medical treatises of Galenic inspiration, of which the last is an *Anatomy*, described in the text as "drawn with figures by Guido, physician of the said King" and conceived as a pendant to the extract from Galen's *On Interior Things* that formed part of the manuscript.[104] Its eighteen paintings were meant to substitute for the experience of viewing a dissection, a practice familiar to Guido from his Italian training, most likely at the University of Bologna, but still viewed with suspicion in northern France. ("Because the Church prohibits performing dissection on human bodies," Guido wrote, "and since the medical art cannot be fully known without knowing anatomy, as Galen says in the first book of *On Interior Things* . . . , I will demonstrate the anatomy of the human body clearly and openly, using well painted figures.")[105] While the series focuses on the dissection of a male cadaver according to the order described in Mondino de' Liuzzi's *Anatomy*, the tenth figure shows the uterus (figure 2.4). This emphasizes two aspects of uterine anatomy mentioned by Mondino: its seven cells, as described in the

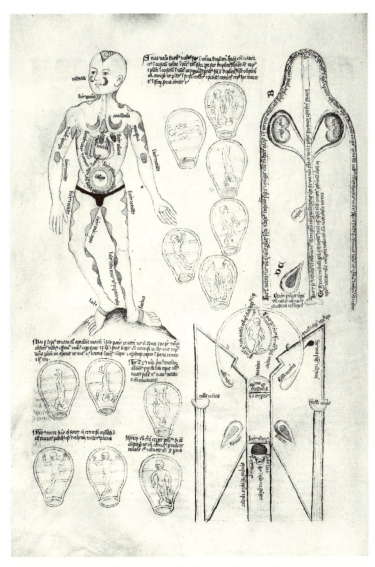

Figure 2.3. Diagram of the female genitals (*lower right*) and of fetal presentations (*center and lower left*). [*Historia incisionis*], London, Wellcome Library for the History and Understanding of Medicine, ms. 49, fol. 35v, German, ca. 1420.

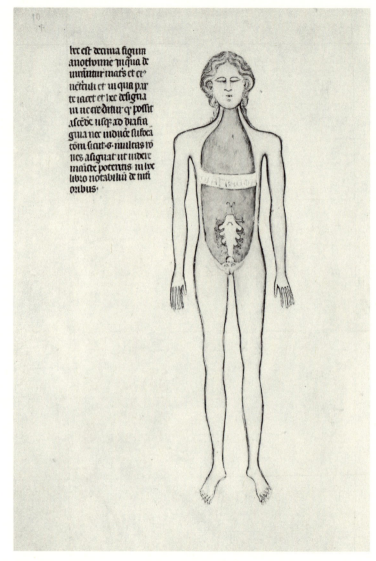

Figure 2.4. Anatomy of the uterus. Guido of Vigevano, *Anothomia Philipi septimi, Francorum regis*, tenth figure, Chantilly, Musée Condé, ms. 344, fol. 267v, French/Italian, 1345.

pseudo-Galenic *On Seed*, and the fact that, as Galen argued in *On Interior Things*, the illness known as suffocation of the uterus does not arise from the upward motion of this organ to compress the diaphragm and stomach, since it is anchored below the navel.[106] (Suffocation of the uterus was a potentially fatal disease of women resulting from retaining seed and menses.)[107] The small, hook-shaped structures visible between the cells in the painting are probably intended to represent the ligaments that hold the uterus in place. Although commissioned by an Italian physician, probably from an Italian artist, this unique image does not appear to have circulated in either Italy or France.

The sparseness of late medieval Italian anatomical illustrations relative to their northern European counterparts may arise, at least in part, from the very situation that makes it so surprising: the currency of human dissection in Italy. As Guido explained in his introduction, he had his own (text-based) images painted to substitute for the viewing of a dissection, because dissection was "prohibited" in France.[108] The same was true of the early fourteenth-century French surgeon Henri of Mondeville, also educated at Bologna, who lectured on anatomy at the University of Montpellier using a series of thirteen images that stood in for the dissected cadaver.[109] Given the expense of creating paintings or images of the dissected human body, in a culture where anatomies were being performed in a variety of contexts, it may have seemed less urgent to include them.

Various elements of the *Fasiculo* woodcut of the uterus show the difficulties involved in creating a tradition of anatomical illustration based on direct observation in the absence of preexisting models (figure 2.1). Most striking are what appear to be gross lapses of communication between the designer, who produced the drawing on which the print was based; the woodblock cutter; and the man responsible for inserting the identifying lettered captions as they had appeared in the 1491 woodcut (figure 2.2).[110] The last apparently failed to notice that the carefully carved lines radiating outward from the female figure's genitals were intended to lead to letters that corresponded to the lettered captions in the first part of the appended text. Taking a wild guess, they interpreted these lines as fumes emitted from the uterus through the vagina, a phenomenon related to suffocation of the uterus as described by Mondino, and labeled them at the bottom of the image as such (*fumosità dela matrice*); the letters referencing the captions (labeled *a* through *ii* in the earlier woodcut) disappeared from both image and text.[111]

More subtle, but also more significant, is the apparent difficulty experienced by the block cutter in interpreting the drawing of the woman's open abdomen, which may reflect the designer's difficulty in imagining a set of graphic conventions to show the incision and the exposed internal organs. The block cutter evidently failed to understand both the importance of reversing the drawing so that, for example, the liver ended up on the right side of the figure, and the spatial relationship between the inside and the outside of the body. The prominent "horns" radiating from the sides of the uterus, which probably represent the supporting ligaments — crucial structures, since they proved that the uterus could not wander within the body, as Mondino and Guido of Vigevano had also emphasized — connect directly to the lines representing the dissector's incision. The relationship between the enormous vagina, with its own "horns," and the incision is also far from clear. The block cutter's difficulties may in turn reflect difficulties on the part of the designer (or designers, since there may have been several intermediate drawings), who found himself faced with a novel challenge: to represent simultaneously the exterior of the body (upper chest, arms, legs, and head); the exposed internal organs of the lower abdomen, with an emphasis on the reproductive organs; and the incision that formed the transition between the two. It took fifty years for Italian artists to develop a stable and effective set of conventions to represent anatomized bodies (see Chapters Four and Five), and the designer of the 1494 *Fasiculo* had no models on which to draw. Given the novelty of his task, he struggled both with making visual sense of the body's interior and with finding ways to convey what he saw, oscillating between schematic two-dimensional and volumetric representations (of the nonreproductive and reproductive organs, respectively).[112]

The shift from the anatomical diagram of the 1491 woodcut to the depiction of a dissected corpse in the 1494 version also transformed the meaning of the texts that accompanied the images in both the Latin and the Italian editions: thirty-five medical recipes and notes concerning the female reproductive system, its functions, and its malfunctions, followed by a short treatise titled *Problems, or Investigations of the Genitals, that is, the Uterus and Testicles, or Secrets of Women*.[113] As the subtitle suggests, these texts formed part of the manuscript tradition of writing on generation and female sexuality that I described in the first part of this chapter. The Italian translation of this work in the printed *Fasiculo* greatly broadened the audience for the material, rendering it accessible not only to the learned doctors that made up the target audience of the original, 1491 edition but also to surgeons and other practitioners without a reading knowledge of Latin, as well as to lay readers. Expanding the audience in this way was clearly the publishers' intention, for they revamped the format to make it more attractive to those with humanist tastes and interests. They reduced the volume's size, replaced double columns of Gothic type with full pages of text set in an attractive Roman font, and commissioned four new images, including the new female figure and a famous depiction of an anatomy lesson based on human dissection (figure 2.5) — the latter specially created to introduce the Italian translation of Mondino's *Anatomy* that was added to the compilation.[114]

This new cultural and bibliographic context — the "publicity" associated with publishing — acted to undercut the putative secrecy of the inside of the human body and of the female reproductive organs in particular. The earliest discussions of the "secrets of women," notably the *De secretis mulierum* attributed to Albertus Magnus, as I have already argued, adopted the point of view of the learned, Latinate, and therefore male reader, for whom knowledge concerning sex, fertility, and childbirth was the exclusive property of women, orally transmitted in the vernacular and intentionally kept from men. In the late fourteenth century, this material was translated back into the vernacular for the benefit of a broader male audience; at the same time, manuscript treatises such as *I secreti delle femine* and *Le segrete cose delle donne* began to transform the old trope of women's secrets, suggesting that the information they contained was a male monopoly — secrets unknown to, rather than possessed by, women. The *Fasiculo de medicina* did much the same thing in the medium of print, supplying not only a third vernacular treatise on the topic (the *Problems*) but also a woodcut that presented information about generation and the uterus in an appealing pictorial form. In this way, it offered its readers privileged access to useful and titillating knowledge concerning the genitals and female sexuality and to the even more exclusive visual experience of viewing a dissection. In this new regime of publicity, the only people excluded in principle from dissection-based knowledge of the female body were nonreaders, which included almost all women.[115]

In this context, the invocation of secrecy in the text associated with the image of the uterus in the 1494 *Fasiculo* seems largely residual; the secrets of women have been unveiled and disseminated by learned doctors in service of public interest and public

116

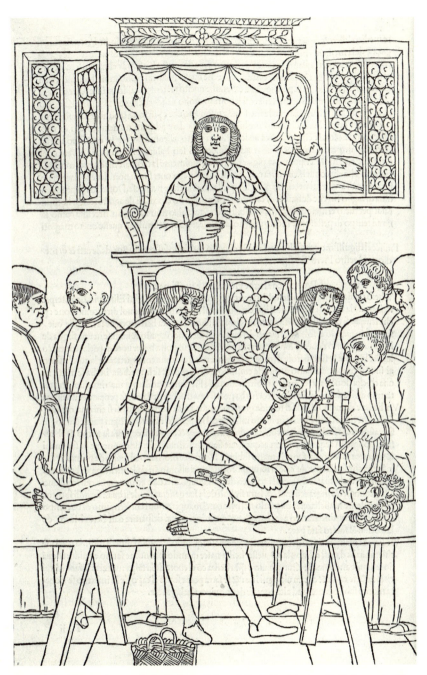

Figure 2.5. Anatomy lesson. *Fasiculo de medicina*, attributed to Johannes de Ketham, sig. f2v.

good. Even the posture of the anatomized woman reinforces the message of contemporary medical writers such as Guaineri and Savonarola, who, as male practitioners aspiring to attract a female clientele, had good reason to downplay issues of female modesty. Her pose is in every respect the opposite of the *pudica*, the standard figure of the naked woman in contemporary art and sculpture.[116] Instead of closing her legs and covering her genitals and breasts with her hands, like Venus in Sandro Botticelli's famous painting, she splays her knees and makes no effort to hide her body; she holds her right knee open with her right hand, as if to facilitate her own display, while her left hand, with its pointed finger, demands the viewer's attention. Indeed, the image draws much of its force from its violation of artistic conventions regarding the portrayal of the naked female body as an object of secrecy and shame, while the figure's upward gaze, characteristic of contemporary portrayals of saints and martyrs, gives the woodcut an aura of incongruous piety that further undercuts associations with women's notorious lustfulness and deception.

Even in the image of the dissected woman, however, it is not clear that the trope of sexuality and generation as "women's secrets" had entirely lost its force. One of the most striking changes between the 1491 and 1494 versions of the woodcut is that the latter shows the uterus as opaque rather than transparent. As a result, the viewer cannot easily tell if the woman is pregnant, particularly given the uncertain scale of other elements in the woodcut. It is unclear why the designer, editor, and publishers of this dramatically innovative image chose to represent the uterus as undissected. Most likely they had no access to a pregnant woman, or to a woman far enough long in pregnancy to support the expected image of a nearly full-term fetus. Whatever the reason, however, the effect is to enhance the authority of the female subject — in contrast, for example, to the man who is about to be dis-

Figure 2.6. Birth of Hercules. Ovid, *Ovidio Metamorphoseos vulgare*, trans. Giovanni Rossi (Venice: Giovanni Rossi for Lucantonio Giunta, 1497), fol. 77v.

sected in figure 2.5. While he lies on a makeshift table, his limbs contracted in rigor mortis, she sits calmly on a chair, whose scrolls recall both contemporary Venetian images of birthing stools (figure 2.6) and the anatomical lecturer's *cathedra*. Like the lecturer, she extends her finger in the canonical gesture of authoritative teaching or affirmation, which suggests that knowledge regarding the inside of the female body is not the property of men alone.[117] Although produced by male dissectors for male students and physicians, and disseminated by male editors and printers for lay male readers, it also still belongs to the women who embody it.

Yet this image does not throw us back onto the older model of women's secret knowledge, for this woman chooses to communicate what she knows. Like Chiara of Montefalco and Margherita of Città di Castello, she is simultaneously subject and object, although even in death she adopts a more active stance of self-disclosure than they, revealing her dissected body to the viewer while simultaneously explaining its secrets. In this way, the woodcut rejects both of the meanings of women's secrets that informed earlier writing on the topic — women's secret knowledge about generation and the secret hidden physically inside the female body — to embrace a regime of anatomical publicity in which dissector and dissectee collaborate to reveal the body's truths.

CHAPTER THREE

The Mother's Part

On August 6, 1477, Fiametta di Donato Adimari had a daughter by her husband, Filippo di Matteo Strozzi, a wealthy and powerful businessman from a prominent Florentine family. She had married Filippo in 1466, at the age of sixteen, and over the next decade she assured the continuation of Filippo's lineage by bearing seven children, five daughters and two sons. Childhood and childbearing were risky business in late fifteenth-century Florence, however. When Alessandra was born in 1477, Fiametta and Filippo had already lost a son and a daughter — only three of their children were to survive to adulthood — and several weeks later Fiametta herself fell fatally ill.[1] Her husband noted the event in his *ricordanza*, the book used by male heads of prosperous Florentine households to register family affairs. "On the twenty-third day of the said month," he wrote, "it pleased God to call to himself the blessed soul of my wife Fiametta, and she was buried on the twenty-fourth in Santa Maria Novella, in our usual tomb. Her problem was that she was not well purged after the birth; nonetheless, although she had had a little fever and some pains for several days afterward, she had felt well for four days and was up and around the house." On the evening of the twenty-third, however, she began to complain of intense pain around her heart.

Help was called quickly but in vain; as Filippo put it, "the many efforts of the women and doctors were of no avail."[2] She died two hours later.

Thus far, the outlines of Fiametta's story differed little from that of the many Florentine wives and mothers who bore a series of children in quick succession, only to die in childbirth in their twenties or early thirties.[3] (Fiametta was twenty-five.) More unusual was her husband's decision to have her anatomized. In Filippo's words,

> I had the body opened and among the others there to see it was Master Lodovico [di Maestro Piero dal Pozzo Toscanelli, a prominent Florentine physician],[4] and he later said to me that he had found her uterus full of putrefied blood, and that this caused her death. And in addition, her liver was in very bad shape, together with her lungs, which had begun to attack her kidneys. So that if she had not died of this illness, she would have fallen into consumption [*tixiccho*]. Her death caused me great sorrow, because every day she pleased me with her many good qualities.[5]

He remarried before the end of the year.

Fiametta's anatomy is the first case I have found of a practice that was to become increasingly common among well-to-do citizens in Florence and elsewhere: autopsies in the context of family health care, performed at the request of patients, relatives, and their physicians.[6] For example, ten years later, in 1486, Bernardo di Stoldo Rinieri granted his wife's request to have her opened and examined after her death.[7] Entries describing similar procedures appear in the *ricordanze* of Tribaldo d'Amerigo de' Rossi (regarding the death of his married sister Alessandra in 1493) and Cambio di Manno Petrucci (regarding the death of his wife, Grazia, in 1512).[8]

Domestic autopsies of this sort took their place alongside earlier forms of dissection that dated to the late thirteenth and early fourteenth centuries, including the officially mandated anatomies in the interests of criminal justice, public health, and medical training I described in Chapters One and Two. To begin with, they were rare; accounts of the sort I have quoted appear in only four of the fifty-eight Florentine *ricordanze* from the late fifteenth and early sixteenth centuries surveyed by Christiane Klapisch-Zuber, although there were certainly others for which the documentation has not survived.[9] In this they did not differ markedly from the older types of institutionally mandated anatomies. Even the instructional dissections sponsored by medical faculties and (beginning in the mid-fourteenth century) by guilds and colleges of physicians or surgeons were relatively infrequent. Although the statutes at the universities of Padua, Bologna, and Florence prescribed the formal, or "public," dissection of one male and one female cadaver a year, it is by no means clear that this was carried out in practice.[10] In part, this infrequency arose from the scarcity of appropriate cadavers — foreign criminals executed in winter. Obtaining female corpses proved particularly difficult, since the principal legal source of cadavers for public dissection was condemned criminals; in 1464 or 1465, the medical faculty of the University of Pavia went so far as to petition Francesco Sforza, Duke of Milan, to commute the sentence of a woman who had been condemned to burn at the stake, so that she might be dissected for the benefit "not only of the university but of the entire world."[11] In part, too, the infrequency of dissections reflected the bookish nature of anatomical study; as Nancy G. Siraisi has pointed out, "the dissected cadaver seemed to add relatively little to the information available from other sources," notably authoritative medical texts.[12]

This situation began to change in the last two decades of the fifteenth century. Learned physicians' new enthusiasm for ancient Greek medicine stimulated their interest in studying anatomy through dissection, on the model of Galen, who considered it fundamental to the art and science of healing (more on this in the next chapter). Physicians in private practice, who had absorbed the new Galenic interest in anatomy in the course of their own university training, increasingly prized the observation of the internal organs as a valuable source of information about disease processes and cause of death, and therefore about prevention and treatment — an attitude that was embraced by their clients, as Fiametta's anatomy shows. Domestic anatomies thus served the interests of both practitioners and patients; they made up for the scarcity of cadavers available for formal public dissection, and they helped to allay the distress and satisfy the curiosity of relatives of people who, like Fiametta, died mysteriously or unexpectedly. Because they could be performed rapidly, privately, and respectfully, without ruining the appearance of the corpse, unlike public dissections, they were fully compatible with family honor and funerary practice.[13]

The best contemporary evidence regarding private anatomies comes from the hand of Antonio di Ser Paolo Benivieni, an eminent Florentine physician. When Benivieni died in 1502, he left an extensive set of manuscript notes that documented more than two hundred noteworthy cases he had heard of or personally encountered in more than three decades of medical practice. About half of these were published posthumously by his brother Girolamo under the title *Some Hidden and Marvelous Causes of Disease and Healing*.[14] Benivieni's full manuscript included references to eighteen anatomies, all but two of which — the opening of a body of a man executed for theft and a forensic inquest — took place in the context of private practice.[15] The patients included people as socially varied as a nun from the monastery of San Donato, who died of an abdominal obstruction, and a male country-dweller, who suffered from a perforated intestine.[16] Most of the rest, however — eleven of sixteen — were members of Florentine patrician lineages such as the Adimari (Fiametta's family), Corbinelli, and Niccolini.

According to Benivieni's notes, the initiative for these procedures came from both families and practitioners. In the case of Michele Niccolini, for example, Benivieni wrote, "his relatives decided to have the dead man cut open, and it was found that almost his entire stomach was hardened, and the cause of his death was discovered."[17] The relatives of Luigi Mancini were moved less by curiosity than by the desire to embarrass the physicians — Benivieni implied he was not among them — who had been unable to cure their kinsman: "They ordered that the dead man be opened, rather to expose the ignorance of the doctors than to know the nature of his disease, for they thought that they had been deceived and mocked by the doctors' words." The result confirmed their suspicions, revealing a large number of gallstones, which the physicians had failed to treat.[18] In other cases, the physicians themselves suggested the autopsy, and only once did Benivieni indicate that the family had refused, in response to "I don't know what superstition."[19] The physicians' motivations were also various. In one case, they wished to prove to the father that the patient's death was inevitable. In another, Benivieni explained of one of his own kinsmen, who proved to have succumbed to heart disease, that "the cadaver of the deceased was cut open for the sake of the public good [*publicae utilitatis*]."[20] Most often, however, anatomies were performed when the attending doctors had disagreed among themselves. (When gravely ill, the well-to-do were usually treated by multiple physicians working in tandem, a practice that was thought to provide additional security for both practitioners and patients.)[21] In cases of this sort, the results either served to confirm the diagnosis of one or another party — mostly, needless to say, Benivieni's[22] — or allowed the doctors to justify and to dispel their collective perplexity. For example, regarding one of Ruggiero Corbinelli's daughters, who died after persistent pain in her right side, Benivieni concluded, "it is hardly surprising if in diseases of this sort, whose causes are uncertain and hidden, the opinions and pronouncements of doctors differ."[23]

One of the most striking things in Benivieni's notes is his re-peated invocation of "hidden," "uncertain," and "concealed" dis-eases and their causes when referring to male and female patients alike — a theme his brother emphasized in the title of the pub-lished book. This is a significant innovation in fifteenth-century medical writing about illness, and it both recalls and expands the notion of "women's secrets" that characterized much fourteenth- and fifteenth-century writing about generation. It is as if the new emphasis on autopsy and dissection in medical training and prac-tice called attention to *all* bodies, not just female bodies, as pos-sessed of an anatomically defined interior that determined how they felt and what they did. This had, of course, been implicit in late medieval writing on medicine, which categorized diseases by the affected part, external and internal. But the importance of individual organs was overshadowed, in the minds of patients and, as a result, of many practitioners, by a model of health and illness based on imprecisely localized notions of complexional balance and imbalance or internal obstruction and flow. Neither of these required intensive anatomical study or analysis in the narrow sense. The late fifteenth-century fascination with the hidden causes of disease evident in the writing of Benivieni thus reflected the new vogue of what Roger French has called "structural" (as opposed to "complexional") anatomy, which recast the interiors of all bodies as stuffed with separate internal organs that required inspection and elucidation.[24]

At the same time, however, there continued to be a notable

focus on the interiors of particular types of bodies — those of mothers and their children. Of the sixteen private autopsies described
by Benivieni, five involved women and six involved children (one
girl and five boys). This emphasis on young and on adult female
bodies is particularly striking when contrasted with the strong
focus on adult males in formal dissections and anatomy texts. It
appears consistently throughout the relatively rare sources that
document this kind of practice: family *ricordanze* and physicians'
case records and medical opinions (*consilia*), such as those of
Benivieni and Bernardo Torni in late fifteenth-century Florence
and Girolamo Cardano in sixteenth-century Milan.[25] The main
exception, beginning in the 1490s, involved the bodies of important male political figures such as Lorenzo di Piero de' Medici
(the "Magnificent"); because of the extraordinary political significance of their deaths, as well as the constant fear of poisoning —
and because their viscera had already been removed for embalming — they were often anatomized as well.[26]

I know of no contemporary visual depictions of private autopsies of this sort from Florence or elsewhere in Italy, but a late
fifteenth-century painting from a French manuscript of Guy
of Chauliac's famous surgical summa portrays the scene in a way
that is generally consistent with the Florentine accounts (figure
3.1).[27] The anatomy takes place in a domestic space, apparently
the woman's bedroom, for the bed in which she died dominates
the area to the left of the window; beside it stands a female ob-
server, probably a relative or servant. The woman's naked body
has been laid out on a temporary trestle table, in front of which a
young assistant holds a bowl, ready to receive her viscera, next to
a stool with the surgeons' tools. At least nine medical men are in
attendance. The two at the far right appear by their clothes to be
senior physicians or teaching masters; one is pointing to and presumably explaining the findings for the benefit of the others, who

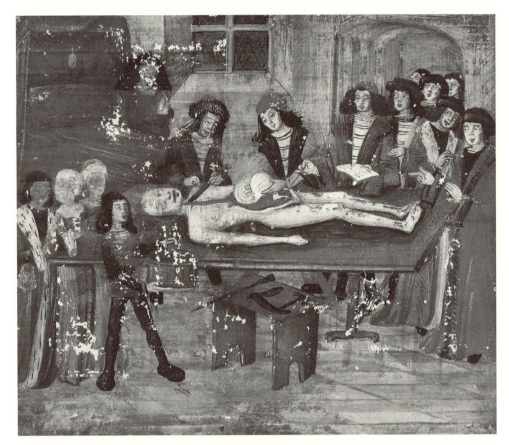

Figure 3.1. Anatomy of a noble mother. Guy of Chauliac, *La grande chirurgie*, Montpellier, Musée Atger, Bibliothèque de la Faculté de Médicine, ms.fr. 184, fol. 14v, French, late fifteenth century.

include several younger men — surgeons and junior physicians or medical students, judging by their dress. The surgeon at the far left of the group is beginning to cut into the woman's chest, while the man next to him removes the uterus from her already open abdomen. It is hard to know how literally to take this image — it has much in common with conventional portrayals of human dissection in contemporary French illustrations (figure 3.2)[28] — but it does reflect key aspects of Florentine practice: the domestic setting of such anatomies, the presence of multiple medical men, the delegation of the duties of cutting to a barber or surgeon, and the special interest in women's bodies. More surprising is the presence at the head of the table of three children, two girls and a boy, presumably the woman's own. Their presence, whether literal or symbolic, together with the focus on her uterus, underscores the centrality of family and generation to the scene.

In the rest of this chapter, I will explore the new use of autopsy in late fifteenth-century Florence to explore the causes of death of patrician wives and mothers, and I will discuss the ways such autopsies reflected contemporary thinking concerning kinship and lineage as well as the proper roles of mother and wife. Fiametta and the other matrons whose bodies were opened in this way were the opposite of the evil, deceptive women who haunted the imaginations of the authors and readers of the treatises on women's secrets I described in the preceding chapter. Those women were above all bad wives and mothers. Putting their private lusts and interests above their families, they cuckolded and dishonored their husbands, exploiting their secret knowledge of generation to abort (or prevent the conception of) the children who threatened to limit their pleasure and freedom. In contrast, good wives and mothers like Fiametta — and like the woman in the French manuscript — opened themselves to scrutiny in the interests of their children and their husband's families, to whom

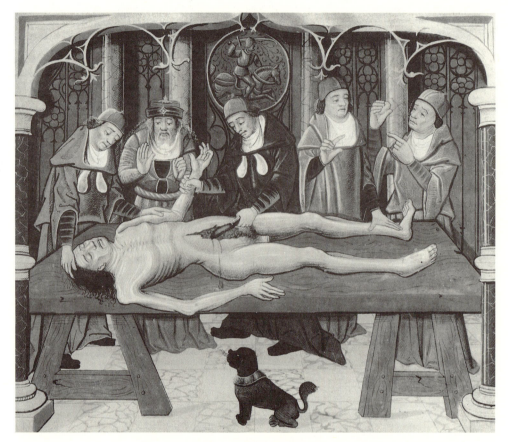

Figure 3.2. Dissection scene. Bartholomaeus Anglicus, *Le propriétaire des choses*, trans. Jehan Corbichon, ed. Pierre Farget, Paris, Bibliothèque Nationale, ms. fr. 218, fol. 56r, France, late fifteenth century.

those children by definition belonged. In this way, they dramatized their corporeal identification with their husband's lineages,
which Florentine patrician ideology defined as composed of males
descended through the paternal line. Wed in their teens to much
older men, these women were supposed to perpetuate the families of their husbands by producing as many male children as their
bodies could bear. Ideally, indeed, the patrician wife was to pre-
decease her husband, leaving him in sole possession of her dowry,
her body, and its progeny.[29]

Like those of Chiara of Montefalco and Margherita of Città di
Castello, however, the anatomies of wives and mothers reflected a
complicated reality, where the different parties to the procedure
might understand the same events in different ways. If the texts I
discussed in the previous chapter reflected the theories concerning women's bodies subscribed to by learned natural philosophers
and medical writers, as well as physicians' increasing claims to
practical expertise in the treatment of women's illnesses, my
focus in this chapter will be the fragmentary and elusive sources
that reflect actual practice, as well as the views of female patients
and their families. These documents make it clear that by the
second half of the fifteenth century physicians were indeed beginning to provide substantial gynecological and obstetric care to
elite women and to play a significant role in managing their pregnancies and reproductive lives. They also suggest that laypeople's
understanding of women's nature and reproductive role did not
correspond neatly to the theories proposed by learned writers on
generation, which tended to downplay the mother's physical contribution to her offspring in favor of the father's, as patrilineal
ideology required. Laypeople, in contrast, saw the anatomy and
physiology of generation as establishing indelible corporeal links
between a mother and her children — the agonizingly fragile bodies
on which the lineage depended.

Men and Women at the Bedside

By the second half of the fifteenth century, Florentine patrician matrons relied on both women practitioners and male physicians to help them conceive, carry, and bear their children. Fiametta Adimari was typical in this respect. The actual birth of Fiametta's last child, Alessandra, was managed by the two female practitioners who appear in every account of childbirth in this period, the midwife and the *guardadonna*. The latter, as her title suggests, was a kind of doula, attending to the mother's needs while the midwife delivered the child. Unlike the midwife, she typically remained in the household for several weeks to look after the mother during the lying-in period.[30] According to her husband's records, Fiametta's *guardadonna*, Maria, seems to have stayed with her until her death three weeks after the birth.[31]

The midwife's function was much more limited; she oversaw the later stages of labor and the delivery itself. I have to date found no evidence that midwives were consulted before or during pregnancy, let alone for any other gynecological or medical condition, or that they were recalled to see mothers for complications after birth. As skilled specialists in delivery, however, they were paid much better than *guardadonne*, with fees depending on the length of labor; for example, the Florentine midwife who delivered Caterina, the wife of the notary Ser Girolamo di Ser Giovanni of Colle, in 1473 received almost as much for slightly less than twenty-four hours of work as did her *guardadonna* for five weeks of service.[32] Although their relatively high fees — on the order of two lire for a single delivery — suggest that midwives were prized for their expertise,[33] there is no evidence that they had any special legal status or recognition. In this respect they differed significantly from their counterparts in many northern European cities, who were beginning to be inspected and licensed by city officials and whose duties were being legally recognized and prescribed.[34] As a result, relatively little is known about the practice of midwifery in Italy — including when it first appeared as an urban occupational identity. (Rural areas could not support this kind of specialized practice.) Monica H. Green argues that this may have occurred much later than usually imagined, possibly as late as the thirteenth century, and that the occupation of midwife, in the sense of someone with specialized skills recognized by the community, emerged only gradually out of the services previously performed by kinswomen and female neighbors.[35] Even in the fifteenth century, midwives learned their trade through experience and informal apprenticeship. There is no evidence that they — or even their relatively well-educated and wealthy clients — read works on women's illnesses, childbirth, or generation. None of the surviving Italian vernacular manuscripts on "women's secrets," for example, gives any indication of female readership, and Michele Savonarola's mid-fifteenth-century treatise on childbirth and the care of infants, ostensibly addressed to the women of Ferrara (including midwives), seems to have had little circulation of any sort.[36]

The relative lack of occupational organization and autonomy on the part of Italian midwives may reflect the early involvement of Italian physicians in treating women in matters related to generation and birth. (Women had long been seen by male doctors for nonobstetric conditions.) This did not represent a "usurpation" of the functions of midwives by physicians, since the activities of midwives were traditionally restricted to the delivery of children. Rather, it seems that male physicians began to provide more medical services to women in matters in which these women had earlier received little or no specialized care of any sort and had relied instead on the advice of female family, friends, and the occasional empirical practitioner.[37]

While physicians seem to have played a small role if any in routine pregnancies and deliveries — except in the case of some women from princely, as opposed to merely patrician, families — they were regularly consulted when problems arose.[38] Indeed, the involvement of male physicians in the reproductive lives of fifteenth-century patrician women appears to have been considerably more extensive than that played by midwives, since they dealt with a much wider range of conditions; they treated infertility and illnesses both during pregnancy and after delivery, and they were regularly called in for life-threatening miscarriages and births.[39] Thus when Fiametta began to suffer "a little fever and some pains" several days following the birth of her daughter, her husband procured the services of an eminent physician, Mosè di Giuseppe Spagnolo, "to treat her in her childbirth."[40] (Master Mosè was the personal physician of Giuliano de' Medici — he had cared for Giuliano's famous lover, Simonetta Vespucci, only the year before — and his presence testifies to the prominence of Jewish doctors in the higher levels of medical practice in late fifteenth-century Florence.)[41] It is not clear how often Master Mosè visited Fiametta or for how long, but he most likely saw her a number of times, since a week before Fiametta died Filippo paid him the handsome sum of two florins and credited him (prematurely) with saving her life.[42] Similarly, when Filippo's daughter-in-law Clarice di Piero de' Medici nearly died during her delivery of twin boys in 1526, she was attended by both doctors and a midwife; after the birth, she was nursed by attendants (*astanti*) but treated by doctors alone.[43]

Male practitioners were often present at the most difficult and tragic form of delivery, when a mother died during labor and her corpse was opened to remove the child for emergency baptism.[44] (Such children almost never survived.) In 1462, for example, the Bolognese master mason Gasparo Nadi lamented the death of his wife, Catalina, in his *ricordanza*: "Since she couldn't give birth, the doctor Master Giovanni of Navarre extracted [the child] from her body; it was a boy, and it pleased God that it was fated that he die after a little more than an hour. I did this because it was impossible for me to save her, since I loved her enormously."[45] Although Nadi's brief description supplies no details regarding the people present at Catalina's deathbed — and does not even say if Master Giovanni was a physician or surgeon — accounts from the first half of the sixteenth century attribute the decision to perform an operation of this sort to the midwife and the female servants, friends, and relatives in attendance, and usually specify that they called a barber or surgeon for the purpose.[46] A doctor was almost certainly present at the death of Francesca Pitti under similar circumstances in Rome in 1477. Her husband, Giovanni di Francesco Tornabuoni, the papal treasurer and manager of the local branch of the Medici bank, described the experience to Lorenzo de' Medici the next day in a letter:

> I am so oppressed by grief and pain on account of the bitterest and unimaginable misfortune that befell my sweetest wife that I don't know where I am. As you will have heard, as it pleased God she passed from this life in childbirth yesterday at the twenty-second hour. Having cut her open, we extracted the dead fetus [*creatura*] from her body, which doubled my grief. I am quite certain that, with your usual mercy, you will excuse me if I do not write to you at length.[47]

The division of labor between midwives and male medical practitioners appears clearly in a contemporary depiction of the death of a woman in childbirth, which has been associated, though insecurely, with Francesca's tomb (figure 3.3). (There are no signs of a Caesarean operation in the relief, which has been attributed to the workshop of the Florentine sculptor Andrea del Verrocchio, and it is not completely clear that the child it depicts is dead.)[48] The right-hand half of the scene (figure 3.4) shows the woman

Figure 3.3. Death of a woman in childbirth. Workshop of Andrea del Verrocchio (attr.), Florence, Museo Nazionale del Bargello, ca. 1477.

slouched over on her couch while the midwife, a woman of evident age and experience, holds her right arm and the other women at the bedside collapse in grief. On the left (figure 3.5) — the gap between the two sections of the relief corresponds to two separate rooms, as is clear from other contemporary images[49] — an older woman, perhaps the same or another midwife, shows the child to the father, who is attended by a number of men, including, on his immediate left, a physician. The midwife focuses on the mother and baby, while the physician advises the father, as complementary and collaborative presences in cases of difficult birth.

Despite the physician's association with the father in this relief, there is no evidence that his presence was the result of masculine interest or pressure. By the second half of the fifteenth century, pregnant and parturient women from well-to-do families not only welcomed but even insisted on the help of male doctors, as shown by a series of letters regarding Clarice Orsini, the young wife of Lorenzo de' Medici, the year after Francesca Pitti's shocking and well-publicized death. In early September 1478, when she was more than eight months pregnant with her first child, Clarice began to feel unwell. Lorenzo's friend and protégé Angelo Poliziano wrote him describing her nervousness: "She is afraid of mis-

Figure 3.4. Detail of figure 3.3 showing the dead mother, midwife, and mourning women.

Figure 3.5. Detail of figure 3.3 showing the father and associates outside the birth room being shown the (dead?) child.

carrying, or of having the same problem as the wife of Giovanni Tornabuoni," he wrote in a letter dated September 7.[50] Clarice's panic is evident in the flurry of letters written by her and others on that and the following day. The women around her, including the mother of Andrea Panciatichi, whom Poliziano described as "very knowledgeable," told her that she was in no danger, but their authority in this matter paled beside that of Master Stefano della Torre, the physician who was quickly called to treat her. As Clarice herself wrote to her husband, "at this point Master Stefano came, and by the grace of God he found me feeling much better, and he gave me great comfort. He will write you with more information about how he sees the situation."[51] Although della Torre remained in charge of the case, other physicians were called in to reassure the young woman, as is clear from Clarice's letter of the following day. "Master Stefano let me know that I am not in danger, as did those other doctors," she wrote to Lorenzo, and Master Stefano followed up with two letters of his own.[52]

Clarice's granddaughter Clarice di Piero (Filippo di Matteo Strozzi's daughter-in-law) echoed her grandmother's attitudes during her life-threatening delivery in January 1526. As her husband wrote his brother a week later, "She has such faith in Master Marcantonio that she wants him to come as soon as possible, as she is not satisfied with either her doctors or [female] attendants. We would like him to be here in four days."[53] Documents of this sort testify to the fear that childbirth inspired in women in this period — fear that was fully justified by the mortality statistics, which show that roughly half of the women who predeceased their generally much older husbands died in childbed, three times as many as died of disease, even in the relatively unhealthy period following the Black Death of 1348.[54] They also show that women did not necessarily accord exclusive authority to others of their sex in the area of reproductive health and that learned physicians had managed firmly to establish their expertise in such matters. Midwives, *guardadonne*, and other women (mothers, friends, and attendants) may have had the advantage of experience in difficult pregnancies and deliveries, but physicians, with their degrees, their fur-lined robes, and the dignity of their position, seem to have been accepted by those who could afford them as the gold standard for the care of women's health.[55]

Women's trust in male physicians does not mean, however, that they delegated decisions regarding their own health care to their male relatives, for the birth chamber, the sickroom, and the deathbed — at least in the case of female patients — continued to be dominated by women. Innumerable letters and *ricordanza* entries list the women who nursed and watched over the seriously ill. These were for the most part female relatives: the sisters, in-laws, and, especially, mothers of female patients.[56] These women apparently made most of the decisions concerning treatment and the eventual disposition of the corpse, including decisions regarding its autopsy. The *ricordanza* entries that describe the anatomies of Florentine female patricians illustrate the central role women played in these events. Although Filippo di Matteo Strozzi described himself as the one who made the decision to open Fiametta in 1476, later cases suggest a different pattern. Bartolomea Dietisalvi asked her husband to have her body opened in 1486,[57] while Tribaldo d'Amerigo de' Rossi attributed the idea of anatomizing his sister Alessandra to the women at her deathbed in 1493: "They buried her the same day ... in San Lorenzo in the tomb belonging to Piero Ripetti [her husband]. Her uterus and liver and lungs were damaged, since he [Piero] had her opened because the women said to."[58] Cambio di Manno Petrucci's account of the death of his wife, Grazia, in 1512 was similar. "There were at her death Alessandra, wife of the late Lorenzo d'Alessandro Buondelmonti, her sister, and Lucrezia, wife of the late Marco di Tolosino de' Medici," he wrote, "and they had her opened, and her illness involved the uterus and white [illegible: spleen?]."[59]

Given that the birth chamber was restricted to women, one of the apparent paradoxes involved in these autopsies of wives and mothers is the degree to which they would seem to violate the modesty, or *onestà*, that was understood to be the foundation of female virtue and masculine honor in patrician families (cf. figure 3.1). Prescriptive sources from this period emphasize adult women's duty to avoid being seen — much less touched — by men unrelated to them; they were to stay at home as much as possible and were not to appear in public except hidden under voluminous layers of cloth.[60] These ideas were motivated in large part by sexual concerns. Visible women ran the risk of inspiring lust in the men who saw them and of responding to male advances with lust of their own. The power of these prescriptions was so strong that historians have cited them as sufficient evidence that male physicians could not possibly have involved themselves in the treatment of female illnesses, at least insofar as this required inspecting and touching the genitals of respectable women.

In practice, however, the situation was somewhat different. As Alessandro Valori has argued, patrician women tended to interpret the demands of *onestà* with flexibility, focusing on the substance rather than the form. For them, the core of modesty lay less in avoiding public activity than in their competence as wives and mothers and their fidelity to their husbands' social, economic, and dynastic interests. This involved managing large households, receiving their husbands' associates, and looking after the welfare not only of their own children but also of the children of friends, kinfolk, and domestics. It also required frequent trips to the houses of relatives, neighbors, and men with whom their husbands had business relationships, as well as attendance at religious festivals and weddings — a subject of recurring contention between husbands and wives.[61] A similar principle appears to have governed women's choices regarding their gynecological and obstetric care. Precisely because this aspect of their lives touched their husband's interests so closely, in terms of their ability to produce healthy children, it was an obvious area in which men and women might agree to waive the technical requirements of modesty in the interests of the lineage.

As I will argue in the next section, the autopsies of Florentine patrician matrons indeed responded to male dynastic interests. At the same time, however, they reflected an understanding of generation that attributed great importance to the mother's role in reproduction, which existed in tension with the theories of generation proposed by learned medical writers, as well as with contemporary constructions of kinship, both of which emphasized the child's relationship to its father, while denying or effacing its relationship to its mother. Thus where medical writers concerned with infertility were careful to stress that the problem was at least as likely to lie with the father as with the mother, lay measures to encourage conception, from medicines to magical rituals to prayers and pilgrimages, focused overwhelmingly on the mother.[62] The same was true of attempts to prevent certain constitutional forms of hereditary disease in children.[63] In both cases, despite the stress laid by learned writers on the importance of the father's contribution to generation — its failures as well as its successes — laypeople mostly emphasized only the mother's part.

Mothers, Fathers, and Generation

As Joan Cadden has shown in her study of ideas of sex difference in medieval Europe, learned medical writers attributed greater activity and power in generation to men than to women, describing paternity in terms of creating and begetting and maternity in terms of birthing and nurture. They were not always in accord on the details. Some, following Aristotle, tended dramatically to downplay the female role in procreation, denying not only the efficacy but even the existence of a female contribution to the fetus in the form of female seed. Others, looking to Galen and his medieval Islamic followers, such as Avicenna and Haly Abbas, attributed a more active role to women, including the production of their own seed, which, while by no means equal to the father's in importance, nonetheless played an ancillary role in genera- tion.[64] But all agreed that women's primary function was to serve as a receptacle for the growth of the fetus and to supply the uter- ine blood that would nourish this growth. This function was char- acterized in purely passive terms; the male seed stamped the father's impression on the mother's menses like a seal on soft wax. In this way, in line with the Aristotelian precept that any- thing reproducing itself tries to create the most exact likeness of itself possible, the father reproduced himself, literally (at least in theory), using the mother's body as his tool.[65]

The matter of children's resemblance to their parents compli- cated learned discussions of generation, however, since under- standing this as a process by which the father reproduced himself using the mother as a nurturing receptacle made it difficult to account for the birth of daughters and of children who resembled their ancestors in the female line. The model of the seven-celled uterus, which I described in the preceding chapter, addressed this problem; children conceived in the three left-hand cells were fe- male, while those conceived in the three right-hand cells were male (cf. figure 2.4). By the late fourteenth and the fifteenth century, however, this theory was in retreat. Medical writers increasingly described generation as an agonistic process, involving a battle between the primarily active contribution of the father and the primarily passive contribution of the mother (the menses, whether it was assimilated to female seed or not); the latter put up a kind of passive resistance, which was often described in terms of "disobedience" or "indisposition." The outcome of this conflict determined the child's sex and appearance.[66] In the worst-case scenario, the process produced daughters who resembled their mothers; more desirable outcomes of the attempt of the father's seed to master the mother's matter included, in ascending order, daughters who resembled their fathers, sons who resembled their mothers, and — the ideal case, rarely achieved, which reflected the full submission of the female principle — a son who resembled his father in every respect. Whatever the appearance of the child, however, learned writers described it as receiving its life and human identity (in other words, its human soul) from its father rather than its mother, so that the children were always, in a metaphysical sense, more his than hers.

The idea that women's role in generation was essentially passive can be found in lay as well as learned texts, particularly prescriptive lay texts with a misogynist bent. One set of Florentine legislative deliberations from 1433, aimed at restraining women's "bestial" vanity and profligacy in the matter of dress, invokes their duty "to carry [the children] procreated by their husbands — they who, like a little sack, hold the perfect natural seed of their husbands, so that people will be born."[67] Lay authors also described the ideal course of generation as the exact reproduction of the father. In his dialogue on the family, written in the 1430s, Leon Battista Alberti described the father's "delight in seeing [his sons (*figliuoli*)] express his very image and likeness."[68] Writing to his brothers in 1455, Marsilio Ficino also commented on the close resemblance between father and sons: "The son is very like the father, being produced of his substance and close to him in complexion," he wrote. He therefore owes him "gratitude, reverence, and obedience."[69] In this view, the process of generation mapped onto the proper hierarchy of gender in the patriarchal family; just as the ideal household was one in which the wife (and children) were completely obedient to the husband, so the ideal course of reproduction was one in which the mother's mostly material contribution to the fetus accommodated itself perfectly to the formal impetus of the father's seed to produce a son identical to himself, avoiding any contamination of the male principle by the female.[70] This view of generation naturalized the deeply and increasingly patrilineal notions of kinship in late fifteenth-century Florentine patrician culture, in which, as Christiane Klapisch-Zuber put it, "men *are* and *make* the lineages [*maisons*]" and women are but "passing guests."[71]

While this male-oriented model of the relations between parents and children dominated learned writing on generation, it did not exhaust lay views on the topic. Alongside notions of generation that emphasized the father's contribution, lay writers expressed a complementary set of ideas that stressed the importance of the mother's role in shaping the child — ideas that in many respects accorded better with the observed facts of gestation and birth. The tension between these two sets of views found at least a partial resolution in the distinction between soul and body. If children received their souls — their human life principles — from the paternal seed, their mothers shaped them in their flesh. Even though the mother's relationship to her child was rooted in the physicality of gestation rather than in the metaphysical realm of identity, it nonetheless had strong effects. For the nine months during which it shared its mother's body, the fetus took its corporeal form and condition from the uterus that sheltered it; its mother could alter its disposition by her behavior — by how she held herself and what she ate or drank.

Additionally, as I have already described in connection with the production of miraculous objects inside the bodies of early fourteenth-century holy women, it was generally accepted that mothers could physically shape their children by impressing on them, intentionally or unintentionally, images of external objects they had seen during pregancy. A craving for strawberries might produce a strawberry mark, for example, or meditation on a painting of John the Baptist might produce a child covered with hair.[72] This theory of maternal impression continued to influence birth practices in patrician families, where part of the ritual of pregnancy involved giving women wooden salvers known as birth trays (*deschi da parto*), which might be painted with images of beautiful baby boys (figure 3.6). (There are no corresponding images of female children.) By gazing attentively at these objects, mothers could literally shape their offspring, raising their chances of producing a well-formed son.[73] This theory also permitted less salutary forms of female intervention in the process of generation; as medical writers were fond of pointing out, women who had become pregnant in the course of an adulterous relationship might escape detection by contemplating their husband's face or portrait, thereby impressing his likeness on a bastard child.

As important in many respects as the child's sex or appearance was its state of health, for a weak or sickly child, whether male or female, might prove unmarriageable, if it even lived to marriageable age. In this area, too, the mother's overwhelming physical influence was a source of concern, as is clear from frequent references to what might be called constitutional disease, in the sense of chronic disability or illness — or susceptibility to chronic disability or illness — that ran in the family or was present from birth. Medical writers emphasized that conditions of this sort were transmitted by both father and mother. As the anatomist Mondino de' Liuzzi explained in his commentary on Avicenna's chapter on the generation of the embryo, even though the father's seed did not contribute to the matter of the fetus, it contained formally or virtually a variety of what he called "habitual corporeal dispositions"

Figure 3.6. Healthy male child. Wooden childbirth tray, Boston, Museum of Fine Arts, Ferrara, ca. 1460.

seated in the heat of the paternal seed; these might include illnesses such as leprosy (*lepra*), chronic scabies, ringworm (*tinea*),
and gout (*podagra*).[74] The mother's contribution to consistutional
disease was more material, the product of an indisposition of the
uterus or — in the case of leprosy — exposure to impure menstrual
blood at the moment of conception.[75]

Lay writers, on the other hand, focused overwhelmingly on the
mother's influence in constitutional illness, often in connection
with the man's need to choose a wife free from physical taint. In
his *Book of Good Customs* (ca. 1370), for example, the moralist
Paolo of Certaldo included the following recommendation: "Make
very sure that the wife you take has not been born of the stock
of lepers or of consumptives or of people afflicted with scrofula,
madness, ringworm, or gout, because it will often happen that
some or all of the children who are born from her will have some
of the said defects and conditions."[76] These concerns were more
than theoretical, as contemporary marriage negotiations show.
One notorious case involved the successive betrothals of Galeazzo
Maria Sforza, the son and heir of Francesco Sforza, Duke of
Milan. Francesco had promised Galeazzo Maria to Susanna, one of
the daughters of Ludovico Gonzaga, Marquis of Mantua.[77] The
engagement was broken off in 1457, when Susanna showed signs
of developing a hunchback, and Ludovico proposed to substitute
his younger daughter Dorotea for her sister. The Sforzas agreed,
but they inserted a clause into the agreement that would annul it if
she showed signs of her sister's condition. In 1463, when Dorotea
was approaching marriageable age, Duke Francesco demanded
that she be examined by doctors. The Gonzagas balked, citing
Dorotea's modesty. After heated negotiations — and a scheduled
medical examination by four eminent physicians that was aborted
at the last minute — Dorotea ultimately ended up being examined
by her prospective mother-in-law, Bianca Maria Visconti. Although

Bianca Maria seems to have determined that Dorotea's back was normal, the marriage was nonetheless called off. Despite his mother's pressure, Galeazzo Maria flatly refused to marry a Gonzaga girl, because, as he explained to her, "from these women, born of the blood of hunchbacks, would be born other hunchbacks or lepers, as Avicenna says."[78]

This rich and sometimes contradictory mix of ideas concerning generation and constitutional illness helps to parse the elliptical descriptions of women's autopsies in fifteenth-century Florentine *ricordanze* with which I began this chapter. Although the women involved died under different circumstances, in all four cases the uterus was involved. Fiametta Adimari had failed to purge herself fully after childbirth; in her husband's words, the doctors "had found her uterus full of putrefied blood, and this caused her death."[79] Both Alessandra de' Rossi's and Grazia's autopsies also revealed damaged wombs.[80] The most interesting case, however, was that of Bartolomea Dietisalvi. Unlike Fiametta, Bartolomea had survived her childbearing years — her last baby had been born six years earlier, when she was in her mid-thirties — and had succumbed to a chronic uterine ailment. In the words of her husband, Bernardo,

> she died of an illness of the womb that caused a flux from her belly that lasted about eighteen months, and the physicians never found a remedy. She left me six children born from her, four girls and two boys. And because she asked me to have her opened to see if her illness was consumption [*tisicho*] and to give medicine to her daughters or others, I did so. And her illness was found to be of such a nature that her uterus was so hardened that it couldn't be cut with a razor.[81]

Bernardo's account recalls the autopsy of Fiametta, who suffered from a related set of conditions; her failure to purge properly

148

after the birth of her daughter led to a buildup of toxic blood in the uterus, and the physicians who opened her body noted that, had she lived, she, too, "might have contracted consumption." But the description of Bartolomea's case contains an additional element, since she was concerned not only that her chronic uterine illness might be related to consumption but also that she might have transmitted this weakness to her daughters, presumably *in utero*.[82] Her final act of maternal responsibility was to call for her own anatomy, out of concern for their health.

Similar fears regarding constitutional illness seem also to have motivated the anatomies of patrician children, as is evident from a *consilium* prepared in 1496 by Bernardo Torni, professor of medicine at the Florentine university in Pisa. Torni had opened the body of Battista, the teenage son of the Florentine wool merchant Francesco di Antonio di Taddeo, when Battista died after a sickly childhood. The anatomy revealed a massive obstruction of the chilic vein (vena cava), which Torni interpreted as contracted "either from birth or in the course of time."[83] If Battista's illness was congenital, it could pose a danger to his siblings, as Torni made clear. "I grieve over your situation," he wrote to the father, "for it is bad to lose a child, worse still to lose a son, and worst of all to lose him by a disease not yet fully understood by doctors. But truly, for the sake of your other, remaining children, I think it will be of the greatest utility to have seen his internal organs." On the basis of his findings, Torni recommended that Francesco protect "any of [his] sons of the same complexion until the age of twelve with the usual medicines," for which he provided a prescription, and concluded with a promise to visit the family often, in order to monitor the children's health.[84]

These anatomies of Florentine matrons and their children reflect not only a sense of women's fragility, especially as regards their reproductive organs, but also a broadly held lay appreciation

149

of the corporeal link between mother and child. Under carefully controlled circumstances, by gazing on images of beautiful male infants, women might intervene consciously and intentionally in the course of generation, to improve their husband's stock. In other cases, though, the process was unpredictable and less benign; instead of developing into a mirror-image of its father, the fetus might be contaminated by a host of poorly understood factors relating to the messy female nature. An accidental impression might mark or deform the child. Maternal illness could ruin a fetus's complexion, while heredity or conception during menstruation might create a disposition to constitutional disease. The mother's seed might successfully resist the father's, producing a daughter — a form of constitutional illness in itself. Like everything else relating to generation, misfortunes of this sort were largely uncontrollable and unpredictable, since their causes and effects lay hidden within the unruly female frame.

Myths of the Maternal Body

This persistent concern about the mother's part in generation informs two striking stories about the opening of maternal bodies. Both stories, which are found in a number of fifteenth-century Florentine manuscripts, belonged to a single narrative: the history of the Roman emperors, embodied in the lineage of Julius Caesar. As related in vernacular histories of the Roman Empire that circulated widely in late medieval France and Italy, this narrative began and ended with a mother's exposed and opened womb.[85] According to one such history, a short vernacular chronicle of the Roman emperors usually appended to the much longer *Imperial Book*, the Roman Empire began with the birth of Gaius Julius Caesar out of the corpse of his mother, the eponymous "Caesarean" birth:

150

Julius Caesar was the first emperor of Rome, and he was the son of a nobleman named Catullus. Poets and authors say that Caesar descended from the Trojan Aeneas and was a Trojan [himself]. And it was he who alone had all the honors of the world.... He was called Caesar because his mother died before he was born, and she was cut open and he was pulled from her body alive."[86]

This scene was illustrated in fourteenth- and fifteenth-century manuscripts of another influential work on Roman imperial history, *Deeds of the Romans* (see figure 1.9),[87] as well as in a woodcut from a Venetian edition of Suetonius' *Lives of the Twelve Caesars* published in 1506, which shows the midwife extracting the child through an incision made by the surgeon (figure 3.7).[88]

The history of the Roman Empire, as embodied in the Julian line of emperors, traditionally ended with Nero's opening of his mother, Agrippina (figure 3.8). According to the chronicle I have already cited,

the emperor Nero reigned for fourteen years, six months, and thirty days after the death of Claudius, and he was a very bad man.... He had his mother killed to see the place where he had been, that is, he had her opened.... Then he ordered the philosophers to make him pregnant, so that he could bear a son who would resemble him, because he doubted that his wife would bear him legitimate children. And the philosophers did this by putting a frog in his body. Finally, he killed himself a half-mile outside of Rome and his flesh was carried away by wolves and dogs.[89]

These two stories of Roman emperors and their mothers — of the birth of the heroic founder of the Julian line and the death of the evil emperor that destroyed it — reflect many of the issues raised by the autopsies of fifteenth-century patrician wives and

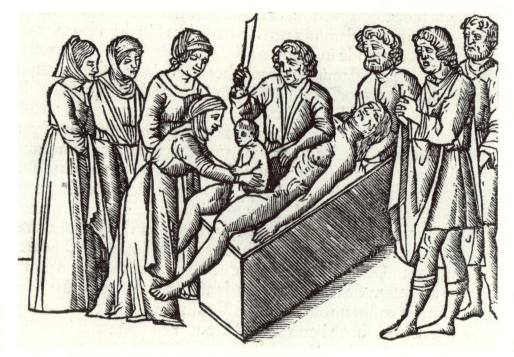

Figure 3.7. Birth of Julius Caesar. Suetonius, *Suetonius Tranquillus [De vita duodecim caesarum]...* , ed. Filippo Beroaldo and Marco Antonio Sabellico (Venice: Giovanni Rossi, 1506), fol. [1r].

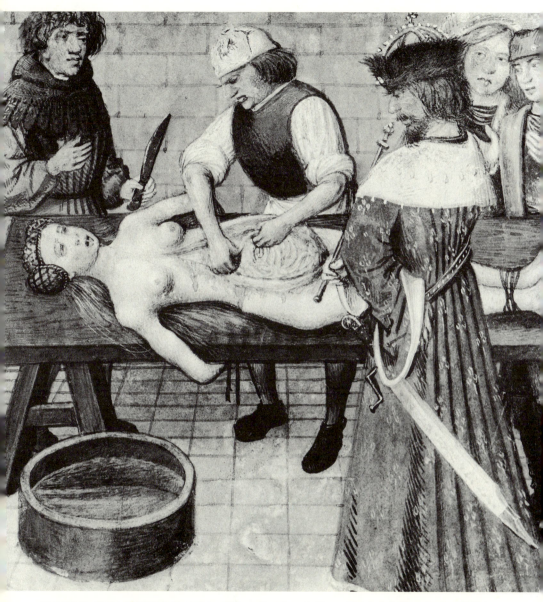

Figure 3.8. Nero supervising the anatomy of his mother. Jean de Meun, *Roman de la Rose*, London, British Library: Harley ms. 4425, fol. 59r, Flemish, ca. 1500.

mothers. As far as the story of Caesar goes, it is important to know that, while there is no evidence that the Caesarean operation was ever actually performed in ancient Rome, births of this sort were touted as highly auspicious, an indication of a heroic future for the child. (I am not familiar with any corresponding stories about female children.) In the words of the Roman natural history writer Pliny, "It is an excellent omen when the mother dies in giving birth to the child; instances are the birth of the elder Scipio Africanus and of the first of the Caesars, who got that name from the cut-open uterus of his mother [*caeso matris utero*]."[90] Nor was this kind of birth confined to human heroes; Asclepius was also delivered in this way, as Ovid recounted in the *Metamorphoses*.[91]

Janet Adelman has discussed the meanings of Caesarean birth in Renaissance Europe, in connection with Shakespeare's play *Macbeth*. Macbeth is told that "none of woman born" can harm him, eventually learning that his conquering nemesis, Macduff, was born by Caesarean section — "from his mother's womb untimely ripped."[92] Such a man, Adelman argues, appears to be free of the weakness implied by being birthed by and therefore dependent on a woman: like Caesar or Scipio Africanus, he becomes the bearer of what she calls "the shared fantasy that secure male community depends on the prowess of the man not born of woman, the man who can carve his own passage out, the man whose maleness is the mark of his exemption from maternal origin and the vulnerabilities that are its consequences."[93] Caesarean birth solves "the problem of masculinity by eliminating the female" as a generative force; although the mother is not altogether absent, she functions primarily as a receptacle and incubator, living only long enough to receive her husband's seed and bring her son to term, in line with the theories of kinship and generation that minimized mothers' contribution to their children.[94]

The fantasy of the dead mother found a more extreme expression in the story of Nero, whose meditations on the open uterus of his mother, Agrippina, made him want to conceive and bear his own child. Nero's project of parthenogenesis had two related goals. First, he believed this was the only way to ensure his own paternity; in the words of the imperial chronicle, "he doubted that his wife would bear him legitimate children." In other words, he embodied pervasive contemporary fears concerning women's sexual infidelity and the fragility of the link between father and child. Second, he wished to bear what the chronicle called "a son who would resemble him" — a phrase that recalls ideas concerning the recalcitrance of female matter and its inclination to resist the perfect impression of the father's form, as I discussed above. In seeking to produce a child in his own image, he reflected the patrilineal hope of eliminating the contamination introduced by mothers into the process of generation. At the same time, however, Nero's story sets clear limits on this fantasy, marking it as evil and deranged. As the imperial chronicle notes, he was a "very bad man." The depravity of his project is underscored in one of the most detailed and influential versions of his story, which appears in Jacobus de Voragine's *Golden Legend*, a mid-thirteenth-century collection of saint's lives that circulated widely in Renaissance Italy.

Jacobus' version of the story of Nero is so striking that I will quote it at length:

> Obsessed by an evil madness, [Nero] ordered his mother killed and cut open so that he could see how it had been for him in her womb. The physicians, calling him to task over his mother's death, said: "Our laws prohibit it, and divine law forbids a son to kill his mother, who gave birth to him with such pain and nurtured him with so much toil and trouble." Nero said to them: "Make me pregnant with a child and then make me give birth, so that I may know how much

155

pain it cost my mother!"...They said to him: "That is not possible because it is contrary to nature, nor is it thinkable because it is contrary to reason." At this Nero said to them: "Make me pregnant and make me give birth, or I will have every one of you die a cruel death!" So the doctors made up a potion in which they put a frog and gave it to the emperor to drink. Then they used their skills to make the frog grow in his belly, and his belly, rebelling against this unnatural invasion, swelled up so that Nero thought he was carrying a child.... At last, unable to stand the pain, he told the doctors: "Hasten the delivery, because I am so exhausted with this childbearing that I can hardly get my breath!" So they gave him a drink that made him vomit, and out came a frog horrible to see, full of vile humors and covered with blood. Nero, looking at what he had brought forth, shrank from it and wondered why it was such a monster, but the physicians told him that he had produced a deformed fetus because he had not been willing to wait the full [nine-month] term.[95]

Nero's experiment with parthenogenesis was a failure, as the conclusion of Jacobus de Voragine's account makes clear. The emperor was unable to bring the fetus to term, and after he had vomited up the bloody frog, his immediate reaction was one of horror: "Is this what I was like when I came out from my mother's womb?"[96] Shocked by the deformity of his offspring, "he commanded that the fetus be fed and kept in a domed chamber with stones in it" — a final act of cruelty that inspired the Roman mob to chase him from the city into a nearby field, where he committed suicide, thus ending the Julian imperial line. In this version of the story, the physicians speak for both reason and nature, neither of which condones a son's murder and evisceration of his mother, or a father's efforts to bear his own child.

These stories of mothers and sons are part of the mythologized Roman history that played an important role in shaping and transmitting Florentine patrician culture and identity. Fifteenth-century Florentines used episodes of this history to illustrate the values they saw as underpinning the order of family and state. They told and retold it in Latin histories and orations and in vernacular poetry and chronicles such as the one that I cited above. They also decorated their private houses with scenes that exemplified important virtues, including the proper relationships between men and women and the proper roles of husbands and wives. Favorite themes, found on wedding chests and decorative panels (*spalliere*), included the rape and suicide of Lucretia, a parable of married female chastity, and the rape and reconciliation of the Sabine women, a fable of the political and dynastic benefits of marriage and the production of legitimate heirs.[97] Stories of this sort, which hinged on sexualized violence in connection with generation, emphasized female submissiveness — sometimes to the point of self-annihilation — in the service of the dynasty and the patriarchal family; empires, like lineages, were founded on the compliant, generative bodies of mothers and wives.

The stories of Caesar, Nero, and their mothers form part of the same general complex of themes. The opening of Nero's mother was perverse, contrary to both nature and reason, as his doctors warned. It reflected his selfish curiosity — his wish to see the womb that bore him — rather than any plausible dynastic or generative goal; its sterility was dramatized in his failed attempt to reproduce himself. Caesar's mother, in contrast, was a good wife and mother, who lived only to generate her husband's heir. Her opened uterus marked her acceptance of and compliance with his dynastic interests, and it was echoed in the autopsies of patrician women such as Fiametta Adimari and Bartolomea Dietisalvi. In Fiametta's case, her husband's curiosity regarding her cause of death was legitimate and natural, a mark of affection and respect for his wife and the mother of his heirs. Bartolomea's case was more complicated; in requesting her own anatomy, she invoked her daughters' interests above all, and her emphasis on her female children suggests a somewhat more expansive understanding of family than the all-male patriline. Husbands used their wives to produce sons, but wives on occasion used their husbands to produce daughters; female corporeal submissiveness could never be complete.

Thus there was no single model of generation in fifteenth-century Florence. Theories of procreation, like theories of kinship, reflected contemporary assumptions concerning gender, and they inevitably raised the question of the nature and relative importance of the contributions of mother and father to the formation and growth of the child. Even within the learned tradition, there was considerable variation among writers on this important topic. In the minds of laypeople, the tension between assigning the dominant role to the father (which naturalized the patrilineal family) and stressing the mother's ability to conceive and shape the fetus was even more pronounced. These two ideas coexisted in patrician culture, like the battling male and female seeds in the uterus, or agnatic and cognatic models of descent.[98] They found expression in a whole host of stories and practices, including those centered on the opened maternal womb. Maternal autopsy is one of the most striking of these practices; performed by university-trained physicians, usually with the assistance of a male surgeon or barber, who did the actual cutting, it raises questions regarding not only lay notions of generation but also the expanding role of male doctors in gynecology and childbirth. By the second half of the fifteenth century, physicians were treating women for infertility and potential miscarriage, as well as the sequelae of childbirth — functions midwives, as specialists in delivery, appear not to have performed. Doctors' inspections of patrician mothers' wombs should be seen in the context of this increasing attention to women's reproductive health. Thus male physicians entered the domestic life of patrician families in new and important ways, bringing professional ideas concerning generation into contact with lay ones. This multiplied the resources available to lay people to understand and to try to control their reproductive lives. At the same time, however, it highlighted tensions and discrepancies concerning the mother's role in procreation that continued to play themselves out not only in the spaces of the patrician household but also in the anatomy theater.

CHAPTER FOUR

The Evidence of the Senses

The death of Elena Duglioli had been heralded by portents, including a comet that appeared over the house of one of her followers.[1] A woman from a respected citizen family in Bologna, Elena had gained local renown for her gifts of vision and prophecy. After she died, in September 1520, people flocked to see her corpse. According to her anonymous *vita*, written about ten years later, "they adored it and kissed it like a precious relic, sprinkling it all that morning with fragrant herbs."[2] That evening, it was transferred to the church of San Giovanni in Monte, where some years earlier Elena had commissioned a chapel dedicated to Saint Cecilia.[3] There it underwent a series of what the author of the *vita* described as "experiments" (*experientie*) initiated by a group of pious followers who were eager to promote her cult. Although Elena had initially gained fame for her chastity — fifteen years earlier, she had revealed that, like Saint Cecilia and Saint Valerian, she and her considerably older husband, Benedetto Dall'Olio, had lived together in complete continence since their marriage — the experimenters were more interested in other, more fleshly manifestations of her holiness. She told her confessor, Pietro Ritta of Lucca, that in 1507 Jesus had permanently removed her heart from her body, and three years later, when she was in her late thirties, she had begun to lactate and menstruate; Jesus had subsequently informed her that "the milk in her virginal breasts would last until the end of the world."[4]

The postmortem experiments conducted on Elena's corpse focused initially on her breasts.[5] The first two nights following her death, while her body was lying in the church of San Giovanni in Monte, Ritta uncovered her chest in the presence of a number of laypeople and clerics and pressed her breasts with his hands, producing a gush of "pure and white milk."[6] Several of her most devoted supporters suckled from Elena's corpse, which was then eviscerated and embalmed by two local surgeons. The third night Ritta again expressed milk from her breasts. At that point, recalling her claim that Jesus had taken her heart to be with him in heaven, Ritta ordered an inspection of her remains by the two surgeons who had embalmed her the previous evening, Battista of Bologna and Damiano. (The latter was the nephew of the renowned surgeon and anatomist Jacopo Berengario of Carpi, of whom more below.)[7] Also present was Girolamo of Firenzuola, a professor of medicine at the University of Bologna and one of Elena's most ardent followers.[8] The inspection of Elena's viscera yielded extraordinary results. Her heart was missing, having been replaced by a pale, flat, flaccid mass, which the author of her *vita* described as being "like a piece of soft liver." Indeed, he noted, it was "so unlike a heart that a man familiar with that human organ would never have recognized it as a heart. And the doctors in attendance said that it was a very strange thing and that they had never heard of anything like it."[9]

So far, the manipulations of Elena's corpse recall the anatomies of Chiara of Montefalco and Margherita of Città di Castello, which I described in Chapter One. Like them, Elena was first eviscerated for embalming; only afterward were her viscera inspected, revealing anomalous structures that her supporters took as signs of holiness. The author of her *vita* made the parallel explicit, citing as a precedent the events following Chiara's death.[10] But the procedure in Elena's case was different, for, once buried, her body was not allowed to rest in peace. In order to convince powerful local skeptics, and specifically to confirm her prophecy that her milk would last until the end of the world, it was exhumed not once but twice over the following three months. In the course of the inspections following these exhumations, it was examined for signs of corruption, and its breasts were cut open and assessed by eight additional experts in medicine and natural philosophy, mostly teaching masters at the universities of Bologna and Padua, to determine whether they were indeed incorrupt and full of milk. Both times, the experts failed to reach consensus, which did not aid those pushing for Elena's immediate canonization; she was not beatified until 1828.

One of the most striking things about the accounts of these events in the hagiographic materials Elena's supporters compiled is the importance attributed to medical and natural philosophical expertise. This makes them very different from the stories of Chiara and Margherita told two hundred years earlier, in the early fourteenth century, in which learned experts had played a decidedly minor part. Chiara's body was opened not by medical men but by five of her fellow nuns, who found the crucifix and other holy objects hidden inside her heart and gallbladder; these were inspected only after the fact by physicians and natural philosophers, whose opinions, if they were ever recorded, did not survive.[11] Although doctors played a more prominent role in the opening of Margherita's body, their intervention was exclusively technical; they eviscerated it for embalming. It was the Dominican friar entrusted with cutting apart her entrails in order to transfer them to a reliquary who found the three miraculously figured stones in her heart. The several doctors present at this second event served only as incidental witnesses among many others, and the *vitae* do not even record their reactions, let alone attribute to them any special authority.[12]

In the case of Elena Duglioli, in contrast, medical men were present in ever-increasing numbers at each of the several inspections of her corpse, and the accounts repeatedly invoke their competence in matters concerning the body. Not only did two surgeons open and embalm her — according to standard practice — but her supporters, including Ritta, also took it as self-evident that her corpse should be evaluated by men with medical and natural philosophical qualifications, in order to determine whether its peculiarities were the result of supernatural or natural processes. Furthermore, those peculiarities differed from the ones that had characterized the bodies of Chiara of Montefalco and Margherita of Città di Castello. Where the corpses of those early fourteenth-century holy women had produced objects, separable from their internal organs and suitable for veneration (a crucifix, the instruments of the Passion, marvelous figured stones), Elena's produced signs (a flaccid heart, an appearance of relative incorruption, breasts filled with a white substance that smelled like rancid butter). The miraculous nature of the phenomena associated with Chiara's and Margherita's bodies was apparent even to lay observers. The phenomena associated with Elena's, in contrast, were ambiguous, requiring expert interpretation in order to determine if they were miraculous at all.

The differences between the early fourteenth-century anatomies of Chiara and Margherita and the early sixteenth-century anatomies of Elena reflect changes in both the ecclesiastical and the medical realms. The intervening period had seen a marked increase in clerical suspicion of visionary women. This suspicion was manifested by a growing concern to eliminate both natural and demonic causes as explanations for these women's unusual abilities and behaviors (thus demonstrating that they were, in fact, miraculous), as well as a growing conviction that one of the most effective ways of differentiating among the sick, the demonically possessed, and the truly holy was by expert reading of their bodily signs.[13] At the same time, over the course of the fourteenth and fifteenth centuries, the consolidation of natural philosophy and medicine as university disciplines had reinforced the authority of their practitioners as experts on the body.[14] As a result, while miracles, morals, and local reputation remained important in assessing the holiness of prospective saints, both promoters and opponents of holy women increasingly invoked the opinions of learned men, especially those of physicians and surgeons.

Anatomical evidence became more significant in understanding these sixteenth-century cases. As I have already described in connection with the work of Antonio Benivieni in Florence, anatomy acquired new prestige after 1490, largely as a result of new work on Galenic sources by learned physicians and other scholars influenced by the intellectual movement of humanism, with its emphasis on the study of ancient Greek authors.[15] Fourteenth- and fifteenth-century medical writers had relied for the most part on the relatively brief anatomical passages in Avicenna's *Canon* and on abbreviations and adaptations of the Galenic texts that circulated under the titles *On Interior Things* and *On the Uses of the Members (De juvamentis membrorum)*.[16] After the publication of the first edition of Galen's works in 1490, however, medical writers had easy access to the Latin translation of the more detailed treatise on which the former was based, *On Affected Places*, as well as more specialized Galenic treatises, such as *On the Anatomy of the Uterus*.[17] At the same time, Niccolò of Reggio's early fourteenth-century translation of the full text of Galen's *On the Use of Parts (De usu partium)*, of which *De juvamentis* was an abbreviation, began to circulate more widely in manuscript, although it was not printed until 1528. In addition to providing late fifteenth- and early sixteenth-century anatomists with more material, these newly recovered — or newly disseminated — Galenic works infused the discipline of anatomy and the practice of dissection with intellectual energy and cultural prestige. Like Galen, the medical writers who followed his lead emphasized the importance of direct experience, of learning from the body itself. The result was a dramatic surge in the study of anatomy in late fifteenth- and early sixteenth-century Italian universities, as well as in municipal colleges of physicians and surgeons, which paralled and reinforced the contemporary interest in domestic autopsies such as those of Fiametta Adimari and Bartolomea Dietisalvi.[18]

Inspired by these Galenic treatises, Italian medical writers began to compose their own specialized anatomical works. Before 1490, the only original Italian treatise on the topic was the *Anatomy* of Mondino de' Liuzzi, who had taught medicine at the University of Bologna in the early fourteenth century. In contrast, the years between 1490 and 1543, the publication date of Andreas Vesalius' celebrated work *On the Fabric of the Human Body*, saw the composition of at least eleven new anatomical treatises by Italian authors — including three by Vesalius — several of which appeared in multiple editions.[19] These reflected a renewed sense, traceable in large part to Galen's influence, that anatomy was fundamental to medical knowledge and practice and that dissection was fundamental to anatomy.

Alessandro Benedetti, who lectured on anatomy in Venice (and possibly Padua) in the late fifteenth and early sixteenth centuries and was renowned for his commitment to Greek sources, stressed the latter point in the final chapter of his *Anatomy*, which he called "In Praise of Dissection." Here he exhorted all medical practitioners, whether they were "students or experienced physi-

cians and surgeons, to frequent anatomical dissections, which should be held at least once a year, because in them we see the truth; we contemplate things open [in front of us], so that the works of nature lie before our eyes as if alive."[20] According to Benedetti, direct experience of this sort was indispensable. Although texts and images stimulate the memory and "disperse the shadows of the mind," they are only "simulacra," like maps — substitutes for and approximations of the phenomena they record. For this reason, he noted (citing Plato and echoing Galen), "they often err who, trusting written texts without examining objects, do not reflect on the things themselves."[21] Such powerful ancient testimonials increased the cultural prestige of dissection, attracting the interest even of nonprofessional elites. Girolamo Manfredi, a professor of medicine and astrology at the University of Bologna and the author of the first original anatomical treatise since Mondino's, dedicated his 1490 *Anatomy* to Giovanni Bentivoglio, Lord of Bologna, who had attended one of his anatomical demonstrations.[22] Even more ambitiously, Benedetti addressed his own work to the Holy Roman emperor Maximilian I and framed it as a kind of utopian fantasy, in the form of a multistage dissection to which he "invited" a host of eminent men of letters, patricians, and local dignitaries, whom he invoked by name.[23]

As the audiences for the annual formal dissections held by medical faculties in temporary "theaters" grew, the procedures functioned increasingly as civic spectacles, which dramatized the cultural achievements of the city and its university for the benefit of both locals and foreign visitors.[24] At the same time, as these events became more spectacular, and the standards for anatomical teaching more ambitious, anatomical writers began to dismiss their utility as serious teaching exercises, supplementing them with dissections organized for small groups of serious students or colleagues. Many such dissections were performed on animals in the master's own home — the medical student Ippolito of Monte-reale produced a detailed account of one such exercise in Perugia in 1519 — while others took place in hospitals or in the houses of private patients.[25] One of the earliest proponents of dissections of this sort was Jacopo Berengario of Carpi, who taught surgery and anatomy at the University of Bologna and who (with his nephew Damiano) was one of the medical experts involved in the anato-mies of Elena Duglioli in 1520. In his *Commentaries on the Anatomy of Mondino*, published the next year, Berengario stressed the im-portance of frequent dissections and the need to study many bod-ies, animal and human, of both sexes and varying ages, conditions, and body types, in order to accumulate accurate knowledge con-cerning the variety of the human frame. He was his own best example, for in his *Short Introduction to Anatomy* of 1522 he claimed to have personally anatomized several hundred corpses — an extraordinary number relative to even the most active of his fifteenth-century predecessors.[26]

The work of Benedetti, Berengario, and their contemporaries transformed the nature of anatomy as a learned discipline. Exten-sive experience with dissection allowed them to identify errors made by previous anatomists, such as Mondino and even Galen, and to create new anatomical knowledge of their own. (Among the discoveries Berengario claimed to have made were the exis-tence of voluntary muscles inside the nose and the fact that male fetuses urinate through the urethtra, not the umbilical cord.)[27] It also promised to reveal the truth concerning countless matters on which earlier writers disagreed: for example, the identity of the vessels that serve the gallbladder, the location of the veins in the arms, and the structure and function of the female reproductive organs.[28] The last were of great interest to all anatomists, not only because of their practical importance, but also because they were the site of several classic debates among medical authorities, most notably the dispute over the nature and existence of male and female seed and the respective contributions of fathers and mothers to their children.[29] In addition, as Berengario noted, the uterus was the subject of one of what were thought to be only two specialized anatomical treatises written by Galen (the other was on the eye), which gave the topic great intellectual cachet.[30] Finally, the female body — and the uterus in particular — retained its symbolic importance for Berengario and his contemporaries as the ultimate natural secret. These various considerations led him to devote more than a fifth of his massive *Commentaries on the Anatomy of Mondino* to the female genitals: 106 folios, whereas the heart, for example, merited just 27.[31]

As I will argue in this chapter, the increasing prestige of anatomy as one of the fields in which contemporaries might build on and even correct the work of ancient authors, as well as the emergence of specialists such as Benedetti and Berengario, whose expertise was based on extensive experience with dissection of both humans and animals, transformed the study of the female body, giving it confidence, texture, and detail. While earlier writers on women's anatomy, including Antonio Guaineri and Michele Savonarola, had confined themselves, for the most part, to simple and schematic descriptions of the genitals and fetal development, supplemented by brief discussions of the variable size of the uterus and uterine suffocation, Italian anatomists of the late fifteenth and, especially, sixteenth centuries could use first-hand knowledge — of Galenic texts and women's bodies — to make detailed determinations concerning the general physiology of generation and the formation of the fetus. The same knowledge allowed them to discuss and treat specific medical conditions that afflicted women, particularly those relating to childbearing, and to evaluate contemporary claims regarding the miraculous bodily powers and experiences of female ecstatics and visionaries such as

169

Elena Duglioli. In anatomical works of this period, women appear increasingly as the objects not only of solicitude but also of direct study on the part of male experts.

At the time of Elena Duglioli's death in 1520, Berengario was arguably the most knowledgeable such expert. He had practiced surgery and taught anatomy for many years in Bologna, and his *Commentaries on the Anatomy of Mondino*, which would appear the following year, established him as the most innovative and (with the possible exception of Gabriele Zerbi) the most prolific Italian writer on female anatomy before Vesalius. Thus it is hardly surprising to find him among the experts invited to evaluate Elena's corpse. In this chapter, I use the posthumous encounter between Elena and Berengario, the virgin and the surgeon, to explore how the new interest in dissection in late fifteenth- and early sixteenth-century Italy recast discussions of both sainthood and generation. At the same time, I focus on the development of anatomical illustration in learned medical writing, which echoed the new commitment to knowing through direct experience. Berengario's *Commentaries* was remarkable not only for its imposing length (528 folios) and its exhaustive treatment of the female genitals, but also for its twenty woodcuts; the first profusely illustrated anatomical text to appear in print, it served as both model and foil for later writers, including Vesalius. The woodcut images of women's anatomy in Berengario's *Commentaries* demonstrate the anatomist's ability to fathom hidden realities concerning human bodies, while they also model the compliance required of those bodies in the service of anatomical and physiological truth.

The Anatomy of Sanctity

Elena was not the first holy woman to attract the attention of medical men in the period after 1490. In 1501, two decades before Elena's death, the corpse of the visionary and prophet Colomba of

Rieti, a Dominican tertiary, was anatomized in Perugia. According to her *vita* — the work of the Dominican friar Sebastiano Bontempi of Perugia, who had served for some time as her confessor — Colomba was opened for embalming by a "most experienced physician [*peritissimo physico*]," who simultaneously performed an anatomy.[32] Two notable findings emerged: first, that Colomba's stomach and intestines were almost empty, confirming reports that she lived exclusively from the Eucharist, and second, that her heart was bathed in uncongealed blood, which Bontempi described as "liquid, full of life, bright, and pure as if it had flowed from the throat of a living dove."[33] Throughout his eyewitness description of Colomba's anatomy, Bontempi referred repeatedly and respectfully to the professional expertise of the anonymous physician who performed the procedure, calling him an "eminent master physician [*egregius Magister physicus*]" and even a "most experienced master anatomist [*peritissimus Magister anatomista*]."

This was not the first time that Colomba had come under the scrutiny of professionals. According to Bontempi, sometime before she died, she had been examined by "certain illustrious men, learned in the disciplines of philosophy and of nature, and not unfamiliar with holy scripture."[34] Three groups of anonymous experts — physicians, natural philosophers, and theologians — were asked to analyze and assess from the perspective of their own disciplines Colomba's famous behaviors and abilities: her almost complete abstinence from food and drink, her prophetic gifts, and her frequent ecstasies, in which she sometimes lay rigid and unresponsive for several days. Their brief was to consider all possible explanations for these remarkable capacities, in order to determine whether they might result from natural causes or demonic intervention. Only if these two causes were definitively excluded could a strong case be made for her canonization. Physicians thus subjected Colomba to a series of intensive medical examinations described approvingly by Bontempi: they "asked for her horoscope" and "tested . . . her nails and hair. They considered her sweat and odor, as well as her menstruation [*passione muliebri*] and her excretions. They observed her teeth while she spoke, the color of her face, and the pupils of her eyes."[35] Based on these observations, and after extensive deliberation, they agreed that Colomba was not suffering from a physical illness — specifically melancholy, which might produce delusions and other forms of unregulated behavior that mimicked religious ecstasies — or from demonic possession, which would have manifested itself in additional somatic and behavioral signs. They also confirmed that her abstinence from food was genuine, which would explain the fact that she did not menstruate, as her prioress certified.[36]

Like the postmortem exploration of her corpse, the study of Colomba's living body shows the degree to which both her supporters and her detractors acknowledged the authority of medical men and natural philosophers as experts on human physiology and the causal processes that governed the material world. At the same time, her case illustrates the uncertainties that inevitably arose when the learned were consulted. While it initially appeared to make sense to mobilize as many specialists as possible in hopes of generating the strongest possible case, it turned out that multiplying specialists also multiplied the opportunities for argument — or, as Bontempi euphemistically put it, when "men astute of sense and perspicacious of intellect came together, among whom the discussion of doctrines flourished, there was vigorous debate."[37] In other words, the more experts were consulted, the more likely they were to disagree.

In Colomba's case, as it so happened, the physicians failed to reach consensus regarding her pulse and respiration, which became almost imperceptible during her ecstasies. This matter was crucial to determining the nature of her raptures: whether they were supernatural, whether they involved a temporary but natural suspension of her vital powers, or whether they resulted from some much more unusual process — specifically, the receptivity to the impressions of distant events or the control over physical bodies attributed to occasional remarkable souls in magical texts such as the *Asclepius* and some of the newly translated works attributed to Hermes Trismegistus.[38] The experts' disagreement was unsurprising not only because technical disputes of this sort were the basis of contemporary university pedagogy, and hence the meat and drink of academically trained physicians,[39] but also because the experts in this particular case were being asked to resolve issues that lay at the far boundaries of the natural — in a period, furthermore, during which pressure on those boundaries was intense. Simultaneous with the rise of anatomy, the late fifteenth and sixteenth centuries saw a dramatic expansion of interest in what contemporary naturalists sometimes referred to as "preternatural" phenomena, which included both demonic activity and the rare and unusual physical events that lay outside the ordinary course of nature but resulted from natural, albeit unfamiliar, causes, such as subtle vapors, celestial influences, or the powers of extraordinary souls.[40] These topics were of obvious relevance to the assessment of potential saints, and it is worth noting that one of the experts consulted in 1520 regarding the corpse of Elena Duglioli, Pietro Pomponazzi, a professor of natural philosophy in Bologna, composed an entire treatise describing and explaining preternatural phenomena that same year.[41]

The pitfalls of invoking expert medical testimony are even more obvious in the case of Elena's multiple anatomies. Elena's clerical champions, like Colomba's, had obvious confidence that the physicians' methods — the careful assessment of corporeal signs in search of underlying causes — would vindicate as supernatural Elena's remarkable bodily experiences: her virginal lactation and

Jesus' extraction of her heart. This led them to describe their initial examinations of her corpse in the contemporary language of natural inquiry and demonstration, as their use of the term "experiment" makes clear. But the invocation of medical expertise proved to be a double-edged sword. Having secured testimony regarding the flaccid appearance of her heart from the two surgeons who, with Girolamo of Firenzuola, a professor of medicine at the University of Bologna, extracted her viscera, Ritta would probably have been willing to leave Elena's remains in peace. However, other Church officials, including the archbishop of Bologna, were skeptical regarding Elena's case,[42] and they pressured her supporters to continue the anatomical initiatives that Ritta had ordered, mobilizing the vast reservoir of medical and natural philosophical knowledge for which the city was renowned.

The anonymous author of Elena's *vita* described in detail the subsequent events. The first of the two supplementary anatomies, forty-five days after her death in early November, was conducted by three eminently qualified medical men: the surgeon Angelo of Parma, who was directed to open her left breast, and two professors of medicine at the university, Girolamo of Firenzuola, who was present at the initial autopsy, and Ludovico Leone. The result of this investigation was "schism and difference of opinion"[43] regarding both the degree of incorruption of her breasts — for all agreed that the rest of her corpse was already rotting — and the reasons for their relatively good state of preservation. While Elena's follower Girolamo insisted on their incorruption, Leone and the surgeon attributed this to natural causes and to the fact that the body had been embalmed: the breast, "being full of nerves and arteries, [was] slower to putrefy than the rest of the body, which . . . was aided by the myrrh, aloes, and alum that . . . had been placed in the eviscerated body." Furthermore, they argued, since it was well known that human milk was composed of menstrual blood that had been purified and concocted in the breasts by the body's natural heat, the white substance in the breast had to be flesh, not milk, "since the causes of this — the menses and natural heat — had wholly ceased, the sign of which was that it did not appear to flow."[44] On this point, the experts' task was no doubt made more difficult by the loss of Elena's uterus, which seems to have been discarded after her evisceration.

As these remarks suggest, the problems raised by Elena's breasts were complicated and technical, involving issues that went beyond human anatomy and physiology to engage more general natural processes, specifically putrefaction (the process at work on Elena's flesh) and concoction (the process responsible for the production of milk).[45] Although both putrefaction and concoction depended on the body's natural heat, which slowly dissipated after death, as Aristotle had described in the fourth book of the *Meteorology*, their interaction was complex and unpredictable; putrefaction involved an initial increase in heat, which might allow lactation to continue naturally after death, assuming that the requisite raw material, menstrual blood, was present. In order to address these more general questions, the principal promoters of Elena's cult, the priors of the church of San Giovanni, consulted two eminent natural philosophers at the University of Bologna who had special expertise in these matters: Ludovico Boccadiferro and Pomponazzi, both of whom later produced commentaries on *Meteorology* 4.[46] At the same time, to make sure they had covered all the physiological and anatomical bases, they called in two additional professors of medicine, Niccolò de' Passeri of Genoa, from the University of Padua, and Jacopo Berengario of Carpi, from the University of Bologna, who was in the process of preparing his massive *Commentaries on the Anatomy of Mondino* for publication.

Again the experts split, with one medical man and one philosopher on each side. Passeri and Pomponazzi — surprisingly,

given the latter's well-documented naturalism in many other matters — argued unambiguously for the presence of milk in Elena's
breasts and for their supernatural preservation, noting that "wom-
en's breasts are among the parts of the body that rot most easily
... on account both of the moisture in them and of the tenderness
and softness of this glandular flesh."[47] (Pomponazzi's caution in
this case may have been a reaction to the intense public controversy generated by his treatise *On the Immortality of the Soul*,
which had led to a papal warning in 1518.) Berengario and Boccadiferro, on the other hand, while agreeing that Elena's breast
contained congealed milk, "nonetheless identified this as a work
of nature without anything miraculous about it and attributed it
to the natural heat preserved in [the breast], saying that it was
ridiculous to say that there was no heat there, ... nor was it credible that [the milk] would flow, since it was preserved in its proper
place, and to assert that there was not milk there, because it did
not flow, was not fitting for a man of learning."[48]

After the failure of the November anatomy to generate a consensus, Elena's body was exhumed yet again, in late December.
This time, the forty witnesses to the opening of her right breast
included six doctors — the three who had been present at the
November procedure, together with Berengario, the physician
Virgilio of Modena,[49] and an unnamed colleague — with the same,
predictable results: ambiguity and indecision. In the words of the
frustrated author of the *vita*, "the medical men, who are always
enemies of miracles and have recourse to the works of nature,"
held that the breast had begun to putrefy, although not nearly to
the degree expected three months after death, and they proposed
to send a piece of it to the Pope, "so that the Roman doctors
could pass judgment regarding the prophecy of the preserved
milk."[50] The *vita* does not record if this suggestion was ever
carried out.

Comparing these events to the fourteenth-century anatomies of Chiara of Montefalco and Margherita of Città di Castello, it is easy to see that much had changed, and that the revival of an earlier practice in a new intellectual and institutional context produced unintended and unwelcome results, at least in the view of Elena's and Colomba's supporters. The clerics who initiated these women's anatomies had the examples of Chiara and Margherita explicitly in mind.[51] As in Chiara's and Margherita's cases, they believed they had found unambiguous evidence of sanctity in their hearts, which in their literal softness and impressionability revealed receptivity to God's word; while Elena's heart had been replaced by something that resembled soft liver, Colomba's was surrounded by fluid blood and had the consistency of wax, a fact her hagiographer related to Psalm 22.[52] In other respects, however, the stories of Elena's and Colomba's anatomies diverged dramatically from those of Chiara and Margherita, for the latter's clerical supporters, working in small towns, in an environment in which the authority of academic medicine was inchoate and the practice of dissection in its infancy, maintained full control over the interpretation of the results. In the cases of Elena and Colomba, however, the anatomies were performed in university cities by medical men well versed in medical theory and surgical and anatomical practice, in a period when the latter was a focus of intense professional study. Colomba's anatomy was relatively successful from Bontempi's point of view; the only physician present at her anatomy was the man who performed the dissection, and he was already a professed believer. Elena's case, however, involved multiple medical experts, and the process of interpretation proved impossible to control. Although they were by no means all hostile to Elena, they disagreed on the meaning of almost every particular of her case.

Elena's anatomies and Colomba's examination by physicians,

natural philosophers, and theologians became opportunities for learned men not only to promote their own interpretations and reputations but also to advance medical and natural philosophical knowledge — occasions for first-hand research. Through these women's anatomies, they could explore new issues, such as the operation of natural heat in the interlaced processes of putrefaction and lactation or the fashionable topic of the imaginative powers of elevated souls. Although I have not been able to find any mention of Elena in the works of any of the medical writers and natural philosophers involved in her anatomies (specifically Berengario, Pomponazzi, and Boccadiferro), the anatomist Alessandro Benedetti referred to Colomba in his medical reflections on fasting, and Berengario included in his *Commentaries* a detailed anatomical defense of the miraculous nature of the flow of water and blood from the side of the crucified Christ.[53]

In this new context, the notion that bodies revealed their holiness self-evidently, through signs that were readable by any layperson (stigmata, lactation, incorruption, the internal generation of holy objects), could no longer be sustained. As anyone with academic medical training knew, corporeal signs were equivocal and multivalent; to read them correctly required great erudition, expert judgment, and long experience.[54] Although the traditional medical doctrine of signs focused on external manifestations of internal states, as in the doctors' examination of Colomba's living body, the rapid development of anatomy and dissection in late fifteenth- and sixteenth-century Italy opened up a whole new domain for semiological exploration, as Elena's anatomies make clear. Thus whereas the bodies of Chiara and Margherita produced objects — a crucifix, the instruments of the Passion — the bodies of Elena and Colomba produced evidence. This evidence might take the form of purely natural signs (Colomba's emaciation and empty stomach) or result from supernatural processes (Elena's uncorrupted and milk-filled breasts, the bright, liquid blood in Colomba's chest, and the peculiar appearance of both women's hearts).

The production of objects, as I argued in Chapter One, was conceived of in gendered terms; women's bodies were defined by their ability to generate objects inside themselves, so the discovery of the crucifix inside Chiara's corpse was easily understood in terms of the birth of Christ in the heart and even in terms of the contemporary practice of extracting children from the wombs of their dead mothers. Although this spiritual process was applicable in theory to both men and women, it seems to have manifested itself materially only in the bodies of two holy women, Chiara and Margherita — although St. Francis wrote eloquently concerning the birth of Christ in the soul, his body manifested only *external* signs, in the form of stigmata — and it may well be that its association with female reproductive anatomy and physiology initially discouraged the internal exploration of the corpses of holy men. The shift from wonder-working objects to anatomical evidence, which we see in the openings of Colomba and Elena, effectively removed the gendered associations of holy anatomy, making it equally appropriate and available for the assessment of male bodies. And indeed, by the middle of the sixteenth century, the practice had been extended to saintly men. The most famous such autopsies in the first century of the Counter-Reformation were those of Ignatius Loyola (d. 1556), who was inspected by the famed anatomist Realdo Colombo; Carlo Borromeo (d. 1584); and Filippo Neri (d. 1595).[55] This seems to have created some pressure to rewrite the history of the practice, to provide a male precedent for the general practice of holy anatomy as well as for the specific discovery in the heart of signs of divine habitation. Explaining his decision to open Neri's heart in 1597, the physician Antonio Porto recalled not only the example of Chiara of Montefalco but also that of Bernardino of Siena (d. 1444), in whose heart "was found the good Jesus, because [Bernardino] never spoke of anything else."[56] Although Bernardino was embalmed, I know of no evidence that his viscera were ever inspected, let alone that anything unusual was found in his heart.[57] Thus Porto's offhand remark suggests the degree to which, by the end of the sixteenth century, holy anatomy had become detached from the female body and accepted as common practice, appropriate for the bodies of both women and men.

The Anatomy of Generation

Despite the interesting issues it raised regarding the limits of natural causation, the anatomy and physiology of holy virgins such as Colomba and Elena remained a recondite topic. Although medical men were available to consult on such matters, they remained, for obvious reasons, far more invested in the internal structure and workings of the bodies of more typical women, which they for the most part reduced to their reproductive organs. The two late fifteenth- and early sixteenth-century Italian anatomical writers who devoted the most time and attention to female bodies in this connection were the anatomist and surgeon Jacopo Berengario of Carpi and Gabriele Zerbi, who had taught medicine at the University of Bologna a generation before Berengario.[58] The two were very different in temperament and intellectual orientation. Zerbi's massive *Book of the Anatomy of the Human Body and of its Individual Members* (1502) was principally a work of erudition, based on the exhaustive comparison of textual authorities, while Berengario's *Commentaries on the Anatomy of Mondino* (1521) stressed sensory experience and the practice of dissection. Despite these differences, however, both discussed the female reproductive organs at length, including both the pregnant and the nonpregnant uterus.[59]

Zerbi's and, especially, Berengario's disproportionate emphasis on the uterus relative to the other members of the human body, including the heart and brain, reflects the prestige that accreted to the topic as a result of the discovery that Galen had written a specialized treatise on its anatomy. Furthermore, not only were women's reproductive functions crucially important in the lives of the men and women who made up the physicians' clientele, but they were acknowledged to be the most complicated, mysterious, and difficult to understand of all the body's processes. In large part, this was because of the amazing variability of the uterus, not only in size, as Mondino had already noted, but also in texture and form. As Berengario put it,

> it is sometimes small, in accordance with the age and size of the body in which it is located, and sometimes large; and it is sometimes swollen with menses and sometimes thin, since it becomes finer close to the fetus or fetuses...; and sometimes it is shortened and wrinkled and becomes quite dense and thick, as if it were a fleshy organ; and sometimes it is filled with a fetus or growth, and sometimes empty, and sometimes rough, and sometimes soft and smooth, and sometimes closed, and sometimes open.[60]

This mutability meant that the anatomist had to dissect many more uteruses than any other organ in order to gain even an approximate sense of its nature and functions. This problem was particularly acute for the pregnant uterus, which represented the most complicated and difficult case, since the fetus itself also grew and changed over the course of its gestation.

Because of its importance, its complexity, and its inaccessibility, the uterus continued to symbolize the ultimate frontier of the physician's understanding. In Berengario's hands, however, it also acquired new meaning, as an emblem of the peculiar epistemological status of anatomy, where — more than in any other branch of medicine — experience trumped erudition and verbal description. "And let no one believe that this discipline may be learned through oral instruction or books alone," he wrote in the preface to the *Commentaries*, "since it requires sight and touch."[61] One testimony to Berengario's belief in the importance of the senses to anatomical study is his decision to illustrate his book with woodcuts. The use of images to describe nature was controversial in this period, since they could be seen as poor substitutes for direct experience, as Galen (and, after him, Alessandro Benedetti) had argued.[62] This was certainly one reason why none of the late fifteenth- and early sixteenth-century anatomists before Berengario had seen fit to illustrate his work, except Leonardo da Vinci, who was an artist by training.[63] Berengario himself collected ancient and contemporary art — among other things, he seems to have owned a painting of John the Baptist attributed to Raphael and a marble torso from an antique sculpture, and he was acquainted with a number of artists, including Benvenuto Cellini — which may explain his willingness to experiment with visual description.[64]

The most methodologically articulate of the *Commentaries'* twenty woodcuts is in fact devoted to the female genitals (figure 4.1). It shows a standing woman, her abdomen opened to reveal her dissected uterus, pointing toward a second uterus, lying on a plinth. Her gesture indicates that the author's knowledge of her genitals comes from direct inspection, rather than from reading earlier texts; these lie beneath her left foot, serving as both the foundation and the discarded precursors of his work — an apt metaphor for Berengario's relationship to earlier writers on anatomy. The caption underscores this point, with its repeated references to sight. It reads (in part),

> You have in addition an everted uterus outside her belly, which is the figure on which you see the index finger of the present figure of the

Figure 4.1.   Anatomy of the uterus, third figure. Jacopo Berengario of Carpi, *Carpi Commentaria cum amplissimis additionibus super Anatomia Mundini...* (Bologna: Girolamo de' Benedetti, 1521), fol. 226v.

woman. And at the base of the uterus there is a certain depression, as you see, which is what distinguishes the right hollow from the left. Nor is there any other division in the uterus. And the black dots are the cotyledons. And you see how the vagina [*collum matricis*] lacks cotyledons. And you see how the vagina resembles the male penis.[65]

Berengario's caption exemplifies the practice for which he is famous: using observations made in the course of his own dissections to "determine" the answer to a disputed question, in the language of contemporary scholastic debate — in other words, to draw the correct conclusion regarding matters on which textual authorities disagreed, such as whether the uterus had two or seven cells and whether there were cotyledons on the walls of the vagina as well as the uterus. (Cotyledons were believed to be the orifices of the seminal, or spermatic, vessels, which carried female seed, the menses, and the blood that nourished the fetus. In fact, cotyledons, which are contact points between the placenta and the wall of uterus, are found only in ruminants, which confirms Berengario's statement that he, like his contemporaries, relied heavily on animal dissection to study the female genitals.) Although recourse to first-hand experience may seem to modern readers like a straightforward and obvious way to resolve such disagreements, it was rare in medical and natural philosophical writing of the late Middle Ages, where truth was assumed to lie within the text. The resolution of disputed questions, as Jole Agrimi and Chiara Crisciani have noted, was typically an exercise in hermeneutics; this involved reading beneath the surface of the conflicting texts to find a level on which they agreed, so that apparent contradictions were reduced to differences in emphasis or semantics.[66] Even when the determiner, as occasionally happened, squarely rejected the position of one of the relevant authorities, he usually did so on grounds relating to the accuracy of the Latin translation of the Arabic or Greek text in question or to the text's authenticity.

Zerbi's *Book of Anatomy* of 1502 still belonged to this earlier world of intellectual practice, as is clear from his discussion of the womb in Book 2. For example, after noting that various authorities (Galen, Haly Abbas, Avicenna, Gentile of Foligno) disagreed regarding the number of tunics or coverings of the uterus — some claimed one, others, two — he reconciled their differences by suggesting that there were in fact two tunics, but they were attached so tightly that they looked like one.[67] Similarly, in addressing the number of cells or concavities in the uterus, he noted that the "common agreement of the illustrious" was that there were only two. In both cases, his reasoning centered on texts rather than corpses. On the latter point, for example, he argued (appealing to the fourteenth-century medical writer Pietro of Abano) that although Galen appeared to have argued for seven cells in his *On Seed* (*De spermate*), this work was very possibly inauthentic.[68]

Berengario's procedure was completely different. Although his *Commentaries*, as the title indicates, took Mondino's early fourteenth-century *Anatomy* as its point of departure, Berengario's ultimate reference, at least in theory, was his own observations. As the woodcut of the pointing woman indicates, he aimed to resolve contradictions by looking at bodies, not books. For example, he determined — or, more accurately, inveighed — on the basis of self-evident appearances that the uterus had only two hollows, barely divided by the "depression" referred to in the caption of figure 4.1, characterizing Mondino's claim that it had seven cells (due most likely to his ignorance of Galen's *On the Anatomy of the Uterus*) as an unforgivable howler.[69] Nor was Berengario loath to criticize even Galen. On the matter of the cotyledons, visible in the same woodcut as black dots on the interior of the uterus, he invoked his own experience to side with Mondino against Galen, criticizing Galen for contending, in *On the Anatomy of the Uterus*, that the seminal vessels terminated in both the vagina and the uterus, when they were in fact confined to the latter. He argued as a result that the text of *On the Anatomy of the Uterus* was corrupt and that the true doctrine of Galen on this particular point was contained in *On Seed*, which he, unlike Zerbi, took (wrongly) to be authentic.[70]

Berengario agreed entirely with Galen, however, on the relationship between the male and female genitals, which he understood as inversions of each other; in the words of the caption to figure 4.1, "you see how the vagina resembles the male penis." As Galen had argued in *On the Use of Parts*, women, being of colder complexion, are incomplete and imperfect versions of men, with their genitals still hidden within their bodies rather than fully expressed on the outside. If men's genitals were folded inward, they would resemble women's, with the scrotum corresponding to the uterus, the male testes to the female ones (our "ovaries"), the vagina to the penis, and the foreskin to the labia.[71] Berengario noted that this resemblance is easily visible in women who have just given birth — something he personally witnessed in the course of his surgical practice — since the uterus is open, which makes the relationship between the female and male genitals extremely clear.[72]

Although much has been made of such expressions of the homology of the male and female genitals, they should not be taken as evidence of a "one-sex" model of the human body that supposedly characterized the period from Galen through the eighteenth century, as argued most famously by Thomas Laqueur.[73] Anatomists' commitment to this homology correlates strongly with their interest in *On the Use of Parts*, the work in which Galen described it; as I have already mentioned, this text had relatively little circulation in Latin Europe before the late fifteenth century and was not published until 1528. For this reason, references to the homology appear most clearly in the works of Renaissance anatomists such as Zerbi, Berengario, and, later, Vesalius, who had direct access to this work.[74] But these two generations of anatomists do not exhaust the fourteen hundred years that separated Galen from his sixteenth-century followers. References to the homology between the male and female genitals were conspicuously absent from medieval anatomical texts and images before the thirteenth century, when they began infrequently to appear as a result of the influence of Avicenna's *Canon*.[75] However, the vast majority of medieval writers on anatomy (including Mondino), who had never read *On the Use of Parts* and relied instead on its abbreviated version, *On the Uses of the Members*, supplemented by *On Interior Things*, made little or no mention of the homology. As Mondino put it in his *Anatomy*, "the members of generation in men and women are similar in some respects and different in others" — the most crucial difference being the inability of the female testicles to produce real seed.[76] And although the idea of genital homology enjoyed a real vogue in learned medicine in the first half of the sixteenth century — a vogue that persisted considerably longer in vernacular sources — it soon came under attack by anatomists and learned physicians.[77]

For Berengario and Zerbi, however, the homology was real, with real-life consequences — Berengario referred to the possibility that women might turn into men when their internal genitals everted themselves[78] — and important physiological implications, which helped to resolve the persistent debate about the relative contributions of mothers and fathers to their children, which I described in the preceding chapter. Aristotle had argued for a radical difference between males and females; this was expressed, among other ways, in the fact that men alone contributed animating seed to the fetus, while women supplied only the menstrual blood that nourished it and the uterus within which it grew. Galen and Hippocrates, on the other hand, had held that both males and females produced seed that communicated form and motion to the fetus and influenced its character and appearance. Faced with these two contradictory authorities, medical writers of the fourteenth and fifteenth centuries (such as Mondino) for the most part split the difference; while acknowledging the existence of female seed, they attributed to it at best an ancillary role in generation, privileging paternity over maternity, even though this flew in the face of lay intuitions and observations concerning the resemblance between mothers and children and the transmission of hereditary illness through the female line.[79] For a sixteenth-century Galenist like Berengario, however, accepting that women's genitals were a literal inversion of men's — a fact he considered manifest to the senses — entailed not only the incontestable existence of female seed (albeit subtler, colder, and emitted in smaller quantities than its male equivalent) but also the basic parity of the mother's and father's physiological relationship to the fetus. Both parents contributed matter and form, although mothers were responsible for more of the former and fathers for more of the latter; as a result, "fetuses appear evidently to resemble both [parents]."[80] In particular, Berengario described hereditary illness as transmitted in the seed, without drawing any distinction between the parents in this respect.

Berengario founded these conclusions on his own observations, in line with his often-repeated commitment to the truths of sight and touch. As he noted, the opinion of Galen and Hippocrates on the mother's contribution to the fetus was "more consistent with sense [experience]."[81] Yet anatomists had great difficulty in obtaining pregnant cadavers for dissection. Women tended to die in or shortly after childbirth, rather than before it, which restricted the supply of bodies for private anatomy, and convicted criminals were rarely executed until after they had given birth. For these reasons, as Zerbi noted, it continued to be "more convenient to conduct this research on monkeys and pigs and other animals similar to women."[82] Berengario referred clearly and unambiguously to only two anatomies of pregnant women: one of an executed criminal, witnessed by "almost five hundred students at the university of Bologna, as well as many citizens"; and one, in 1520, of a private citizen, who died in the ninth month of her pregnancy "without being able to give birth."[83] He conducted the latter in his own home in the presence of "many trustworthy scholars [doctoribus] and students." He was able to supplement these frustratingly rare experiences with observations collected in the course of his extensive surgical practice. I have already described his discovery of an extrauterine fetus in the course of a failed attempt to extract a living child from the uterus of its dead mother.[84] He had also treated a pregnant woman with an enormous swelling in her womb and participated in at least two successful operations to ligate and excise prolapsed uteruses — one as an assistant to his father and the second on his own, in May 1507.[85] Furthermore, having operated several times on "noble Bolognese girls" whose hymens were too thick to permit intercourse, he could confidently state, against Avicenna and Mondino, that the hymen lay nearer to the mouth of the vagina than to the cervix.[86]

In this and similar passages, Berengario's *Commentaries* confirms that women in late fifteenth- and sixteenth-century Italy, at least in elite urban households, looked to male medical practitioners for the treatment of gynecological conditions. Surgeons excised overgrown hymens and manipulated prolapsed uteruses, while physicians dealt with a wide range of illnesses, from infertility to uterine suffocation. At the same time, childbirth itself remained the sole responsibility of midwives, female friends and relatives, and attendants, except when the mother died in child-birth and it was decided to extract the baby for baptism — an infrequent operation that seems generally to have been entrusted to barbers or surgeons like Berengario.[87] This situation limited anatomists' research, since it meant that they had no direct access to miscarried fetuses — a situation that Berengario deplored. "Experience is the mistress we must believe in this matter," he wrote, "but this experience is extremely difficult . . . , since we see miscarried [*abortivos*] fetuses only very rarely and with difficulty, on account of the custom of our women, who do not wish to show them, except with difficulty and by theft. And wishing to see this, I managed to see it with difficulty, by secretly giving money to midwives."[88] For the same reasons, anatomists found it difficult to test ideas widely held by laypeople but not found in ancient texts: for example, that the number of "knots" in the umbilical cord predicts the number of children a woman will bear.[89] Regarding this matter, Berengario wrote, "I asked many old and very experienced midwives about this, and they said that it was a simple lie, and I believe this to be true, since every umbilicus has rollings and twistings of veins and arteries, which make up the aforesaid knots."[90] In one case, his own experience allowed him to explode what was literally an old wives' tale. Noting that many midwives believed that a baby delivered with the umbilical cord around its neck was destined to die by hanging, he remarked that "whenever I dissected a pregnant uterus, I always found the umbilical cord wrapped around the neck" — a contrivance "made by nature," to protect the umbilicus when the fetus moved.[91]

Such remarks confirm the unique difficulties presented to anatomists by "women's secrets": the female genitals and their functions, which included fetal as well as female anatomy. At the same time, they show that the dissemination of more and better information, resulting from both the recent recovery and translation of ancient anatomical texts and the new currency of dissection in the late fifteenth and the sixteenth centuries, had allowed considerable inroads in this domain. Experience was indeed the mistress of knowledge, as in all branches of anatomy, but experience of this sort was no longer restricted exclusively to women; pregnant cadavers were scarce but not unattainable, and there were other ways of gaining direct access to human fetuses, even if they involved bribery and theft. It was obvious to male practitioners that research on the genitals and generation was ultimately in women's best interest — an opinion shared by at least some of their female patients — and that cultivating their compliance served the general good.

Self-display

Berengario influenced learned anatomy in Italy not only through his texts, with their emphasis on the hands-on work of the practicing surgeon, but also through his use of illustrations. The twenty woodcuts of the *Commentaries* are scattered unevenly throughout the treatise. In addition to depictions of the abdominal muscles; the superficial muscles of the front, back, and sides; the veins of the arms and legs (important for bloodletting); the skeleton; and the bones of the hands and feet — all of which show male bodies — these include three illustrations of uterine anatomy (figures 4.1, 4.2, and 4.3).[92]

Although nothing specific is known about Berengario's role in the design of the woodcuts, it is clear that he was closely involved. The drawings for the woodcuts were the work of several artists active in Bologna, including Amico Aspertini and possibly Ugo of Carpi, and the blocks may have been cut by Berengario's printer, Girolamo Benedetti, so it would have been easy for the anatomist to collaborate in their design and production.[93] The captions of several of the *écorchés* foreground the utility of anatomy to artists

Figure 4.2. Anatomy of the uterus, first figure. Jacopo Berengario of Carpi, *Commentaria*, fol. 225v.

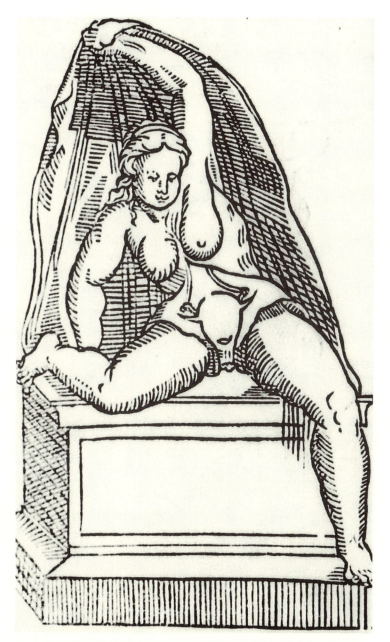

Figure 4.3. Anatomy of the uterus, second figure. Jacopo Berengario of Carpi, *Commentaria*, fol. 226r.

as well as medical men, which further suggests that Berengario advised the draftsmen of the images.[94] While these illustrations are not integrated into the text to the same degree as the ones in Vesalius' *Fabrica*, their captions and their style reflect specific points in the anatomist's argument — such as the number of subdivisions in the uterus and the homology between the male and female genitals — as well as his strong interest in surgery and his classicizing tastes in art. At least one of the figures demonstrating the muscles in the *Commentaries* (figure 4.4) was modeled on a specific contemporary sculpture, Michelangelo's *David*.[95]

The ideal of cooperation between women and medical men in the unraveling of the mysteries of the uterus had already been modeled in the woodcut of the dissected female figure in the *Fasiculo de medicina* of 1494 (figures I.6 and 2.1), which showed her instructing the anatomist (and the reader) regarding the details of generation, rather than keeping them secret, as women were thought to do. The three woodcut illustrations of female anatomy in Berengario's *Commentaries* elaborate this theme of self-revelation. The women in all three, like the woman in the *Fasiculo* illustration, spread their legs so as to give an unimpeded view of their genitals. Figures 4.1 and 4.3 show their subjects actively inviting the anatomist's inspection by unveiling their dissected bodies for the viewer, a gesture also alluded to by the half-drawn curtain in figure 4.2. But the new rhetoric of visual demonstration means that the participation of Berengario's women in the process of discovery is more limited than that of the woman in the *Fasiculo* woodcut, in the sense that they no longer speak but merely show. This is in accord with Berengario's rejection of Zerbi's style of anatomy as a verbal discipline, based on the oral explication of texts, for an anatomy based on the evidence of the senses; while the woman in the *Fasiculo* raises her finger in the gesture associated with lecturing, the woman in figure 4.1 points

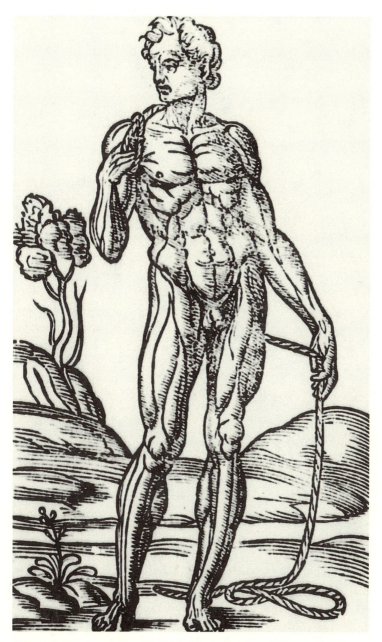

Figure 4.4. Exterior muscles of the front of the body. Jacopo Berengario of Carpi, *Commentaria*, fol. 519r.

mutely to the uterus on its plinth to illustrate the author's comments, conveniently summarized in the caption, on the location of the cotyledons and the homology between the male and female genitals.[96] The truth of the anatomist's statement is self-evident; more words would be beside the point.

The women in figures 4.2 and 4.3 have even less to do. They present themselves without engaging in any other gestures of demonstration; their lessons are wholly displaced to the captions, where the author explains the uterine diagrams that are superimposed awkwardly on their dissected bodies and gives basic information about the names and relationships of the uterus and adjoining organs (including the "horns" of the uterus described by Galen).[97] Although their eyes are cast down in modesty, this can also be read as curiosity; in opening their bodies, Berengario has revealed to them information they never knew. The effect of all three figures is to transfer the last vestiges of authority from the woman to the anatomist, in the sense that he needs her no longer to instruct him, but only to permit him to observe the body he has dissected; he is the expert, while she is the object of his expertise.

The relationship between the anatomist and his female subjects in the text and images of the *Commentaries* recalls in some respects that of holy women and their confessors, as in the cases of Colomba of Rieti and Elena Duglioli of Bologna, with whom I began this chapter. Sebastiano Bontempi and Pietro Ritta, Colomba's and Elena's confessors, were responsible for bringing to light the interior secrets of their protégées, cultivating their visionary vocations, amassing evidence concerning their prophetic gifts — including arranging for their anatomies — and publicizing their holy acts to others during their lives and after their deaths, both orally and in writing. (Bontempi was the author of Colomba's *vita*, and while the long *Legend* devoted to Elena is anonymous, it is closely related to Ritta's more compact *Narrative* as well as the letter he wrote regarding her to Pope Leo X.)[98] These men served as the women's sounding boards, informants, and mouthpieces, supported them through painful and frightening experiences, guided them to culturally recognizable and effective expressions of holiness, and spoke for them in an environment where women had no public voice. In turn, the women revealed to their confessors the most intimate details of their internal life, including, in Colomba's case, cooperating in a highly intrusive medical examination.

As John Coakley has argued, this relationship was truly collaborative; the confessors of holy women drew powerful inspiration from their protégées' spiritual lives. Immersed in the mundane world of teaching, preaching, and administration, they were attracted by the way these women's separateness from the world of ecclesiastical responsibility and learning — indeed, their very lack of literacy — allowed them to turn their backs on words and books to achieve direct contact with God. In Coakley's words, "the power of the women to represent the divine goes hand in hand, therefore, with their very differentness from the men, their strangeness of experience, their lack of office and schooling, their oppositeness of gender."[99]

This collaborative dynamic clearly structured Ritta's relationship to Elena. She apprenticed herself to him in 1505 or 1506, when she first revealed her married virginity, and she must have spoken with him almost daily until her death.[100] He was the first to hear of her visions and had the privilege of revealing them at his own discretion to the wider world. One of the earliest and most powerful of these visions came during her wedding, when she was only fifteen or sixteen, long before she associated herself with Ritta. In the middle of the festivities, she suddenly found herself entranced by "celestial choirs, like another Cecilia, among songs and sounds from above."[101] (In addition to being a married

197

virgin, like Elena, Cecilia was the patron saint of music.) Repeated experiences of this sort eventually left her unable to bear the sound of earthly instruments and voices,[102] as shown in the painting that Elena and her disciple Antonio Pucci commissioned from Raphael for the altar of her chapel in San Giovanni in Monte, probably in 1515 (figure 4.5). Composed following a program most likely devised by Pucci, this depicts Cecilia (closely identified with Elena), flanked by the saints Paul, John the Evangelist, Augustine, and Mary Magdalene; ignoring the delapidated portable organ in her hands and the musical instruments scattered on the ground in front of her, she attends only to the singing of angels that appear in the sky.[103]

This painting resembles in obvious ways the third female figure in Berengario's *Commentaries* (figure 4.1), which was published only one year after the author had participated in an anatomy of Elena's corpse in the church in which the painting hung. Both images use the figure of a standing woman to establish a hierarchical relationship between two ways of knowing: spiritual experience, as opposed to the experience of the external senses, in the case of the painting; and sensory experience, as opposed to textual knowledge, in the case of the woodcut.[104] Both reject the latter in favor of the former and represent this rejection by placing objects embodying the rejected form of knowing (musical instruments, books) underfoot.[105] In both cases, the woman represents the epistemological power of unmediated experience: it is better to hear celestial harmonies directly than mediated by metal, wood, and leather; it is better to study anatomy through dissection than mediated by paper and ink. At the same time, however, both images imply the presence of a crucial figure, who appears in neither but stands outside the frames of both: the man whose expertise has made meaning of the revelation in question, using all the resources of intellect and sense. In the woodcut, this is the anatomist Berengario, who has

Figure 4.5. Raphael, *The Ecstasy of Saint Cecilia*, Bologna, Pinacoteca Nazionale, ca. 1515.

opened the woman's body to reveal her uterus. In the painting, it is the confessor Ritta, who has recorded and interpreted Elena's revelations, publicizing them as seemed appropriate. Each man has taken the raw materials of a woman's inner self and staged them as public knowledge — a staging that took dramatic, physical form in the crowded postmortem examinations of Elena's corpse, which find an echo in the dissected body of the woman in Berengario's woodcut.

Like the relationship between confessor and confessee, the collaborative relationship between dissector and dissectee in Berengario's illustrations is not intrinsically gendered. Male visionaries had their own confessors, and several of Berengario's male figures, like the female ones, cooperate with the anatomist, holding their bodies open to display their abdominal muscles (for example, figure 4.6). The effect is somewhat different, in predictable ways: the women's gestures are sexualized, while the men's are not. The latter open their abdomens, while the former spread their legs. The women's downcast eyes contrast markedly both with the forthright gazes of the male figures and with their own extravagantly immodest poses — a tension that reflects the message of contemporary medical men who wished to reassure their female patients that they could reveal their bodies to their physicians while still preserving their virtue and respectability.

But the women in Berengario's woodcuts are nonetheless full of vitality. Although their postures have something in common with the female figures in contemporary erotic prints, such as the famous *Positions* conceived shortly afterward by Giulio Romano, their frankness is in fact far less sensual than the half-hearted gestures of modesty associated with naked women in ancient art and its classicizing imitators, who cover their breasts and genitals with their hands (figure 4.7).[106] Aside from their splayed-leg poses, they have little in common with the much more explicitly

Figure 4.6. Anatomy of the abdominal muscles. Jacopo Berengario of Carpi, *Commentaria*, fol. 6v.

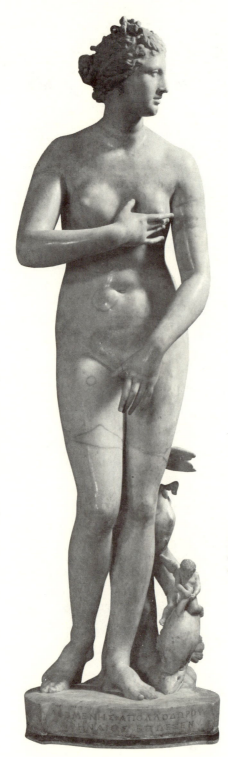

Figure 4.7. *Medici Venus*, Florence, Galleria degli Uffizi, Greek or Roman, first century BCE.

sexual engravings created in the early 1530s to illustrate the French medical writer Charles Estienne's *On the Dissection of the Parts of the Human Body* (for example, figure 4.8), which were modeled on Jacopo Caraglio's *Loves of the Gods*.[107] In place of Estienne's pornography of death, Berengario presents active, energetic women, whose postures are much closer to those of the male figures of Hellenistic statues such as the *Laocoön* and the Belvedere Torso and the *ignudi* on Michelangelo's ceiling of the Sistine Chapel.[108]

Whatever their merits as artistic statements, the three woodcuts of dissected women in the *Commentaries* did not satisfy Berengario. He included them in the first edition of his *Short Introduction to Anatomy*, 1522), an abbreviated version of the *Commentaries* for less advanced readers, but eliminated them from *Short Introduction's* second edition, which came out the following year. In their place, he offered a single figure, which he described as "drawn more elegantly than before" (figure 4.9). Although this illustration refers to the three earlier images' use of drapery, the splayed legs of its female figure, and allusion to the natural world outside the window, it departs significantly from their pattern. Berengario's illustration depicts an obviously lifeless corpse, arms and legs dangling, draped on an elaborate chair (which resembles a piece of studio furniture), in a kind of simultaneous parody of the female figure in the *Fasiculo*, the ur-image of Italian uterine anatomy (figures I.6 and 2.1), and of that book's famous dissection scene (figure 2.5). In this way, it implies that Berengario has superseded both the bookish knowledge of the university lecturer (on his chair) and the secrets known to the woman (on hers). The anatomist no longer needs his female subject's cooperation, let alone instruction. This is but one of the many aspects of Berengario's illustrations that Andreas Vesalius will imitate and elaborate.

Figure 4.8. "The substance of the uterus cut through the middle." Charles Estienne, *De dissectione partium corporis humani* (Paris: Simon de Colines, 1545), p. 285.

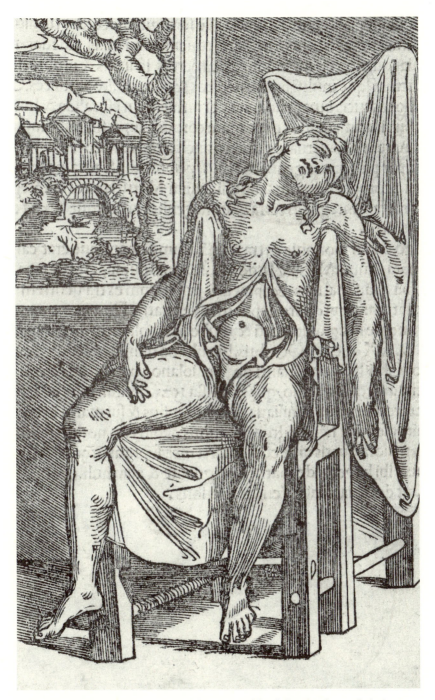

Figure 4.9. Anatomy of the uterus. Jacopo Berengario of Carpi, *Isagogae breves*, 2nd ed. (Bologna: Benedetto di Ettore, 1523), fol. 23v.

CHAPTER FIVE

The Empire of Anatomy

Almost nothing is known about the woman whose corpse is the focus of my last chapter, although she is more familiar by far to historians of medicine than any of the women I have thus far discussed. Her name, occupation, and family status are a mystery, as is the date of her death in the northern Italian city of Padua, most likely occurring in the winter of 1541–1542. Her fame rests on the verbal and visual descriptions of her corpse in *On the Fabric of the Human Body*, published in 1543 by Andreas Vesalius of Brussels, then an ambitious young professor of anatomy at the local university.[1] She was the subject of two of the four illustrations of women's bodies in the *Fabrica* (figures 5.1 and 5.2) and the basis for the cadaver depicted on its title page. This image reveals an idealized and fictionalized version of a dissection scene presided over by Vesalius (standing just to the left of the dissecting table), who is lecturing to a large and highly engaged audience (figure 5.3). The woodcut has been interpreted by generations of historians as an icon of the reformed anatomy of the sixteenth century, which elevated the truths of the body as revealed by human dissection over textual descriptions.[2] Like Jacopo Berengario of Carpi, whose *Commentaries on the Anatomy of Mondino* of 1521 was an important model, Vesalius recruited local artists for his ambitious

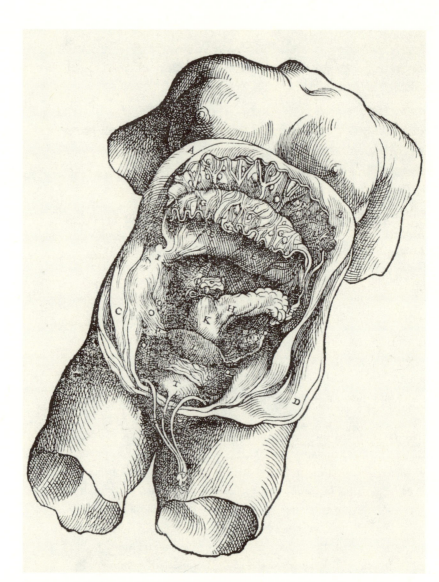

Figure 5.1. Dissected body of a woman showing the position of the uterus (L) and bladder (T).
Andreas Vesalius, *De humani corporis fabrica* (Basel: Joannes Oporinus, 1543), bk. 5, figure 24.

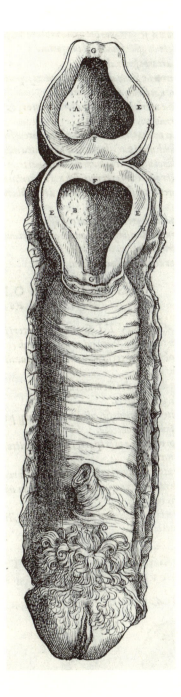

Figure 5.2. Dissected genitals of the woman shown in figure 5.1. Andreas Vesalius, *De humani corporis fabrica*, bk. 5, figure 27.

Figure 5.3. Vesalius dissecting the body of a female criminal. Andreas Vesalius, *De humani corporis fabrica*, title page.

project, in this case from the workshop of Titian in nearby Venice. Although their identities cannot be definitively confirmed — they very likely included Domenico Campagnola and a Fleming, Jan van Calcar — it is nonetheless clear that the *Fabrica* was the product of an intimate collaboration between the anatomist and various artists, both draftsmen and woodblock cutters, and that Vesalius was involved at every stage in the preparation of the woodcuts for which the book is famous.[3]

Vesalius recounted everything we know about this woman at the end of his chapter on the uterus, in Book 5, where he addressed the location of the spermatic vessels, which Berengario had vehemently discussed in his own treatise.[4] According to Vesalius, the woman had been condemned to death for an unspecified crime and had attempted unsuccessfully to stave off execution by claiming to be pregnant. However, "when interrogated by midwives under orders from the podestà — they asserted that she was not pregnant — she was never willing to indicate how long she had gone without menstruating, however diligently we tried to ascertain this."[5] Vesalius' dissection of her uterus confirmed the accuracy of the midwives' judgment. The only additional details available regarding her come from the caption to figure 5.2, which describes her as "a woman of very tall stature who had often given birth," and from revisions to the second edition of the *Fabrica*, published twelve years later, that refer to her as "a woman of rare size and of middle age."[6]

Any narrative of the events leading up to this woman's anatomy is therefore conjectural. She was probably not a native of Padua or its environs, for the university statutes that governed the practice of public dissection exempted any criminal who was born in Paduan territory or was a Venetian citizen. (Padua had come under Venetian control in the early fifteenth century.) Restrictions of this sort, common throughout northern Italy, were intended to protect respectable local families from the horror and shame involved in the public display of the naked and mutilated corpse of a kinsperson or neighbor.[7] Vesalius' silence about the woman's origins suggests that she came from an inconsequential social background, as does the fact that she was hanged, a penalty applied only exceptionally to those of patrician birth. She had most likely been convicted of murder — along with robbery, the most common capital crime in the Venetian territories — although this too is by no means certain.[8] Once condemned, she would most probably have been hanged at the Camposanto, Padua's principal execution grounds, outside the Porta San Giovanni, in the city's western wall.

Assuming she was treated like other criminals, she would have been attended the evening before her death by two members of the confraternity of San Giovanni Evangelista della Morte, a company of pious laypeople whose special task was to assist criminals on the brink of execution.[9] Their mission was to prepare her physically and spiritually for this ordeal and eventually to bury her corpse.[10] The confraternity statutes required two brothers to feed her — dissection of an executed man's stomach, they noted, had revealed that the "great agony" preceding this kind of death speeded up digestion — and to provide for her confession.[11] More than anything else, however, the two brothers aimed, like members of other such confraternities throughout northern and central Italy, to reconcile her to her punishment, so that she not only accepted but welcomed it as a manifestation of divine justice. To this end, they encouraged her to identify with Jesus and the Christian martyrs, who went willingly to their deaths. In this woman's case, because she was to be hanged, they would have asked her to meditate specifically on her similitude to the crucified Christ, "who hung on the wood of the cross."[12] (Those who were about to be decapitated were reminded of Saint Paul; those who were about to be quartered, of Saint Hippolytus.) If she accepted her death with true contrition, they would assure her that "God would remit all [her] penance, so that [she] could go straight to heaven," rather than expiating her sins in purgatory or hell, and "everyone would have compassion for [her]."[13]

The next day, all the brothers would have put on their habits (white capes marked with a black cross) and accompanied her in the extended public procession that delivered her to the Camposanto. After she had died, they would have removed her body from the gallows and, under normal circumstances, they would have buried her in the cemetery there.[14] In this case, however, because she was to be anatomized publicly, her corpse would have been consigned to representatives of the university and carried back inside the city, to a temporary wooden theater constructed for the occasion, perhaps in a local church, courtyard, or classroom. After the anatomy was completed, they would have gathered up the various pieces of her body for burial in a Christian grave.[15] This last act of piety lay at the heart of their mission; before the foundation of confraternities of this sort, beginning in the fourteenth century, the bodies of criminals would have been left to rot in the fields outside the city, as the ultimate symbol of their exclusion from the Christian community.

The woman on the title page of the *Fabrica*, in other words, was among those scholars usually identify as the subjects of human dissection in late medieval and early modern Europe: executed criminals. From the early days of human dissection as a pedagogical practice in the years around 1300, anatomies performed in medical faculties and (somewhat later) in colleges of physicians or surgeons had been largely confined to this group, first by custom and subsequently by decree.[16] In Padua, for example, the statutes of the faculty of arts and medicine that governed Vesalius' teaching prescribed an annual public anatomy of "the cadaver of some criminal provided by the judges after execution, that is to say, of one man and one woman or at least one of [the two] each year."[17] Relying on provisions of this sort, historians of anatomy and the body have made much of the intimate connection between the practice of human dissection and the criminal justice system, either to underscore the supposedly polluting or "taboo" nature of dissection, which pressed anatomists to distance themselves from the abject provenance of their corpses by the use of elaborate visual and verbal strategies, or to show the implication of medical learning and authority in the punitive apparatus of state power.[18] There is indeed ample evidence that medical scholars relied on princes and judges not only for anatomical material but also for subjects of occasional toxicological experiments;[19] nevertheless, historians have significantly overstated the degree to which the criminal justice system was implicated in anatomical study and vice versa. They have asserted that dissection was considered an integral part of the criminal's punishment, despite the almost complete absence of contemporary testimony to this effect, and inflated the relative importance of ceremonial public dissections by downplaying the more intimate anatomies of private and hospital patients — not to mention animals — that lay at the heart of late fifteenth- and sixteenth-century anatomical teaching and research.[20] The latter rarely involved the corpses of condemned criminals precisely because these were so difficult to obtain, given both the infrequency of executions of foreign criminals and the difficulty of coordinating them with the academic schedule. (For obvious reasons, dissections were generally conducted during the cold season, in January or February.)[21] The reliance on private individuals was even greater in the case of female cadavers, since women were rarely convicted of capital crimes, a subject of perennial comment and complaint by medical faculties.[22] Thus, of the various female bodies Berengario referred to in his *Commentaries*, only two belonged to executed criminals; most of the others came from his own practice.

Vesalius was an exception to this rule, however; compared to previous anatomical writers, he seems to have been unusually dependent for material on judges and other secular authorities. Not only was the scope of his research project unprecedented, but he also lacked private patients, as a result of both his youth — he was only twenty-six when he began preparing the *Fabrica* — and his single-minded focus on anatomical teaching and research. (When Berengario published his *Commentaries*, in contrast, he had been working as a surgeon for over forty years.) Nonetheless, despite Vesalius' relatively heavy reliance on the corpses of executed criminals, there is no convincing evidence that he (or any other fifteenth- or sixteenth-century anatomist, for that matter) attempted to obscure their provenance in order to distance himself from their abject origin. On the contrary: the text of the *Fabrica*, like that of Berengario's *Commentaries*, is matter-of-fact about which bodies came from the gallows and which did not. In fact, Vesalius (unlike Berengario) seems to have gone out of his way to underscore the distasteful and occasionally illicit origins of his cadavers, recounting gory details with gleeful satisfaction, as in the cases of the skeleton he stole piecemeal from the gibbet outside Louvain, the "still-beating heart" he extracted from the corpse of a man who had just been quartered, and the body of a woman procured at the initiative of some of his students. These students

pulled from her tomb the attractive whore of a monk at the church of St. Anthony here, who had died suddenly, as if from suffocation of the uterus or some other fulminating illness. They brought her in for public dissection, having removed all the skin from her body with amazing industry, so that she would not be recognized by the monk, who with her relatives was complaining to the Podestà that the corpse had been stolen from the tomb.[23]

A number of the smaller images in the *Fabrica* are similarly explicit. The woodcut initial *I* (figure 5.4) shows a nighttime scene in which playful putti, presumably representing Vesalius' students, pull what appears to be a female corpse from its tomb — an apparent reference to the story of the monk's mistress. In the initial *L* (figure 5.5), two putti let down a corpse from a gallows, watched over by a priest and a person wearing the robe and hood of a confraternity for the comfort of condemned criminals; the initial *N* depicts a putto receiving the head of a decapitated criminal.[24] And of course Vesalius' innovative choice to focus his title-page woodcut on the corpse of a female criminal, surrounded by a large and rowdy male crowd and positioned for maximum sexual effect, also underscores the potentially transgressive elements in anatomical dissection.

Vesalius was by no means the first to exploit the obvious sexual element in the anatomy of the genitals for the entertainment of his students. This practice was probably as old as anatomy itself. Certainly there are references to it as long ago as the early fourteenth century, when the person who recorded Mondino's lectures on the generation of the fetus (probably a student) included a coarse remark in Italian about sex between women at the point where Mondino noted that women could experience pleasure and emit seed in the absence of a male orgasm.[25] In 1540, Matteo Corti, a professor of medicine at the University of Bologna, earned the disapproval of one of the German students who attended his lectures on anatomy (performed with Vesalius' assistance) when he joked repeatedly about the impotence of old men, the naturalness of the search for sexual pleasure, and the familiarity of students with illnesses of the penis. "Here all the

Figure 5.4. Graverobbing scene. Andreas Vesalius, *De humani corporis fabrica*, p. 55.

Figure 5.5. Removing a body from the gallows for dissection. Andreas Vesalius, *De humani corporis fabrica*, p. 168.

Italians laughed, including Corti," the German noted tartly, re-
marking elsewhere that sex was licit only in the context of a legitimate marriage.[26] This brand of humor also appeared in images of
dissection; the late fifteenth-century manuscript painting reproduced in figure 3.2 shows a learned physician (on the right) holding up a cadaver's penis while the man next to him comments on
its size. In this sense, Vesalius' emphasis on the physical and moral
violence involved in anatomical study is a fairly sedate version of
what Michael Sappol has called the "homosocial meaning of anatomical mayhem" — the creation of a sense of mastery and cama-
raderie through rough, often sexualized behavior around corpses.[27]
Where Vesalius differed from his predecessors was in placing this
element front and center on the title page and incorporating it
into his published Latin text, rather than restricting it to oral
comments in the classroom.

In general, Vesalius emphasized the transgressive aspects of
anatomy to support his extraordinary commitment to discovering
truths regarding the human frame. His gratuitous stress on the re-
pugnant, sometimes shocking features of his work allowed him to
demonstrate his personal dedication to his subject and to claim
special epistemological authority gained from immersing himself in
the dirty stuff of nature, body, and soul.[28] Unlike more squeamish
teachers of anatomy, who contented themselves with manipulating
books and words rather than dead bodies, he wrested knowledge
from matter itself. Further, since he was not particularly active in
private practice and not anatomizing his own patients, he had no
particular reason to dwell on their sensibilities and those of their
families.

Vesalius' lack of experience as a medical practitioner and his
unusually heavy reliance on the corpses of executed criminals
shaped his work in important ways. In particular, it limited his
knowledge of women's bodies, so that his account of the female

genitals in the first edition of the *Fabrica* is vague, brief, and inaccurate, and does not begin to compare with Berengario's in richness and detail. Like Berengario and Gabriele Zerbi, Vesalius contemptuously rejected the vulgar account of the uterus as having seven cells and accepted Galen's contention in *On the Use of Parts*, which had finally been published in 1528, that women's genitals were internalized and inverted versions of men's — an idea illustrated to surprising effect in figure 5.2.[29] Like Berengario and virtually every other learned medical writer of his day, he accepted the obvious corollary, agreeing with Galen (against Aristotle) that women, like men, produced seed, albeit of a scarcer, colder, and wetter sort.[30] Beyond these commonplaces, however, he had relatively little useful to say about the female genitals. By his own admission, he had never dissected a menstruating woman, so he could not determine which vessels carried menstrual blood to the genitals and whether these entered the uterus, the vagina, or both, a topic on which Berengario had held forth at length and illustrated in figure 4.1.[31] Vesalius also had no experience with pregnant women, since criminals known to be pregnant were almost always allowed to give birth before execution — hence the (unsuccessful) claim of the woman who was the model for the title page[32] — and unlike Berengario, he did not appear to have purchased miscarried fetuses from midwives, perhaps because his limited medical practice permitted few contacts with this group. Thus he noted that his treatment of the coverings of the human fetus would be based on the writings of others:

> Although I have learned certain things based on the dissection of fetuses and pregnant uteruses and have conducted several demonstrations at the university, I have used only animals, nor have I [studied these organs] with the necessary care, for to this point I have been able to obtain very few women for dissection.[33]

As a result, Vesalius was constrained to substitute a description of canine anatomy, characterized by an annular placenta, which the accompanying illustration attached, incongruously, to a human fetus.[34]

Vesalius rectified many of these problems in the second edition of the *Fabrica* (1555), dramatically expanding and revising the relevant chapters on the basis of his later experience. (He had left his academic post even before the first edition was published, for a position as court physician to the Holy Roman emperor Charles V, to whom he had dedicated the *Fabrica*; this provided considerably more opportunities to inspect the bodies of private patients and to conduct additional anatomies.)[35] Nonetheless, as far as the first edition is concerned, his account of the female organs of generation, in the estimation of his great biographer and admirer Charles D. O'Malley, was the weakest section of the book.[36]

Why then, given Vesalius' own awareness of the inadequacy of his discussion of the uterus — and given that virtually all previous depictions of dissection scenes in books of anatomy and medicine, like that in the *Fasiculo de medicina* (figure 2.5), had shown male cadavers[37] — did he choose to spotlight precisely that topic by placing it at the title page's center? One obvious answer lies in the imagined appeal of this image to a male audience; Vesalius aimed to increase his reputation (not to mention recoup his large investment in artists' fees), and he would certainly have expected this dramatic and novel image to help sell his book. Scholars have identified other important messages communicated by the woman's body. The obvious counterpoint between the skeleton behind the dissecting table and the opened uterus reinforces the function of the image as a *memento mori*, according to the Latin motto "When we are born we die; our end is but the pendant of our beginning."[38] The choice of cadaver also illustrated one of the *Fabrica*'s principal themes, the superiority of Vesalius' account to

Galen's, based on his own access to human cadavers, whereas (he alleged) Galen had to make do with animals.[39] Despite his inexperience with female bodies, Vesalius had dissected enough women to be able to make this point confidently. In his chapter on the (nonpregnant) uterus, he contemptuously rejected Galen's description of its supposed horns, noting that "Galen never inspected the uterus of a woman, unless it was in a dream, but only those of cows, goats, and sheep."[40] Finally, Vesalius certainly intended the scene to refer to the long tradition of using female genitals to symbolize the most recondite aspects of human anatomy and physiology and those least accessible to male readers and writers.

In addition to these general associations, however, Vesalius had more specific models in mind, which become clearer when his title page is compared with other iconographic traditions. The first of these, which focuses on the story of Saint Anthony and the miser's heart, foregrounds not the cadaver's sex but her status as criminal and sinner. The second illustrates the popular version of Roman imperial history that I described in Chapter Three, which began with the extraction of Julius Caesar from his dead mother and ended with Nero's inspection of Agrippina's womb. Placing the title page in the context of these stories, I will argue that the sex of Vesalius' cadaver is central to the author's presentation of himself as the founder of a reformed anatomy and that the image reflects an emergent idea of experience, which revised earlier notions of the relationship between viewer and corpse; this emphasized the distance between the (male) anatomist and his (female) object, using gender difference to loosen the bonds of reciprocity and identification that had informed medieval models of vision.

Saints and Sinners

One index of the new cultural visibility of anatomy in Italian cities was its appearance in devotional images. These included, for

221

Figure 5.6. Death of the pilgrim Larghato. *Fior di virtù historiale* (Florence: Jacopo di Carlo?, 1491).

example, Sandro Botticelli's painting of the opening of Saint Ignatius of Antioch on the predella of the altarpiece he painted for the church of San Barnaba in Florence (ca. 1489) and the woodcut illustration in the didactic volume the *Flower of Virtue through Stories* (1491), which showed the anatomy of the pilgrim Larghato, who had died of joy on visiting Jerusalem (figure 5.6). Although the story of Saint Ignatius lacked a medical referent — he was tortured and executed by order of Emperor Trajan, after which he was found to have the name of Jesus on his heart in golden letters[41] — the fifteenth-century *Flower of Virtue* describes a private anatomy to determine the cause of death, similar to those discussed in Chapter Three:

> Those present, believing [Larghato] was dead, sent for the doctors, and when the doctors saw him, they immediately said that he was dead. And having the body opened up the middle, they found written on his heart, "My sweet love Jesus Christ." In this way the doctors, understanding his devout condition and happy complexion, judged that he had died of the joy he had received from seeing the Holy Sepulcher, since people die more quickly from joy than melancholy.[42]

The best-known story to use dissection in the service of piety, however, concerned the miracle of the miser's heart and Saint Anthony of Padua.[43] The most famous depiction of this scene is the great bronze relief created in 1447 by the Florentine sculptor Donatello for the front of the high altar of Saint Anthony's basilica, only a few blocks from the university where Vesalius performed his anatomies almost a century later (figures 5.7–5.9).[44] Saint Anthony was a renowned miracle worker and the focus of enormous popular devotion in Padua.[45] The version of the story illustrated by Donatello appears in the *vita* written by Sicco Ricci Polentone around 1435. According to Polentone,

Figure 5.7. Donatello, *Miracle of Saint Anthony and the Miser's Heart*, bronze relief from the high altar of the Chiesa del Santo, Padua, 1447.

In Tuscany, which is a large province in Italy, Saint Anthony (as was customary) happened to attend the funeral of a rich and wealthy man. Suddenly, moved by the fervor of the spirit, he exclaimed, "The dead man should not be buried in a holy place, but outside the city walls, like a dog, since his soul is damned to hell and he does not have a heart in his body, for in the words attributed to the Lord by the holy evangelist Luke, 'Where his treasure is, there also is his heart.'" On hearing this, as is easy to believe, everyone became very excited, and various opinions were voiced. Finally, those who had been ordered to inspect the body opened up his chest. They did not find the heart inside it, but rather, as the saint had predicted, in the place where his money was. For this reason, the city praised both God and the saint, and the dead man was not buried in the tomb that had been prepared for him, but dragged to the countryside and buried like an ass.[46]

Donatello's relief captures the horror of Saint Anthony's revelation, as described by Polentone, and the public consternation

224

Figure 5.8. Detail of figure 5.7 showing the miser's heart being pulled from his moneybox.

Figure 5.9. Detail of figure 5.7 showing the opening of the miser's body after the revelation by Saint Anthony (*far right*).

provoked by the discovery of the missing heart. The funeral procession is drawn up in the background, complete with the banner of the confraternity to which the miser hypocritically belonged, in front of the church whose doors are now eternally shut to him. The crowd in the foreground is racked by intense and varied emotions; some bystanders recoil in fear, some adore the saint (who stands to the right of the corpse), and others strain to see inside the dead man's open body, which lies on his bier in the street. The disposition of the central figures — the corpse on its platform being opened by one man while another underscores the lesson with a pointed finger (figure 5.9) — echoes contemporary anatomical practice; a surgeon prepares to open the corpse, while a demonstrator stands by to point out the relevant structures, as in the woodcut of the *Fasiculo de medicina* of 1494 (figure 2.5).[47] The parallels are so striking that it is hard to believe that Donatello had not himself witnessed a dissection at the university during his sojourn in Padua, particularly given the interest in anatomy among contemporary Florentine artists.[48] Not only is the arrangement of the figures suggestive, but the corpse is also shown almost naked, which is not true of previous depictions of the miracle (compare figure 5.10, Francesco Pesellino's version, which served as a model for both Botticelli and the designer of the *Flower of Virtue* woodcut), and the heart being extracted from the money-box at the far left of the relief has a certain amount of anatomical detail (figure 5.8).

Vesalius' title-page woodcut appears to refer to Donatello's relief.[49] The scene takes place outdoors — note the pebbles at the bottom of the image — against an imaginary, classicizing architectural background, rendered with self-conscious attention to perspective. The central figures are surrounded by a large crowd of people with individualized and highly expressive physiognomies, in a variety of attitudes; a few turn away, others are inattentive,

Figure 5.10. Francesco Pesellino, *Miracle of Saint Anthony and the Miser's Heart*, predella panel from Filippo Lippi's altarpiece for the Chapel of the Novitiate in the church of Santa Croce in Florence, Florence, Galleria degli Uffizi, ca. 1445.

but most display intense curiosity. While many stand, several are kneeling in the front. Even the puzzling naked figure clutching a column to the left of the title page has a (clothed) counterpart in the relief.[50] His hand outstretched, Vesalius stands beside the cadaver of the female criminal — like the miser, an abject sinner — as the saint stands next to the miser's corpse. The parallel between the images is broken, however, in the arrangement of the central figures. In the title page, the corpse has been turned with its feet to the reader, and two of Donatello's figures, the man opening the body and the pointing saint, have been subsumed into a single personage: the figure of the anatomist, Vesalius himself. In this way, the woodcut reinforces one of the *Fabrica*'s main contentions: that anatomy can progress as a discipline only if anatomists perform their own dissections, rather than leaving the work to others (as in figure 2.5). Because the Vesalian anatomist opens the cadaver in addition to demonstrating its structures, the surgeon (in front of the table) has been relegated to the ancillary role of sharpening the razors.[51]

One of the effects of Donatello's relief, juxtaposed with Vesalius' title page, is to call attention to the counterpoint between sin and sanctity in the rituals and ideologies relating to the execution of criminals in fifteenth- and sixteenth-century Italy and to emphasize its relationship to anatomical dissection.[52] Donatello's miser is a sinner consigned to eternal damnation, his corpse destined to be abandoned outside the city walls, as executed criminals were before the foundation of confraternities devoted to burying them in hallowed ground. His opened body assimilates him to the cadavers of criminals granted to medical faculties for dissection.[53] At the same time, however, there are intimations of sanctity in Donatello's portrayal of the scene. The miser's lolling head and dangling arm recalls images of the deposition of Christ from the cross, while his bier, set parallel to the church door, recalls an altar. As in many other contemporary devotional images, most notably those of the Crucifixion, martyrdom and criminality are provocatively juxtaposed (for example, figure 5.11).[54] One of the aims of such portrayals was to elicit compassion in the viewer: the compassion ("co-suffering") promised, perhaps disingenuously, to condemned criminals by the confraternity of San Giovanni Evangelista della Morte in Padua and modeled by Mary at the foot of the cross.[55] The viewer was expected to identify with the desperate state of the sinner, taking the martyr and the contrite criminal as models to be emulated, while the criminal who had not repented was to be pitied and abhorred.

The association between executed criminal and martyr — both subjected to spectacular suffering at the hands of secular authorities — is dramatized in several of the woodcuts in Berengario's 1521 Commentaries. Two of these, which demonstrate the exterior muscles of the front and back of the body, are posed as écorchés against a rural background in the posture of saints holding the instruments of their martyrdom, the executioner's rope and ax (figures 4.4 and 5.12). The most striking woodcut of all, however, was the one chosen by Berengario to illustrate the muscular anatomy of the arms: a flayed figure of the crucified Christ (figure 5.13).[56] Here the hanged criminal is identified with Jesus, as confraternities for the comfort of the condemned prescribed. This edgy image may have gone a bit too far for contemporary readers, for Berengario eliminated it from his Short Introduction to Anatomy, published the following year.

Given the associations between criminality, martyrdom, sanctity, and anatomy in contemporary artworks, as well as its relationship to Donatello's relief, the title page of the Fabrica is striking for the degree to which it downplays the religious element. The dissecting table is turned perpendicular to the picture plane, which diminishes its resemblance to an altar; neither the skeleton nor the

Figure 5.11. Andrea Mantegna, *Crucifixion*, predella panel from Mantegna's altarpiece for the church of San Zeno in Verona, Paris, Musée du Louvre, 1457–1459.

architectural surroundings that replaced the church in the background of the relief has an obvious Christian referent. (The skeleton as a death symbol can be found in ancient Roman art.)[57] The same is true of the illustrations embedded in the text, with the exception of the author portrait, which I will discuss below, and the playful woodcut initials that show priests and confraternity members attending executions; the latter have none of the pathos that characterizes Berengario's illustrations of male cadavers and contemporary images of martyrdom. While the title page's foreshortened cadaver may appear to recall Andrea Mantegna's *Dead Christ*, that figure is draped from the waist down and its legs are not splayed, which softens the emphasis on the internal and external genitals that is such a striking feature of the image that opens the *Fabrica*.[58] Finally, Vesalius' cadaver is female, the only woman but one — at the back, between the second and third columns from the right — in a sea of male viewers. This, together with her undignified posture, discourages identification with her, let alone compassion, on the part of either the audience in the title page woodcut or the presumably male reader of the difficult Latin text.

The *Fabrica*'s title page thus marks a significant step in what Andrea Carlino has called the "desacralization" of the anatomical cadaver.[59] As Carlino points out, this process was gradual; Christian references continued to appear in some anatomical illustrations and depictions of dissections well after Vesalius.[60] Nonetheless, the secular cast of the images in the *Fabrica* is striking; the Christian elements in the woodcuts of Berengario's *Commentaries* have been eliminated and the classical ones multiplied and emphasized, as in the famous series of musclemen of Book 2 or the trope of sculptural fragments that informs Book 5 (for example, figure 5.1). Insofar as Christian resonances persist, they are confined to the frontmatter: the author portrait and the title page, with its quotation of Donatello's relief. But even in the latter, the *Fabrica*

Figure 5.12. Exterior muscles of the back of the body. Jacopo Berengario of Carpi, *Commentaria*, fol. 520v.

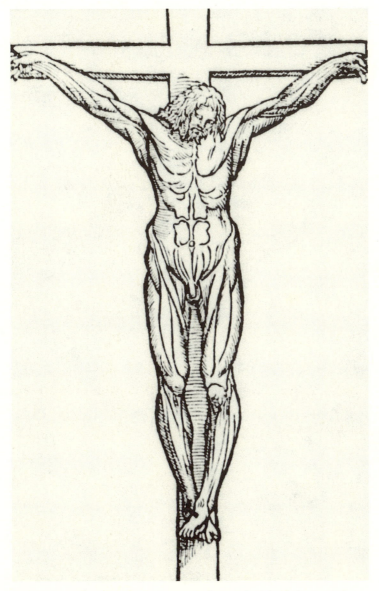

Figure 5.13. The muscles of the arms. Jacopo Berengario of Carpi, *Commentaria*, fol. 19v.

has dramatically shifted the frame of reference: whereas contemporary ideology identified the saint with the cadaver (through the identification of both with the executed criminal), the title page identified the saint with the dissector: Vesalius' position and gesture echo the attitude of Saint Anthony in the relief; each man interprets the corpse for the instruction of his listeners. This daring move is part of Vesalius' transformation of the anatomist from a sober lecturer in an academic gown, a stock figure on many sixteenth-century anatomical title pages, into a heroic, even transgressive, figure as part of his campaign to attract the patronage of the Holy Roman emperor Charles V.

Translations of Empire

Vesalius dedicated the *Fabrica* to the emperor in hopes (soon realized) of obtaining a post as imperial physician. Like the dedicatory preface, the title-page woodcut reflects key themes in Hapsburg imperial ideology, which identified the Hapsburg empire as the successor to the ancient Roman Empire and the Hapsburg dynasty as the heirs of the Julio-Claudian line.[61] In particular, it evokes two episodes in Roman imperial mythography, the birth of Caesar and Nero's anatomy of Agrippina, which I analyzed in Chapter Three as ways of thinking about genealogy, paternity, and the relative importance of paternal and maternal contributions to the child. As I related there, the version of Roman history that circulated widely in vernacular writings began and ended with a woman's open womb; Caesar, the first emperor, was cut from the uterus of his dead mother, while Nero, the last in Caesar's lineage, killed his own mother, Agrippina, to see the place where he was conceived. Both stories constitute in different ways — Caesar in a heroic mode and Nero as a cautionary example — fantasies of masculine self-birth and a pure male lineage uncontaminated by the female principle.

234

These two stories had lost none of their appeal in intervening years, especially in the Low Countries and France, where Vesalius was born and educated. There they circulated in a variety of medieval texts: the *Deeds of the Romans* and other vernacular retellings of Roman imperial history (on the birth of Caesar); and Jacobus de Voragine's mid-thirteenth-century *Golden Legend*, Jean de Meun's section of the *Romance of the Rose*, and Laurent de Premierfait's French version of Giovanni Boccaccio's *On the Downfalls of Famous Men* (on the opening of Agrippina). Woodcuts of the two scenes replaced the painted images that decorated many earlier manuscripts of these medieval works (figures 1.9, 3.8, 5.14, and 5.15) and were inserted into new editions of ancient sources; an early sixteenth-century Venetian edition of Suetonius' *Lives of the Twelve Caesars* includes a striking illustration of Julius Caesar's birth (figure 3.7).[62] Learned writers recognized that the story of Nero's anatomy of Agrippina was a medieval interpolation, but it lost little of its force in the popular imagination. Among other things, it was staged in a French play, *The Vengeance of Our Lord*, which appeared in a variety of late fifteenth- and sixteenth-century printed editions; this included a scene in which three devils appeared on stage in physicians' gowns, advising Nero to open his mother alive.[63]

Stories of this sort, like the story of the Sabine women, which rooted imperial dominion in the violent control of female reproductive bodies, were given new life and force in the sixteenth century with the elaboration of theories and practices of absolutist rule. Violence and procreation lay at the heart of the prince's claim to sovereignty; monarchs welcomed texts that celebrated force as the foundation of princely power and surrounded themselves with images that portrayed the coercive and unequal relationship between the princes and their subjects in terms of the relationships between rapists and their victims, or between

Figure 5.14. Death of Agrippina. Guillaume de Lorris and Jean de Meun, *Roman de la rose*, Paris, Bibliothèque Nationale, ms fr. 24392, fol. 49v, French, fifteenth century.

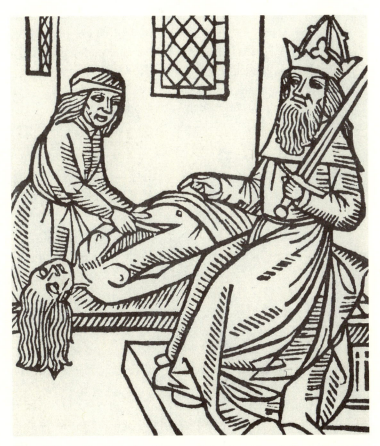

Figure 5.15. Death of Agrippina. Guillaume de Lorris and Jean de Meun, *Roman de la rose* (Paris: Jean Vérard, 1494–1495), fol. 49v.

husbands and wives.[64] The Hapsburgs embraced this patriarchal imagery, using it to downplay the role of women in their geneal-ogy — Charles had in fact inherited the Netherlands through his grandmother Mary of Burgundy — in order to present themselves as a lineage of fathers and sons, on the model of the Julio-Claudian line of emperors, which renewed itself by the all-male process of adoption.[65] Titian, whose workshop produced the *Fabrica* illustra-tions, was one of the masters of such imagery, which he used in his *Danae* (1553–1554) as well as in a series of erotic mythological scenes commissioned by Charles's son Philip, who succeeded his father as emperor in 1556.[66]

Even Nero, an abject exemplum of madness and cruelty, could be incorporated into this framework, as in the *Encomium of Nero*, composed sometime before 1560 by Girolamo Cardano, an emi-nent physician and professor of medicine at the universities of Bologna and Pavia (not to mention an admirer of Vesalius). In this historiographic thought experiment, Cardano defended Nero's notorious actions, including his murder of his mother, as a well-intentioned, if ultimately unsuccessful, attempt to defend the Roman Empire from decline and dissolution.[67] (As a learned writer working with the ancient sources, he made no mention of her apocryphal dissection.) Not only did Agrippina deserve her violent death on account of her crimes, Cardano explained, but Nero's action was also necessary to ensure both his own safety and that of the Roman Empire as a whole. In addition to making this Machiavellian argument, Cardano also justified Nero's cruelty in providential terms, appealing to the idea of translation of em-pire, a staple of Hapsburg ideology, according to which God had engineered human history as a series of cycles in which one regime succeeded another, only to decay and fall in its turn. In this historical schema Nero was merely a link in the chain — a divine scourge and, ultimately, a victim of divine justice.[68] Finally,

Cardano argued, Nero's murder of his mother was excusable, since matricide is a lesser crime than parricide — a position he defended in terms of the physiology of generation. "The laws have established paternal, not maternal authority [*patriam potestatem, non maternam*]," he wrote, "and the father is the principal agent, like seed, while the mother is like the earth; and the plant owes more to the seed than the earth."[69] Because the child is more beholden for its life and existence to its father than to its mother, it owes her less loyalty and respect.

Vesalius had good reason to identify with Hapsburg imperial ideology. He was born into a medical family with several generations of medical service as court physicians and pharmacists, first to the House of Burgundy and then, in the late fifteenth century, to Charles V's grandfather Maximilian Hapsburg, who was the first of his family to become Holy Roman Emperor. Even as professor of anatomy at the University of Padua, Vesalius continued to angle for the position of imperial physician. It is certainly for this reason that he dedicated the *Fabrica* to Charles and the *Epitome* — a smaller, cheaper version of this work, also published in 1543 — to his son. The dedicatory preface of the *Fabrica* resounds with praise for "the divine Charles, the most invincible, greatest Emperor," and his inumerable virtues: "admirable learning, singular prudence, remarkable clemency, keen judgment, untiring generosity, remarkable love for men of letters and scholarship, supreme dispatch in the conduct of business," and so forth.[70] Vesalius' efforts were fruitful, and he left Padua for imperial service in 1543.

Especially in light of Vesalius' aspirations to imperial patronage, the title page of the *Fabrica* has strong affinities with contemporary stories regarding Caesar's birth and Nero's opening of Agrippina's womb.[71] Attentive readers would have noticed not only that it shows a female cadaver but also that the anatomist is in the process of dissecting her uterus, as is clear from a comparison with figure 5.1, the first woodcut of the uterus in Book 5 of the *Fabrica*. Both in its posture and in the disposition of its internal organs, the body of the woman on the title page is a mirror image of the uterine illustration, as happens when the drawing for a print is copied from another image.[72] The anatomist's position beside the corpse evokes those of Nero and of the surgeon who performed the operation on Caesar's mother as shown in contemporary woodcuts (such as figures 5.14 and 3.7). These two scenes were not only well known but also the *only* such scenes (and the only contemporary images) to show the anatomy of a naked, or partially naked, female corpse and the inspection of her opened womb.

Thus the title page of the *Fabrica* was an ambitious exercise in mythography and iconography, designed to function like one of the erudite emblems that were all the rage in European courts. Vesalius' use of this mode of self-presentation in pursuit of princely patronage presages the efforts of a later professor at the University of Padua, Galileo Galilei, the title page of whose *Dialogue Concerning the Two Chief World Systems, Ptolemaic and Copernican* is one of the classics of the emblematic genre.[73] Whether free-standing, part of a collection, or incorporated into a title page, the sixteenth-century emblem combined highly allusive, often allegorical, images with ambiguous texts. The emblematic title page was a kind of rebus or puzzle, intentionally complicated and difficult to decipher. Like other title pages of this sort, that of the *Fabrica* was intended to work on a variety of levels, so that its full meaning emerged only through a careful consideration of both text and image. To decode it required the effort and erudition of its readers and provided the occasion for both social debate and private reflection.[74]

The key to an emblem's meaning was typically encoded in the short Latin motto that accompanied it. In the case of the *Fabrica's* title page, the words that function as the motto appear in capital letters in the shield-shaped cartouche propped against the "stage" on which the dissection takes place: "CVM CAESAREO." While these are the first two words of what is, in fact, the printer's privilege — "WITH the grace and privilege of his IMPERIAL Majesty the King of France and of the Venetian Senate, as contained in their letters of license" — they also provide a clue to how the image should be read. Given the nature of the scene above it, this phrase would have called up the birth of Caesar in the mind of any literate viewer, as well as other births of this sort in classical mythography, notably that of Asclepius.[75]

As the god of medicine, Asclepius is obviously relevant to Vesalius' title page. According to Ovid's influential account in the *Metamorphoses*, he was conceived by the nymph Coronis during an affair with Apollo, who had her killed in a fit of jealous rage. Remorseful, Apollo then cut the infant from its dead mother's womb, functioning simultaneously in the story as the baby's father and as the surgeon who delivered it (figure 5.16).[76] This classical scene formed part of the visual culture of childbirth in sixteenth-century Italy, where it appeared on birth plates given to wives and expectant mothers (which can hardly have been reassuring).[77]

In alluding to Asclepius, Vesalius laid claim to a place in the lineage of distinguished graduates and professors of medicine in sixteenth-century Padua who had adopted Asclepius as part of their personal imagery, including Girolamo della Torre of Verona (d. 1506), della Torre's student and compatriot Girolamo Fracastoro (d. 1553), and — in the generation after Vesalius — the anatomist Girolamo Fabrici (d. 1619, figure 5.17). (The latter two used Asclepius on their portrait medals.)[78] Asclepius also plays an important role in the bronze reliefs on della Torre's elaborate tomb, by the Paduan sculptor Andrea Riccio, which show the master lecturing his students in the presence of Asclepius' father, Apollo, and daughter Hygeia and making a sacrifice to Asclepius, depicted

241

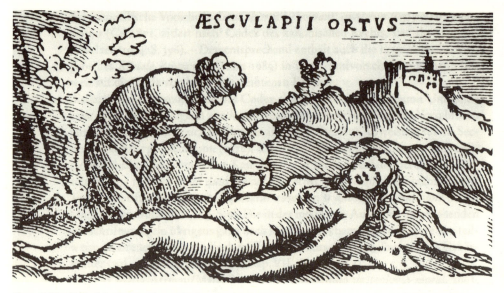

ÆSCVLAPII ORTVS

Figure 5.16. Birth of Asclepius. Alessandro Benedetti, *Habes lector studiose hoc volumine...singulis corporum morbis a capite ad pedes, generatim membratimque remedia....* (Venice: Lucantonio Giunta, 1533), title page detail.

Figure 5.17. Annibale Tosati, medal of Girolamo Fabrici of Aquapendente, London, British Museum, 598.18.3.1.

as a snake (figure 5.18).[79] There are striking compositional similarities between the tomb reliefs and Vesalius' title page; the one
devoted to the death of Girolamo (figure 5.19) is a kind of missing link between the title page and Donatello's relief showing
Saint Anthony and the anatomy of the miser (figures 5.7–5.9).
Riccio modeled Girolamo's bier and naked corpse on Donatello's
figure of the miser and created his own version of the classical
architectural background, the funeral torches, and the intense
reactions of the grieving crowd. Riccio's relief seems to have
served, in turn, as a model for the title page, not least in terms of
the series of arches that frame the crowd, while the sheep awaiting
sacrifice in the foreground of the relief showing the offering to
Asclepius (figure 5.18) are transmuted on the title page into animals awaiting dissection.

The title page of the *Fabrica* plays with these Asclepian associations and with allusions to "Caesarean" birth. On one level, the
scene asks the reader to reimagine it as Apollo's extraction of
Asclepius from the uterus of his dead lover. Here the infant, Vesalius' "son," does not appear on the title page, for it is in fact his
book, the *Fabrica* itself, which, like Asclepius in Ovid's *Metamorphoses*, will give "health and strength to all on earth."[80] Vesalius
emphasized this theme in his dedicatory epistle, where he referred
to medicine as both the "Apolline discipline" and the "art of the
Asclepiads."[81] As Apollo did with Asclepius, Vesalius has both
fathered his book and delivered it to the world.

On another level, the title page evokes the birth of Caesar,
which contemporary popular historiography associated with the
founding of the Roman Empire. In this way, it identifies Vesalius'
publication of the *Fabrica* with the foundation of a new, heroic
empire of anatomy — a "reborn art of dissection," as he called
it in his dedicatory preface, which serves as a manifesto for
this new regime.[82] Vesalius' new approach will replace the old,

degenerate medical order of earlier centuries, riddled with errors and misconceptions regarding the human body that had arisen as physicians increasingly distanced themselves from the world of matter and the body by delegating manual operation, especially surgery, to lower practitioners and retreating to the world of disputation and books. By reintegrating the more philosophical discipline of *physica* (which corresponded roughly to internal medicine) with the manual art of surgery, Vesalius presented himself as superseding the unproductive division of medical labor that supposedly characterized the medieval centuries and reviving medicine as it was practiced in the ancient world. His models were Hippocrates and, especially, Galen, who modeled this integrated approach to medical theory and practice and "frequently repeat[ed] how much he enjoyed working with his hands."[83]

Anatomy was the key to this revival, as the particular discipline that bridged *physica* and surgery, since it simultaneously illuminated the general structure and functioning of the internal organs and helped practitioners improve practical techniques of bloodletting, setting fractures, and the like. According to Vesalius, this discipline, so prized by ancient writers, had since been "ruined" by physicians who focused only on the viscera that concerned internal illness, rather than the bones, muscles, and nerves that were important to the work of surgeons, and who, furthermore, insisted on teaching anatomy without themselves touching the cadaver, leaving it to barbers and surgeons. In lessons of this sort (as illustrated in figures 2.5 and 3.1),

> some perform dissections of the human body while others recite the anatomical information. While the latter in their egregious conceit squawk like jackdaws from their lofty professorial chairs things they have never done but only memorize from the books of others or see written down, the former are so ignorant of languages that they are

Figure 5.18. Girolamo della Torre makes a sacrifice to Asclepius. Riccio (Andrea Briosco), bronze relief from the tomb of Girolamo and Marcantonio della Torre, Verona, San Fermo Maggiore, ca. 1516–1521.

Figure 5.19. Death of Girolamo della Torre. Riccio (Andrea Briosco), bronze relief for tomb of Girolamo and Marcantonio della Torre, as in figure 5.18.

unable to explain dissections to an audience and they butcher the things they are meant to demonstrate.... And as everything is being thus wrongly taught in the universities and as days pass in silly questions, fewer things are placed before the spectators in all that confusion than a butcher in a market could teach a doctor.[84]

In line with the theme of "Caesarean" birth, Vesalius studded his retelling of the history of medicine in the preface with metaphors of renewal and rebirth. In it, he positioned himself as both the restorer of the lost empire of medicine and the founder of a new dynasty of great physicians, on the model of Apollo's establishment of the Asclepian line. This theme resonates cleverly with contemporary Hapsburg mythology, which presented the Hapsburg rulers as heirs and restorers of the Roman Empire, elected by providence as the culminating stage in a sequence of divinely sanctioned regimes that began with the priest-kings of the Old Testament.[85] Just as the gods chose Charles V to preside over a dazzling cultural revival, the preface continues, they selected Vesalius to resurrect the science of anatomy. "Not wishing to be the only one to fall idle while all others are applying themselves with such success to some common topic of interest, or to be unworthy of my ancestors," he wrote, "it was my thought that this branch of natural philosophy should be recalled from the dead."[86] Vesalius has snatched anatomy from the jaws of death, just as Charles resuscitated the Roman Empire, just as the midwife saved the infant Caesar, and just as Apollo rescued Asclepius from Coronis' womb. In his time, thanks in large part to Vesalius' own efforts, the art of dissection has been "reborn."[87]

Vesalius' invocation of "Caesarean" birth, a story of beginnings, resonates with his invocation of the dissection of Agrippina, a story of endings. As the last, deluded emperor of the Julio-Claudian line, Nero represents the old, bad order of anatomy in Vesalius' emblematic scheme; like medieval physicians, divorced from the world of manual practice, he directed others to open Agrippina's body, as described in all the texts and images of this famous scene (for example, figures 5.14 and 5.15). The result was sterility and damnation; Nero's hope of bearing his own child, in the version of the story recorded in the *Golden Legend*, ended in the production of a deformed abortion and ultimately in the emperor's own death (see Chapter Three). In contrast, the renewed or "Apolline" phase of the discipline depended on the willingness of physicians to perform their own dissections, on the model of Apollo's opening of Coronis and of Vesalius' own practice as depicted on the title page. Like Apollo, and unlike Nero, Vesalius has managed successfully to both father and deliver a child, the *Fabrica*, which marks the reinstatement of anatomy as a vital, progressive field of study. The fertility of this new approach to anatomy is emphasized by the size and energy of the crowd on the title page, riveted by Vesalius' words and deeds.

The designer of the *Fabrica*'s title page used a wealth of erudite imagery to fashion an emblem aimed to work on multiple levels. Most obviously, this showed Vesalius with his students and with a dissected cadaver, signaling that this was an anatomy textbook — albeit a textbook with a difference, as the choice of a female corpse and the unconventional disposition of conventional elements make evident; this clearly referred to the old idea of "women's secrets," in which the mysteries of sexuality and generation, embodied in the female organs of generation, stood for the most difficult and inaccessible of the body's truths. On a second level, for those who chose to open the book and read the preface, the title page illustrated Vesalius' specific claim to originality — the fact that he had rejected the medieval separation between the manual art of surgery and the intellectual art of *physica*; dissecting his own corpses allowed him to correct the errors of past anatomists, both ancient and medieval, by seeing for himself. (In this, as in a number of his other claims, Vesalius strategically omitted predecessors such as Jacopo Berengario of Carpi, who had made the same point about their own work.) On a third level, for those familiar with two famous monuments with Paduan connections, Donatello's fifteenth-century altarpiece for the basilica of Saint Anthony and the early sixteenth-century tomb of Girolamo della Torre in Verona, it associated the young Vesalius both with the patron saint of Padua and with one of his eminent predecessors in the medical faculty. Finally, for connoisseurs of Roman history and classical mythology familiar with the stories of the death of Nero and the births of Caesar and Asclepius, the opened uterus of the woman on the title page echoed contemporary images of princely authority and masculine autonomy, evoking translations of empire and reinforcing Vesalius' self-identification as the triumphant founder of a new medical and anatomical regime.

Experience and the Limits of Compassion

In its use of the imagery of gendered violence associated with the ideology of princely power in sixteenth-century Italy, the title page of the *Fabrica* appears to bear out recent interpretations of natural inquiry in early modern Europe as characterized by a growing affective and epistemological distance between subject and object, figured in gendered terms. In *The Body Emblazoned*, for example, Jonathan Sawday describes sixteenth-century anatomy as "a science of *seeing*," represented metaphorically as "a male science which observed and a female subject who was observed . . . , an active male gaze confronting a passive female subject."[88] Sawday argues that the distance between the male viewers (those portrayed on the title page as well as the readers of the *Fabrica*), on the one hand, and the female object of dissection, on the other, is exaggerated not only by the gender of the participants

but also by the fact that the cadaver was that of an executed criminal; as both a woman and a convict, she lies outside the realm of
identification and compassion.[89]

While this reading captures important elements of the image
on the title page, it oversimplifies the case by treating the title
page in isolation from the image with which it was associated in
the frontmatter of the *Fabrica*: Vesalius' author portrait (figure
5.20), which acts as a pendant to the great dissection scene. The
two woodcuts were clearly intended to work together; although
the surroundings are different, each shows the anatomist standing
next to an identically equipped dissecting table, wearing the same
clothes. Like the dedicatory preface, both illustrate Vesalius' emphasis on the manual aspects of medicine and on direct experience, while the classical column or pilaster in the background of
the author portrait recalls the classicizing architecture of the title
page, as well as the preface's classicizing theme. At the same time,
however, the two images exist in counterpoint, embodying two
contrasting models of visual experience.[90] If the title page reflects
the newer model Sawday describes, which disentangled the seer
and the object, repressing the identification between the two and
reserving subjectivity to the former, the author portrait expresses
an older model of vision premised on the interaction and identification of seer and seen, in line with late medieval Christian devotional practice, which I described in Chapter One. Organized
around meditation on images, particularly images of the suffering
Christ, in accordance with the etymology of "compassion," this
practice encouraged Christians to "suffer with" the martyrs in
their agony, Mary at the Crucifixion, and Jesus on the cross.[91]

The author portrait shows Vesalius gazing at the viewer while
holding the partially dissected arm of a large, well-proportioned
male cadaver, whose long hair and draped hips strongly recall
contemporary representations of the crucified Christ (for example,

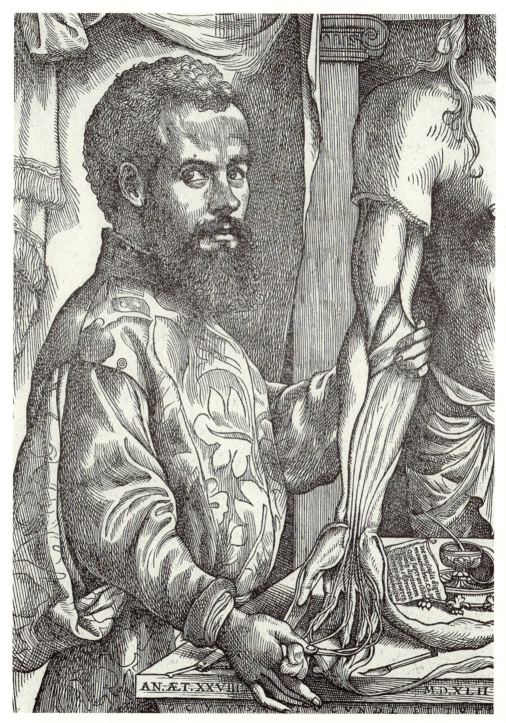

Figure 5.20. Author portrait. Andreas Vesalius, *De humani corporis fabrica*, sig. [6v].

figure 5.11). He gently supports the cadaver's elbow with his left hand, while he holds the tendons of its right hand with his own; this gesture emphasizes the identification between the anatomist and his object and echoes a common motif in images of the deposition of Christ from the cross and the lamentation over his corpse.[92] Like the title page, this image has the hallmarks of an emblem. The inscription on the lip of the table records Vesalius' age, twenty-eight, when the portrait was made in 1542, a common element in author portraits; beneath it, barely visible on the apron of the dissecting table, is the motto "Ocyus, iucunde et tuto," a version of a quotation attributed to Asclepiades by the first-century Roman medical author Celsus, recommending that treatment be performed "safely, quickly, and pleasantly."[93] This recalls the caption of Berengario's woodcut of the crucified Christ, which demonstrated the muscles of the arms and the front of the body; its function, according to Berengario, was to make doctors more knowledgeable and cautious in their operations, so that they "do not become like the torturers of Christ" (figure 5.13).[94] The piece of paper or vellum next to the cadaver's hand on the surface of the table reads, "On the muscles that move the fingers, Chapter 30. Since in the preceding book I investigated the construction of the bones of the five fingers."

As many historians have pointed out, the author portrait embodies Vesalius' manifesto in the preface regarding the need to reintegrate the textual expertise of the physician with the manual expertise of the surgeon. The hand looms large in the preface as the instrument by which this integration will be achieved. Vesalius is equally adept at manipulating the scalpel and the pen, both of which lie on the dissecting table beside him; the scrap of writing next to the inkpot is a version of the opening words of his chapter on the muscles of the hand.[95] Equally important, the author portrait restores the Christian element that was muted in the title page: the association of the anatomical cadaver with the dead Christ, as in figure 5.13. The juxtaposition of the two men's hands, so similar in size and function, also illustrates the identification and mutual interaction of anatomist and cadaver, forcing the reader to confront dissection as the intimate encounter of two human bodies. This intimacy is enhanced not only by the fact that both figures are male but also by the private setting; the drapery behind Vesalius suggests an indoor, domestic space. The visual references to the Christian themes of compassion and redemption emphasize both the piety of the anatomist and the dignity of the corpse. This image in no way obscures the dissection's associations with judgment, execution, and suffering; rather, it relates them to the Passion, surrounding them with a powerful charisma.

The effect of the title page is completely different. Everything in it serves to suppress the viewers' identification with the cadaver. A number of visual devices distance the reader from the scene, including the use of symmetry and perspective — elements that are distinctly muted in the author portrait. While the portrait assumes a viewer close to and on a level with both anatomist and cadaver, the viewpoint of the title page lies well above that of the central group of figures, corresponding to that of a student far from the action, sitting on the top rank of a set of bleachers symmetrical to the one behind the dissecting table. This implied distance is further exaggerated by the female sex of cadaver. Not only is she supine and inert, in contrast to the energetic poses of audience members, but the title page also goes out of its way to emphasize the violation of propriety and female honor involved in the display of a woman's naked body to a crowd of eager men. The corpse is displayed in a way calculated to call maximum attention to her genitals, in the style of contemporary erotic prints (such as figure 5.21).[96] Agency is reserved to the anatomist and his male

253

Figure 5.21. Jacopo Caraglio after Perino del Vaga, *Mercury and Aglaurus*, Rome, Istituto Nazionale per la Grafica, FC 5942, late 1520s.

audience, while the woman's body functions as a source of information and helps create the illusion of depth by carrying the perspectival construction of the space. This use of the female body — which was by no means limited to the *Fabrica* or the sixteenth century — not only deprives her of subjectivity but also, as Linda Hentschel has argued, sexualizes the act of vision itself.[97]

The two models of vision reflected in the author portrait and the title page — one based on identification, the other on (gendered) difference — did not exhaust the forms of seeing available in the mid-sixteenth century. (Other such forms included the decoding of visual emblems for their allegorical meaning and the manipulation of images, either imaginatively, as in the art of "artificial" memory, or manually, as in the case of layered "flap anatomies" such as Vesalius' *Epitome* of 1543.)[98] Nor were they understood as mutually exclusive; the two could coexist, as in the *Fabrica*, doing different types of work as the context required. In this respect, they were reinforced by contemporary views of sex difference, which were also premised simultaneously on similarity and difference. From the point of view of the Galenic theory that framed Vesalius' (and Berengario's) anatomy, women's bodies were anatomically the same as men's, albeit with their genitals inverted, while in other respects — their remarkable reproductive capability, their need to menstruate, the constitutional weakness and susceptibility to disease that constantly compromised their mental and physical well-being — they seemed almost to be another species.[99] In illustrating the female cadaver's genitals, rather than the many parts she shared with male bodies, the title page strictly limits her availability as an object of compassion and of knowing like by like.

The title page's focus on the female genitals recalls the long tradition that identified the mysteries of sexuality and generation as the "secrets of women." In this tradition, women were both knowers and known. In the works of learned natural philosophers and medical writers, "women's" (often "old women's") knowledge stood for all nonliterate, experience-based, and orally transmitted knowledge; women were understood to have a particular understanding of other women's bodies, based on their experience as mothers, matrons, and midwives. Over the course of the fourteenth and fifteenth centuries, as I described in Chapter Two, Italian physicians increasingly concerned themselves with the reproductive care of women — with the management of fertility and pregnancy, though not with childbirth itself. In the process, they recast the secrets of women in terms of the secrecy inherent in the structure of women's bodies, a proper object of learned inquiry, rather than as women's knowledge, inaccessible to men. The story of the title page's female cadaver, which Vesalius told in his chapter on the uterus, echoes this change. A criminal who had pretended to be pregnant in order to avoid execution, she had been examined by midwives by order of the podestà. They concluded that she was not pregnant, a judgment that Vesalius was able to transform into certain knowledge by anatomical demonstration, trumping their conjectural form of knowledge, which was based on haphazard experience.[100]

I have not yet discussed the only other woman to appear on the title page besides the Paduan criminal whose body is the focus of the scene. She stands at the back, one of two figures wedged between the second and third columns from the right; her headcloth, which wraps loosely around her head and neck, unlike the more closely fitted hoods of men in the audience, identifies her as a matron and a working woman — possibly even a birth attendant or midwife (compare figures 2.6, 3.4, and 3.7).[101] Like the men under the portico, she seems to have strayed into the dissecting theater and found herself transfixed by curiosity concerning the scene below. Unlike them, however, she is an incongruous figure — I know of no evidence that women ever attended public anatomies in mid-sixteenth-century Italy — whose presence invites speculation. Perhaps she is like the women in Berengario's illustrations, who peer curiously at their dissected uteruses, eager for the information about their own bodies that only the anatomist can reveal (figures 4.2 and 4.3). Perhaps she is one of the midwives who examined the body of the condemned woman before her death for signs of pregnancy and is waiting for Vesalius' confirmation of her judgment: certainty requires familiarity with the textual tradition and practical experience with human bodies, a set of qualifications that was found only in learned men. While it excluded most men, this model of learned expertise excluded *all* women, depriving them of their traditional authority concerning the the illnesses of the female genitals, as well as the secrets of sexuality and generation — a phenomenon that Monica H. Green calls the "masculine birth of gynecology."[102]

In ending on this note, I do not mean to evoke nostalgia for a lost age when women cared for other women, honing their skills in obstetrics and gynecology and passing that knowledge down to later generations of women, until the anatomists or the judges who presided over the witch trials of the late sixteenth and seventeenth centuries took it all away.[103] This wishful fiction, which continues to dominate much popular historiography, is merely the inverse of the myth that women were the jealous proprietors of secrets concerning generation and female sexuality that they withheld from men; the creation of learned male clerics, this myth reflected a culture of misogyny that identified women with lust and lies. In point of fact, there is very little evidence that even wealthy women received expert prenatal or postnatal care before the late fourteenth or the fifteenth century beyond that provided by kinswomen, servants, and neighbors; medieval midwives seem to have been specialists in the delivery of children,

257

not in obstetrics or gynecology more generally, and it is even difficult to find references to them as an occupational group before the thirteenth century — one of the many areas regarding the history of women's health care in which much more research is needed.[104]

This is not to denigrate the dedication and skill of the women, rich and poor, who offered one another advice and treatment and who certainly had more experience in the care of female bodies than most of the fifteenth- and sixteenth-century physicians who claimed to be authorities on it. It is rather to emphasize, as Green argues, that gynecology and obstetrics were not only an area in which expertise and authority were disjunct in medieval and Renaissance Europe; they were also an area in which the culture precluded women's developing broad-based expertise. Because the vast majority of women were illiterate, especially the working women from whose ranks midwives and female practitioners were drawn, and because Italian midwives were not organized into guilds or the kind of occupational association that would formalize the oral and empirical transmission of knowledge, there was a strict limit on their ability to pass on the knowledge that any woman or group of women might accumulate over a lifetime of practice — especially regarding unusual or problematic conditions — or even to create records for personal use.[105] There were no Martha Ballards in late medieval Italy; levels of female literacy in eighteenth-century Maine far exceeded those in Renaissance Italy.[106] Women were undoubtedly constantly creating new knowledge concerning pregnancy and women's illnesses, but they were unlikely to record it for themselves or to communicate it beyond their immediate friends and collaborators, as Guglielmo of Saliceto pointed out as early as the thirteenth century.[107] Three hundred years later, the woman in the back row of the spectators on the title page of the *Fabrica* may have had a wealth of wisdom concerning the structure, function, and afflictions of the female genitals, far surpassing that of Vesalius, but in his world of literacy and printing, that knowledge was not so much secret as fleeting, lost almost as soon as it was found.

Epilogue

What does it mean to know our bodies? Although my book focuses on the opening of women's corpses in late medieval and Renaissance Italy, it is part of a larger story in which anatomical knowledge gained by exploring the dissected body became a way to think about the self. Literate Europeans had long experienced and described their internal bodily processes in terms of thermal and fluid dynamics: the flow and interplay of vapors and liquids, which the learned parsed as including not only blood and the four humors but also substances such as the radical moisture, airy spirits, and nutritive chyle.[1] The internal organs played a vague and relatively subordinate role in this drama of heat and moisture; in lay writing, as opposed to professional medical works, they appear as generally localized regions — the "heart" as a figure for the self dedicated to the love of God or others, the "stomach" as the site of courage and engagement — without much topological specificity. It is often difficult to determine whether such descriptions were intended to be literal or metaphorical, as in the case of the birth of Christ in the heart. Each organ had its own associations and its own place in descriptions of human experience, but it was rarely placed in a clear relationship with the other organs or visualized as part of an integrated spatial whole.

261

Over the course of the sixteenth century, however, literate men and women increasingly imagined and experienced their personhood in anatomical terms. This resulted in large part from the growing number of vernacular treatises on health and the body written by learned doctors and informed by the work of professional anatomists, as well as the diffuse and often indirect influence of the text and, especially, the illustrations of Vesalius' *On the Fabric of the Human Body* of 1543.[2] Images of this sort (such as figures 4.1 and 5.1) encouraged readers and viewers to imagine the body, in the words of Jonathan Sawday, as "a carefully stowed cabin trunk."[3]

It is difficult for twenty-first-century readers to think of this understanding of the body as having had a beginning. Our culture is saturated with representations that naturalize it: advertisements market everything from migraine pills to antacids using diagrams of internal organs; news programs demonstrate the latest technologies to open clogged arteries and repair defective heart valves with animated images of the relevant parts; we pass around ultrasound images of our fetuses and watch our colonoscopies in real time. But there is little evidence that Europeans before the mid-sixteenth century visualized their bodies' functions and malfunctions in the same way — or indeed visualized them at all.

I have argued here that women's bodies, as they came to be understood during the late Middle Ages and the Renaissance, played an important part in a new conception of the body as defined in part by its internal anatomy. The uterus was central to this development. Inspired by their reading of Galen and writers in the Galenic tradition, late thirteenth-century masters of medicine and surgery in northern Italian schools and universities began to see human anatomy as one of the principal foundations of medical knowledge; following Galen, they believed that this field was best mastered through autopsy and dissection, together with the reading of relevant texts. This idea was fueled and reinforced by new funerary and judicial practices, notably the increasing currency of embalming by evisceration and the use of postmortems to determine the causes of suspicious deaths. From the beginning, female bodies played an important part in this story, a fact that at first appears surprising, since Eve was created out of Adam and the male body was taken as the template for the human form.

If we look outside the setting of medical teaching to the world of practice, however, we find a host of contemporary texts and images that describe the opening of female bodies: the mothers of Caesar and Nero, holy women such as Chiara of Montefalco and Margherita of Città di Castello, and eventually the fifteenth-century Florentine matrons I describe in Chapter Three. Even in anatomical works by academic writers, the dissection of women received special emphasis, whereas male bodies were from the beginning described matter-of-factly. In such works, male interiors appear simultaneously as familiar and as not particularly important for understanding the male self. In contrast, women's bodies — described largely or entirely in terms of their reproductive functions — appear as puzzling and imperfectly known. What was the nature of the seed produced by the female testicles? Did each twin have its own placenta? How was it possible for the uterus to change size so radically during pregancy? Did fetuses breathe and urinate in the womb? Answering these questions required exploring the insides of women's bodies. Thus where the anatomies of men were mostly occasions for demonstrating familiar truths, the anatomies of women could provide opportunities for answering open questions. In this sense, the female reproductive organs became the exemplary object of dissection, demanding visual inspection in order to be known. Male writers used the trope of secrecy to describe this aspect of female identity: the fact that the process of generation took place deep inside the female body made it opaque not only to physicians and natural philosophers but also to family and friends.

It was not until the revival of anatomy as a research field in the early sixteenth century that men's bodies, too, began to be imagined in terms of an imperfectly known interior. It was only as this anatomical perspective became normative for male as well as female bodies, and was extended to the internal organs possessed by both men and women, that these organs came to be an important resource for reflecting on both human identity and the individual self. However men and women thought about the uterus, it never functioned either as a symbol of humanity or as a figure of personal interiority. Although it shaped and often shortened women's lives in the minds of both lay and learned writers, it had always been the organ through which women were conceived of as existing for others: the lineage, the city, and the nation, as well as husbands, daughters, and sons. In this respect, it was unlike the heart, which was not a social organ but which had long stood for those aspects of selfhood that were most personal and particular, a function it did not yield to the brain until the eighteenth century.[4]

One effect of the sixteenth-century explosion of interest in anatomy as a practice and a body of knowledge was to provide Europeans with an expanded repertory of internal organs they could use to think about not only their experiences of illness but also their individual characters and inclinations. Anatomy also became an invitation to reflect on what it meant to be human — to be not only mortal, fragile, and prone to illness but also the product of a divine plan. Anatomies with flaps of skin folded back to reveal the viscera gave readers a way to imagine this process of reflection as one of self-dissection, while the Delphic injunction "Know thyself" appeared in many anatomical prefaces, images, and inscriptions, and was memorably expressed in Juan Valverde's *Anatomy of the Human Body* (1556), published

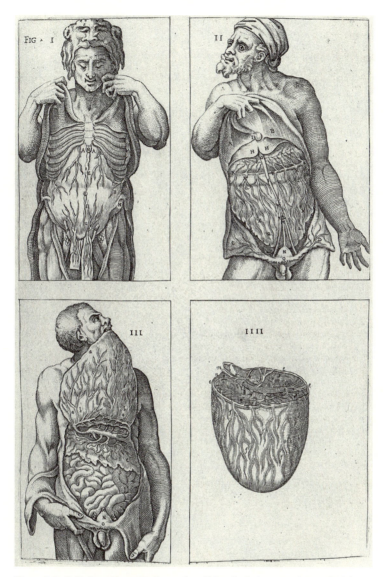

Figure E.1. Self-dissecting male figure. Juan Valverde de Amusco, *Anatomia del corpo umano* (Rome: Antonio Salamanca and Antonio Lafreri, 1560), p. 94.

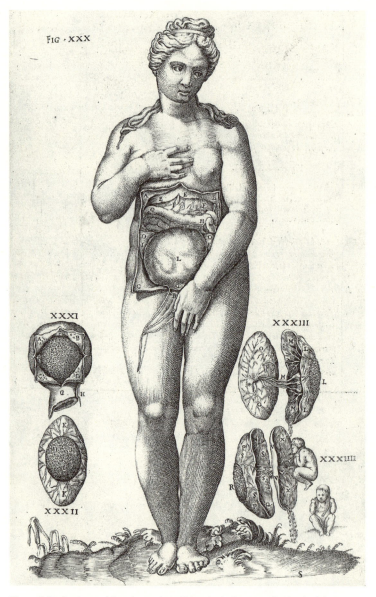

Figure E.2. Anatomy of the uterus. Juan Valverde de Amusco, *Anatomia del corpo umano* (Rome: Antonio Salamanca and Antonio Lafreri, 1560), p. 101.

first in Spanish and then in Italian, which itself depended on Vesalius' *Fabrica*.[5]

Unsurprisingly, this new anatomically inflected ideal of self-knowledge and self-possession reinscribed humanity as male. Several engravings in Valverde's book show figures in novel postures of self-dissection, self-exposure, or self-inspection that illustrate the theme. In one, a flayed man holds his skin and a knife. In another, an anatomized male figure holds open the chest of a cadaver, flanked by details of the thoracic organs; a third shows a trio of men exposing or inspecting their own viscera (figure E.1). In contrast, the only illustration of female anatomy in the text shows a woman with her eyes averted, in the classic posture of self-conscious modesty (figure E.2); she does not participate in the enterprise of dissection, and her gesture covers rather than reveals. Without her hands to hold open the incision, the rectangular window onto her internal organs appears to float awkwardly above her torso, like an anatomical chart. She exists not for herself but for others: the curious reader and the fetus by her side.

Notes

 A C K N O W L E D G M E N T S

1. Katharine Park, "The Criminal and the Saintly Body: Autopsy and Dissection in Renaissance Italy," *Renaissance Quarterly* 47 (1994), pp. 1–33, and "The Life of the Corpse: Division and Dissection in Late Medieval Europe," *Journal of the History of Medicine and Allied Sciences* 50 (1995), pp. 111–32.

2. For good and useful books that exemplify this perspective, see, for example, Andrea Carlino, *Books of the Body: Anatomical Ritual and Renaissance Learning*, trans. John Tedeschi and Anne C. Tedeschi (1994; Chicago: Chicago University Press, 1999), and R.K. French, *Dissection and Vivisection in the European Renaissance* (Aldershot: Ashgate, 1999).

I N T R O D U C T I O N

1. For a good sense of the range and quality of recent work, see Andrea Carlino, *Books of the Body: Anatomical Ritual and Renaissance Learning*, trans. John Tedeschi and Anne C. Tedeschi (1994; Chicago: Chicago University Press, 1999); Jonathan Sawday, *The Body Emblazoned: Dissection and the Human Body in Renaissance Culture* (London: Routledge, 1995); Andrew Cunningham, *The Anatomical Renaissance: The Resurrection of the Anatomical Projects of the Ancients* (Aldershot: Scolar, 1997); and R.K. French, *Dissection and Vivisection in the European Renaissance* (Aldershot: Ashgate, 1999), and the bibliography therein. All four have strengths and weaknesses; for a good historiographical summary and

critical analysis, see José Pardo-Tomás, "L'anatomia rinascimentale: Un soggetto storiografico rinnovato," in Maurizio Rippa Bonati and José Pardo-Tomás (eds.), *Il teatro dei corpi: Le pitture colorate d'anatomia di Girolamo Fabrici d'Acquapendente* (Milan: Mediamed, 2004), pp. 34–43.

2. See Katharine Park, "The Criminal and the Saintly Body: Autopsy and Dissection in Renaissance Italy," *Renaissance Quarterly* 47 (1994), pp. 12–13.

3. On the history of this medieval Christian practice, see Daniel Schäfer, *Geburt aus dem Tod: Der Kaiserschnitt an Verstorbenen in der abendländischen Kultur* (Hürtgenwald: Guido Pressler, 1999), chs. 2–3.

4. On the nature and importance of Italian funerary ritual in this period, see Sharon T. Strocchia, *Death and Ritual in Renaissance Florence* (Baltimore: Johns Hopkins University Press, 1992), esp. chs. 1–2; Giovanni Ricci, *Il principe e la morte: Corpo, cuore, effigie nel Rinascimento* (Bologna: Mulino, 1998); and Agostino Paravicini Bagliani, *The Pope's Body*, trans. David S. Peterson (1994; Chicago: University of Chicago Press, 2000), ch. 5.

5. So-called private (or, more accurately, particular) anatomies (*anatomie particulares*) constituted an intermediate category; used to explore a particular part of the body and to demonstrate its structure to a restricted audience, they were less spectacular and less degrading, and they were sometimes performed on hospital patients. The practice of dissection in Italian hospitals has yet to receive a comprehensive treatment, but see, for Milan, Monica Azzolini, "Leonardo da Vinci's Anatomical Studies in Milan: A Re-Examination of Sites and Sources," in Jean A. Givens, Karen Reeds, and Alain Touwaide (eds.), *Visualizing Medieval Medicine and Natural History, 1200–1550* (Burlington, VT: Ashgate, 2006), pp. 152–54 and 161–67.

6. Quoted in L.R. Lind, *Studies in Pre-Vesalian Anatomy: Biography, Translations, Documents* (Philadelphia: American Philosophical Society, 1975), p. 42 n.1; see also R.K. French, *Canonical Medicine: Gentile da Foligno and Scholasticism* (Leiden: Brill, 2001), pp. 132–34. I have found occasional uses of the term *dissectio* in late fifteenth-century and early sixteenth-century Latin medical texts, but it is uncommon. *Autopsy* and its cognates were first used in the sixteenth century to refer generally to the act of eyewitnessing and did not acquire anatomical

connotations until the mid-seventeenth century; see Harold Cook, "Medicine," in Katharine Park and Lorraine Daston (eds.), *The Cambridge History of Science*, vol. 3, *Early Modern Science* (Cambridge: Cambridge University Press, 2006), pp. 414–15.

7. Like cooking or butchery, these "anatomical" practices required a basic knowledge of the structure of the animal body, which explains why, when no barber or surgeon was available to prepare a human corpse for embalming, the task might fall to a butcher or cook; see Ricci, *Il principe e la morte*, p. 67.

8. Jacopo Berengario da Carpi, *Carpi commentaria cum amplissimis additionibus super Anatomia Mundini*... (Bologna: Girolamo de' Benedetti, 1521), fol. 211v–212r.

9. Ricci, *Il principe e la morte*, p. 62 (on Isabella) and, in general, ch. 5. On Lorenzo de' Medici's death and funeral, see Strocchia, *Death and Ritual*, pp. 215–16; on his autopsy, see Antonio Benivieni, *De regimine sanitatis ad Laurentium Medicem*, ed. Luigi Belloni (Turin: Società Italiana di Patologia, 1951), p. 15 n.4.

10. Berengario, *Commentaria*, fol. 345r.

11. See Chapter One n.64.

12. On the rituals surrounding papal corpses, including ritual nudity, see Paravicini Bagliani, *Pope's Body*, ch. 5. On ritual dispoliation as a practice and hagiographical trope, see Jean-Michel Sallmann, *Naples et ses saints à l'âge baroque: 1540–1750* (Paris: Presses Universitaires de France, 1994), pp. 304–10.

13. Samuel Y. Edgerton Jr., *Pictures and Punishment: Art and Criminal Prosecution during the Florentine Renaissance* (Ithaca, NY: Cornell University Press, 1985), chs. 3–4.

14. Giovanna Ferrari, "Public Anatomy Lessons and the Carnival: The Anatomy Theatre of Bologna," *Past and Present* 117 (1987), pp. 50–106; Cynthia Klestinec, "A History of Anatomy Theaters in Sixteenth-Century Padua," *Journal of the History of Medicine and Allied Sciences* 59 (2004), pp. 404–409.

15. For an important exception, see the sections of Carlino, *Books of the Body*, ch. 2, on sixteenth-century Rome.

16. On the brief appearance of human dissection in ancient Greek culture, see n.25 below.

17. See French, *Dissection and Vivisection*, pp. 14–16, on Salerno, and Nancy G. Siraisi, *Taddeo Alderotti and His Pupils: Two Generations of Italian Medical Learning* (Princeton, NJ: Princeton University Press, 1981), on Bologna.

18. André Vauchez, *Sainthood in the Later Middle Ages*, trans. Jean Birrell (1988; Cambridge: Cambridge University Press, 1997), esp. pp. 183–245; Chiara Frugoni, "The Cities and the 'New' Saints," in Anthony Molho, Kurt Raaflaub, and Julia Emlen (eds.), *City-States in Classical Antiquity and Medieval Italy* (Ann Arbor: University of Michigan Press, 1991), pp. 71–91. Because embalming was an artisanal practice, the first references in medical texts to embalming by evisceration appear only in the mid-fourteenth century; see Paravicini Bagliani, *Pope's Body*, pp. 134–35. The practice was certainly in use earlier, as is clear from Patrice Georges's excellent study, "Mourir c'est pourrir un peu…: Techniques contre la corruption des cadavres à la fin du Moyen Age," *Micrologus* 7 (1999), pp. 359–82, as well as the case studies in Chapter One of this book.

19. Agostino Paravicini Bagliani, *I testamenti dei cardinali del Duecento* (Rome: Società Romana di Storia Patria, 1980), pp. cviii-cxii, and "La papauté du XIIIe siècle et la renaissance de l'anatomie," *Medicina e scienze della natura alla corte dei papi nel Duecento* (Spoleto: Centro Italiano di Studi sull'Alto Medioevo, 1991), pp. 272–73. See, in general, Katharine Park, "The Life of the Corpse: Division and Dissection in Late Medieval Europe," *Journal of the History of Medicine and Allied Sciences* 50 (1995), pp. 111–32.

20. The urtext for this idea, in the Anglo-American tradition, is Andrew Dickson White, *A History of the Warfare of Science with Theology in Christendom* (New York: Appleton, 1897), vol. 2, pp. 49–55, which draws heavily on Jacob Burckhardt's view that fifteenth-century Italians ripped away the "veil of illusion" with which medieval Christianity had obscured the natural world, allowing the "discovery of the world and of man": cf. Burckhardt, *The Civilization of the Renaissance in Italy*, trans. S.G.C. Middlemore (New York: Harper, 1958), vol. 2 (quotations on pp. 279 and 283). On twentieth-century historians' continuing use of the Middle Ages as a foil to a postmedieval period of rational inquiry, see, in general, Edward Grant, *God and Reason in the Middle Ages* (Cambridge: Cambridge University Press, 2001), ch. 7. Michel Foucault's version of

this narrative, which preserves its shape while inverting its moral valence, has greatly influenced historiography on the topic of human dissection since the 1980s; see *The Birth of the Clinic: An Archaeology of Medical Perception,* trans. A.M. Sheridan Smith (1973; New York: Vintage, 1975), ch. 8, which is often read in light of the first few pages of his *Discipline and Punish: The Birth of the Prison,* trans. Alan Sheridan (1975; New York: Pantheon, 1977).

21. For example, Mary Niven Alston, "The Attitude of the Church Towards Dissection before 1500," *Bulletin of the History of Medicine* 16 (1944), pp. 221–38; Charles H. Talbot, *Medicine in Medieval England* (London: Oldbourne, 1967), p. 55; Park, "Criminal and Saintly Body." On the nineteenth-century origins of the myth of Columbus, see Jeffrey Burton Russell, *Inventing the Flat Earth: Columbus and Modern Historians* (New York: Praeger, 1991).

22. On the cultural appeal of this master narrative of the scientific revolution, see Katharine Park and Lorraine Daston, "Introduction," in Park and Daston (eds.), *Early Modern Science,* pp. 15–16.

23. Katharine Park, "Was There a Renaissance Body?" in Allen J. Grieco, Michael Rocke, and Fiorella Gioffredi Superbi (eds.), *The Italian Renaissance in the Twentieth Century* (Florence: Olschki, 2002), esp. pp. 326–31. For some useful reflections on the importance of situating medieval discussions of the body in their own cultural contexts, see Caroline Walker Bynum, "Why All the Fuss about the Body? A Medievalist's Perspective," *Critical Inquiry* 22 (1995), esp. pp. 29–33.

24. Many recent historians of medieval Christianity emphasize the lack of such a belief, which makes it difficult to understand why undocumented "taboos" concerning polluting contact with corpses play such an important role in recent histories of medieval and Renaissance anatomy and medicine; for example, Marie-Christine Pouchelle, *The Body and Surgery in the Middle Ages,* trans. Rosemary Morris (1983; New Brunswick, NJ: Rutgers University Press, 1990), pp. 1, 70–80, 132, 204–205; Jonathan Sawday, "The Fate of Marsyas: Dissecting the Renaissance Body," in Lucy Gent and Nigel Llewellyn (eds.), *Renaissance Bodies: The Human Figure in English Culture, c. 1540–1660* (London: Reaktion, 1990), pp. 126–34; Sawday, *Body Emblazoned,* pp. 81–83; Carlino, *Books of the Body,* pp. 3–7, 160–66, 223–24.

25. See Heinrich von Staden, "The Discovery of the Body: Human Dissection and its Cultural Contexts in Ancient Greece," *Yale Journal of Biology and Medicine* 65 (1992), pp. 223–41; Carlino, *Books of the Body*, pp. 156–69.

26. Peter Brown, *The Cult of the Saints: Its Rise and Function in Latin Christianity* (Chicago: University of Chicago Press, 1981), ch. 1 (quotation on p. 3). See also Philippe Ariès, *The Hour of Our Death*, trans. Helen Weaver (1977; New York: Oxford University Press, 1981), pp. 29–30, and Frederick S. Paxton, *Christianizing Death: The Creation of a Ritual Process in Early Medieval Europe* (Ithaca, NY: Cornell University Press, 1990), pp. 25–27 and references therein.

27. Caroline Walker Bynum, *The Resurrection of the Body in Western Christianity, 200–1336* (New York: Columbia University Press, 1995), pp. 28–32, 43–58, 321–27 (on the absence of fears concerning corpse pollution).

28. Muslim culture does not appear to have had any significant influence on Western Christianity in this regard; embalming was not a part of Muslim funerary ritual, and dissection was neither prohibited nor encouraged, directly or indirectly, by Islam; see Emilie Savage-Smith, "Attitudes toward Dissection in Medieval Islam," *Journal of the History of Medicine and Allied Sciences* 50 (1995), pp. 67–110. Embalming by evisceration seems to have been practiced in the world of Byzantine Christianity, but there is as yet no convincing evidence for anything like the modern practices of autopsy and dissection. On this, see Vivian Nutton and Christine Nutton, "The Archer of Meudon: A Curious Absence of Continuity in the History of Medicine," *Journal of the History of Medicine and Allied Sciences* 58 (2003), pp. 404–405 n.10, regarding the unconvincing character of the texts that have been adduced as evidence by Lawrence J. Bliquez and Alexander Kazhdan, "Four Testimonia to Human Dissection in Byzantine Times," *Bulletin of the History of Medicine* 58 (1984), pp. 554–57; and Robert Browning, "A Further Testimony to Human Dissection in the Byzantine World," *Bulletin of the History of Medicine* 59 (1985), pp. 518–20.

29. On criminals, Edgerton, *Pictures and Punishment*, ch. 5; Giuseppina de Sandre Gasparini, "La confraternità di S. Giovanni Evangelista della Morte in Padova e una 'riforma' ispirata dal vescovo Pietro Barozzi (1502)," in *Miscellanea Gilles Gérard Meersseman* (Padua: Antenore, 1970), vol. 1, p. 809. On visionaries,

Caroline Walker Bynum, "'…and Woman His Humanity': Female Imagery in the Religious Writing of the Later Middle Ages" and "The Female Body and Religious Practice in the Later Middle Ages," *Fragmentation and Redemption: Essays on Gender and the Human Body in Medieval Religion* (New York: Zone Books, 1991), pp. 151–80 and 181–238.

30. See Caroline Walker Bynum, "Material Continuity, Personal Survival and the Resurrection of the Body: A Scholastic Discussion in Its Medieval and Modern Contexts," *Fragmentation and Redemption*, pp. 239–97, and "Violent Imagery in Late Medieval Piety," *Bulletin of the German Historical Institute* 30 (2002), p. 15.

31. Nancy Caciola, "Mystics, Demoniacs, and the Physiology of Spirit Possession in Medieval Europe," *Comparative Studies in Society and History* 42 (2000), pp. 268–306, and *Discerning Spirits: Divine and Demonic Possession in the Middle Ages* (Ithaca, NY: Cornell University Press, 2003), esp. chs. 3–4.

32. See, for example, the essays in Christiane Klapisch-Zuber, *La maison et le nom: Stratégies et rituels dans l'Italie de la Renaissance* (Paris: Editions de l'Ecole des Hautes Etudes en Sciences Sociales, 1990). The relationship between theory and practice in this area was complicated; see the overviews, with references, in Francis William Kent, "La famiglia patrizia fiorentina nel Quattrocento: Nuovi orientamenti nella storiografia recente," in *Palazzo Strozzi, metà millennio, 1489–1989* (Rome: Istituto della Enciclopedia Italiana, 1991), pp. 70–91, and Stanley Chojnacki, "Daughters and Oligarchs: Gender and the Early Renaissance State," in Judith C. Brown and Robert C. Davis (eds.), *Gender and Society in Renaissance Italy* (London: Longman, 1998), ch. 3 and pp. 237–40.

33. Jane Fair Bestor, "Ideas about Procreation and Their Influence on Ancient and Medieval Views of Kinship," in David I. Kertzer and Richard P. Saller (eds.), *The Family in Italy from Antiquity to the Present* (New Haven: Yale University Press, 1991), pp. 150–67; Gianna Pomata, "Legami di sangue, legami di seme: Consanguineità e agnazione nel diritto romano," *Quaderni storici* 86 (1994), pp. 299–334.

34. On the origins of this idea, see Monica H. Green, "From 'Diseases of Women' to 'Secrets of Women': The Transformation of Gynecological Litera-

ture in the Later Middle Ages," *Journal of Medieval and Early Modern Studies* 30 (2000), pp. 5–39.

35. According to Christiane Klapisch-Zuber's research on wives in the *ricor-danze*, or diaries, of Florentine patrician men, almost three times as many women who predeceased their husbands died in childbirth as died from illness; see Klapisch-Zuber, "Le dernier enfant: Fécondité et vieillissement chez le Florentines (XIVe–XVe siècles)," in Jean-Pierre Bardet, François Lebrun, and René Le Mée (eds.), *Mesurer et comprendre: Mélanges offerts a Jacques Dupâquier* (Paris: Presses Universitaires de France, 1993), pp. 281–82, and "Les femmes et la mort à la fin du Moyen Age," in Stéphane Toussaint (ed.), *Ilaria del Carretto e il suo monumento: La donna nell'arte, la cultura et la società del '400* (Lucca: S. Marco, 1995), p. 219.

36. Although I agree with Marie-Christine Pouchelle's intuitions on this matter, I think she is wrong to assume that the desire of medical men to understand women's secrets was necessarily transgressive. Fifteenth-century physicians convincingly presented their interest in these topics as respectful of and beneficial to women; cf. Pouchelle, *Body and Surgery*, pp. 134–37.

37. On the history of this edition (together with its 1494 Italian translation), its unconvincing attribution to Ketham, and its relationship to earlier manuscript versions, see Tiziana Pesenti, "Editoria medica tra Quattro e Cinquecento," in Ezio Riondato, *Trattati scientifici nel Veneto fra il XV e XVI secolo* (Vicenza: Pozza, 1985), pp. 1–28, and *Fasiculo de medicina in volgare: Venezia, Giovanni e Gregorio de Gregori, 1494* vol. 2, *Il "Fasiculus medicinae," ovvero, Le metamorphosi del libro umanistico* (Padua: Università degli Studi, 2001), chs. 1–2.

38. On this image, see Pesenti, *Fasiculo de medicina in volgare*, vol. 2, pp. 104–105 and 143–45, and Chapter Two below.

39. Kenneth D. Keele and Carlo Pedretti, *Leonardo da Vinci: Corpus of the Anatomical Studies in the Collection of Her Majesty the Queen at Windsor Castle* (London: Johnson Reprint, 1979), vol. 1, p. 122r (W.12281r). Leonardo also produced a relatively finished reworking of the bloodletting figure of the *Fasiculo* (figure I.2); see *ibid.*, p. 200 (W.12597r; Keele and Pedretti 36r). On Leonardo's work on uterine anatomy and generation, see Kenneth D. Keele,

Leonardo da Vinci's Elements of the Science of Man (New York: Academic Press, 1983), pp. 348–62. On his work as a dissector, see Azzolini, "Leonardo da Vinci's Anatomical Studies in Milan," and Domenico Laurenza, *Leonardo nella Roma di Leone X (c. 1513–16): Gli studi anatomici, la vita, l'arte* (Vinci: Biblioteca Leonardiana, 2004).

40. Keele and Pedretti, *Leonardo da Vinci*, vol. 1, p. 272 (W.19037v). Leonardo also mentioned beginning with the uterus and fetus on the later sheet (ca. 1506–1508) reproduced as figure I.7: "Your arrangement shall be with the beginning of the formation of the child [*putto*] in the womb, saying which part of it is composed first and so on, and successively putting in its parts according to the duration of the pregnancy until birth, and learning in part from hen's eggs how it is nourished" (translation emended, for clarity and consistency). For an alternative ordering, dating from 1509, see Keele, *Leonardo da Vinci's Elements*, pp. 196–97. Note that some vernacular writers, such as Leonardo, used the same words — here *putto* — to refer to both fetus and child. Many others, and most or all medical writers, however, drew a distinction, referring to a fetus as *f(a)etus*, *embrio*, or *creatura* (*criatura* in Italian); cf. Berengario at n.8 above. On early views of the status of the fetus, see the essays in G.R. Dunstan (ed.), *The Human Embryo: Aristotle and the Arabic and European Traditions* (Exeter: Exeter University Press, 1990), esp. Monica H. Green, "Constantinus Africanus and the Conflict between Religion and Science," pp. 47–69.

41. For a more extensive discussion of this image, see Chapter Five.

42. John Coakley, "Friars, Sanctity, and Gender: Mendicant Encounters with Saints, 1250–1325," in Clare A. Lees (ed.), *Medieval Masculinities: Regarding Men in the Middle Ages* (Minneapolis: University of Minnesota Press, 1994), pp. 91–110.

43. Caciola, *Discerning Spirits*, esp. pp. 207–12.

44. Fundamental early studies include Carolyn Merchant, *The Death of Nature: Women, Ecology, and the Scientific Revolution*, 2nd ed. (1980; New York: HarperSanFrancisco, 1990), and Ludmilla Jordanova, *Sexual Visions: Images of Gender in Science and Medicine between the Eighteenth and Twentieth Centuries* (Madison: University of Wisconsin Press, 1989).

CHAPTER ONE: HOLY ANATOMIES

1. The fresco reflects the description of Chiara's death in her official *vita* (comp. 1315–17): Béranger of Saint-Affrique, *Vita Sanctae Clarae de Cruce: Ordinis Eremitarum S. Augustini*, ed. Alfonso Semenza, *Analecta Augustiniana* 17–18 (1939–1941), pp. 401–402; translated into English as Berengario di Sant' Africano, *Life of Saint Clare of Montefalco*, trans. Matthew J. O'Connell, ed. John E. Rotelle (Villanova, PA: Augustinian Press, 1998), pp. 84–85. On this work and its author, see Claudio Leonardi, "Chiara e Berengario: L'agiografia sulla santa di Montefalco," in Claudio Leonardi and Enrico Menestò (eds.), *S. Chiara da Montefalco e il suo tempo* (Florence: Nuova Italia, 1985), pp. 369–86. On the visual documentation of Chiara's cult, see esp. Silvestro Nessi, *Chiara da Montefalco* (Città di Castello: Edimond, 1999); Giuseppe Zois (ed.), *S. Chiara da Montefalco: Dove ci porta il cuore* (n.p.: Ritter, 1995); and Silvestro Nessi, "Primi appunti sul-l'iconografia clariana dei secoli XIV e XV," in Silvestro Nessi (ed.), *La spiritualità di S. Chiara da Montefalco* (Montefalco: Monastero S. Chiara, 1986), pp. 313–38.

2. Enrico Menestò (ed.), *Il processo di canonizzazione di Chiara da Montefalco* (Florence: Nuova Italia, 1984), p. 339; this and the following quotations are taken from the relevant section of Francesca's testimony on pp. 339–43. This volume also contains an excellent collection of essays and appendixes, which provide essential historical background to Chiara's life and cult. For a detailed discussion of the canonization proceedings, the sources, and the context, see Menestò's introduction, pp. xix–clxxxviii, which is summarized in his "Apostolic Canonization Proceedings of Clare of Montefalco, 1318–1319," in Daniel Bornstein and Roberto Rusconi (eds.), *Women and Religion in Medieval and Renaissance Italy*, trans. Margery J. Schneider (Chicago: University of Chicago Press, 1996), pp. 104–29. The literature on Chiara is vast; for an introduction and a basic bibliography, see Enrico Menestò and Roberto Rusconi, *Umbria sacra e civile* (Turin: Nuova Eri Edizioni Rai, 1989), pp. 137–64 and 233–34. On the opening of Chiara's body, see Katharine Park, "The Criminal and the Saintly Body: Autopsy and Dissection in Renaissance Italy," *Renaissance Quarterly* 47 (1994), pp. 1–4, and "Relics of a Fertile Heart: The 'Autopsy' of Clare of Monte-falco," in Anne L. McClanan and Karen Rosoff Encarnación (eds.), *The Material*

278

Culture of Sex, Procreation, and Marriage in Premodern Europe (New York: Palgrave, 2002), pp. 115–34.

3. On medieval embalming procedures, see Agostino Paravicini Bagliani, *The Pope's Body*, trans. David S. Peterson (1994; Chicago: University of Chicago Press, 2000), pp. 133–36, and Patrice Georges, "Mourir c'est pourrir un peu…: Techniques contre la corruption des cadavres à la fin du Moyen Age," *Micrologus* 7 (1999), pp. 372–79. Important older studies include Ernst von Rudloff, *Über das Konservieren von Leichen im Mittelalter: Ein Beitrag zur Geschichte der Anatomie und des Bestattungswesens* (Freiburg: Karl Henn, 1921), and Dietrich Schäfer, "Mittelalterlicher Brauch bei der Überführung von Leichen," *Sitzungsberichte der Preussischen Akademie der Wissenschaften zu Berlin* 26 (1920), pp. 478–98. For the paleopathological evidence, see Gino Fornaciari, "Renaissance Mummies in Italy," *Medicina nei secoli* 11 (1999), pp. 85–105.

4. Menestò, *Il processo*, p. 428.

5. *Ibid.*, p. 339. Opening Chiara's body "from the back [*ex parte posteriori*]," which would have greatly increased the difficulty of the procedure, was probably chosen as more appropriate to Chiara's modesty and less likely to ruin the appearance of her embalmed corpse, as in the case of a later Umbrian holy woman, Colomba of Rieti (d. 1501), discussed in Chapter Four.

6. *Ibid.*, p. 339.

7. *Ibid.*, p. 344.

8. *Ibid.*, pp. 341–42.

9. Béranger, *Vita Sanctae Clarae*, pp. 446–47; *Life of Saint Clare*, pp. 95–97. The proceedings are full of testimony regarding miracles of wounding and, especially, healing; see Nancy G. Siraisi, *Medieval and Early Renaissance Medicine: An Introduction to Knowledge and Practice* (Chicago: University of Chicago Press, 1990), pp. 39–42.

10. Menestò, *Il processo*, pp. 341–42.

11. Paravicini Bagliani, *Pope's Body*, pp. 135–36; Georges, "Mourir c'est pourrir un peu," pp. 361–62 and 373–74. The nuns emphasized that they had not applied preservatives to Chiara's body until Wednesday, five days after her death; see Menestò, *Il processo*, pp. 84 (testimony of Sister Giovanna di Maestro

Egidio of Montefalco), 153 (testimony of Sister Marina di Maestro Jacopo of Montefalco), and 244 (testimony of Sister Tommasa di Maestro Angelo of Montefalco).

12. Michel Bouvier, "De l'incorruptibilité des corps saints," in Jacques Gélis and Odile Redon (eds.), *Les miracles, miroirs des corps* (Paris: Université de Paris VIII–Vincennes, 1983), pp. 193–221. Margherita was only one of a number of recently deceased local holy men and women whose corpses were found to be partly or totally incorrupt and who might have raised similar expectations and aspirations in the nuns of Montefalco. Others included Zita of Lucca (d. 1278), Giacomo of Bevagna (d. 1301), Niccolò of Tolentino (d. 1305), Pietro of Gubbio (d. 1306), and Angelo of Borgo San Sepolcro (d. 1306). My thanks to Lesley Pattinson for this information.

13. André Vauchez, *Sainthood in the Later Middle Ages*, trans. Jean Birrell (1988; Cambridge: Cambridge University Press, 1997), esp. pp. 183–245; Chiara Frugoni, "The Cities and the 'New' Saints," in Anthony Molho, Kurt Raaflaub, and Julia Emlen (eds.), *City-States in Classical Antiquity and Medieval Italy* (Ann Arbor: University of Michigan Press, 1991), pp. 71–91; and, in general, Aviad M. Kleinberg, *Prophets in Their Own Country: Living Saints and the Making of Sainthood in the Later Middle Ages* (Chicago: University of Chicago Press, 1992). The practice of embalming seems to have spread only later to the corpses of secular leaders, at least in an Italian context; it is not clear when the popes began to be embalmed using evisceration — the first explicit testimony dates from the late fourteenth century — but it may well have been as early as the late thirteenth or early fourteenth century; see Paravicini Bagliani, *Pope's Body*, ch. 5.

14. It is not clear whether Margherita of Cortona's corpse was opened as part of the embalming procedure; the author of her *vita* describes her body only as "preserved with balsam [*balsamo conditum*]." See Giunta Bevegnati, *Legenda de vita et miraculis Beatae Margaritae de Cortona*, ed. Fortunato Iozzelli (Rome: Collegii S. Bonaventurae ad Claras Acquas, 1997), p. 451. On the importance of her embalmed body to her cult, see Daniel Bornstein, "The Uses of the Body: The Church and the Cult of Santa Margherita of Cortona," *Church History* 62 (1993), pp. 163–77.

15. There is still debate about whether Chiara's body, like those of other holy men and women, was mummified naturally or artificially. A chemical analysis of three small specimens taken from her heart, performed in 1985, found no traces of artificial preservative; see Pierluigi Baima Bollone, "Autopsia e conservazione del cuore: 'Un evento unico,'" in Zois, *S. Chiara da Montefalco*, pp. 119–23.

16. Menestò, *Il processo*, pp. 69–70 (testimony of Sister Giovanna di Maestro Egidio) and 294–95 (testimony of Brother Francesco di Damiano of Montefalco, Chiara's brother, a Franciscan friar in the convent of S. Francesco in Montefalco).

17. Béranger, *Vita Sanctae Clarae*, p. 175; *Life of Saint Clare*, p. 35.

18. Complete text of oath in Silvestro Nessi, ed., *Chiara da Montefalco, Badessa del monastero di S. Croce: Le sue testimonianze — i suoi "dicti"* (Montefalco: Associazione dei Quartieri di Montefalco, 1981), pp. 48–57 (quotation p. 52).

19. See Elizabeth A.R. Brown, "Death and the Human Body in the Later Middle Ages: The Legislation of Boniface VIII on the Division of the Corpse," *Viator* 12 (1981), pp. 221–70, and Park, "Criminal and Saintly Body," pp. 10–11. Boniface's supplementary letter of 1303, which forbade any opening of the human body, seems to have had equally little effect.

20. Menestò, *Il processo*, p. 614.

21. *Ibid.*, pp. 156, 244–45, 249, and, in general, 624; see also Siraisi, *Medieval and Early Renaissance Medicine*, pp. 39–42. On the office of the municipal doctor, see Vivian Nutton, "Continuity or Rediscovery? The City Physician in Classical Antiquity and Medieval Italy," in Andrew W. Russell (ed.), *The Town and State Physician in Europe from the Middle Ages to the Enlightenment* (Wolfenbüttel: Herzog August Bibliothek, 1981), pp. 9–46.

22. Menestò, *Il processo*, pp. xxiii and lxiv. The transcript of his testimony has unfortunately not survived. On the role of doctors as expert witnesses in proceedings of this sort, see Joseph Ziegler, "Practitioners and Saints: Medical Men in Canonization Processes in the Thirteenth to Fifteenth Centuries," *Social History of Medicine* 12 (1999), pp. 191–225.

23. Béranger, *Vita Sanctae Clarae*, p. 407; *Life of Saint Clare*, p. 91.

24. Béranger, *Vita Sanctae Clarae*, p. 406; *Life of Saint Clare*, p. 89.

25. See Chiara Frugoni, *Francesco e l'invenzione delle stimmate: Una storia per parole e immagini fino a Bonaventura e Giotto* (Turin: Einaudi, 1993); André Vauchez, "Les stigmates de Saint François et leurs détracteurs dans les derniers siècles du Moyen Age," *Mélanges d'archéologie et d'histoire* 80 (1968), pp. 595–625; and Arnold Davidson, "Miracles of Bodily Transformation, or, How St. Francis Received the Stigmata," in Caroline A. Jones and Peter Galison (eds.), *Picturing Science, Producing Art* (New York: Routledge, 1998), pp. 101–24.

26. "Vita B. Margaritae virginis de Civitate Castelli," *Analecta Bollandiana* 19 (1900), pp. 27–28. This life, written shortly after 1348, seems to be the earlier of Margherita's two surviving Latin *vitae*; see Enrico Menestò, "La 'legenda' di Margherita da Città di Castello," in Roberto Rusconi (ed.), *Il movimento religioso femminile in Umbria nei secoli XIII–XIV* (Florence: Nuova Italia, 1984), pp. 217–37, esp. 223–26. On Margherita, see Menestò and Rusconi, *Umbria sacra e civile*, pp. 167–78 and the bibliography on p. 234.

27. See Jean-Michel Sallmann, *Naples et ses saints à l'âge baroque: 1540–1750* (Paris: Presses Universitaires de France, 1994), pp. 306–309; Katharine Park, "Holy Autopsies: Saintly Bodies and Medical Expertise, 1300–1600," in Julia Hairston and Walter Stephens (eds.), *The Body in Early Modern Italy* (Baltimore, MD: Johns Hopkins University Press, forthcoming 2007); and Chapter Four below.

28. Although Italians were the first to adopt this procedure in a western European context, there may have been Byzantine precedents. William of Malmesbury reports a group of twelfth-century anatomies ordered by the king of Norway while he was in Constantinople; see Ynez Violé O'Neill, "Innocent III and the Evolution of Anatomy," *Medical History* 20 (1976), pp. 429–30 n.5.

29. O'Neill, "Innocent III," pp. 430–31; both inspections exonerated the assailants.

30. See, in general, Edgardo Ortalli, "La perizia medica a Bologna nei secoli XIII e XIV," *Atti e memorie della Deputazione di storia patria per le province di Romagna*, n.s. 17–19 (1969), pp. 223–59, and Eugenio dall'Osso, *L'organizzazione medico-legale a Bologna e a Venezia nei secoli XII–XIV* (Cesena: Addolorata, 1956), pp. 19–40. Joseph Shatzmiller provides a compact summary of this material in

"The Jurisprudence of the Dead Body: Medical Practition [*sic*] at the Service of Civic and Legal Authorities," *Micrologus* 7 (1999), pp. 223–30. The first municipal statutes governing this practice, from 1288, established qualifications for doctors assigned to the pool (age, prestige, wealth, residency) and provided for two names to be drawn at random from the pool whenever medical counsel was needed. Doctors were expected to pronounce on the number and location of wounds and to differentiate between fatal, nonfatal, and postmortem wounds.

31. Documents transcribed in Ladislao Münster, "Alcuni episodi sconosciuti o poco noti sulla vita e sull'attività di Bartolomeo da Varignana," *Castalia: Rivista di storia della medicina* 10 (1954), pp. 210 and 213; for other cases of judicially mandated anatomies in early fourteenth-century Bologna, see Shatzmiller, "Jurisprudence of the Dead Body," p. 229.

32. On this procedure, see Ziegler, "Practitioners and Saints," pp. 193–94; Vauchez, *Sainthood in the Later Middle Ages*, pp. 33–136; and Aviad M. Kleinberg, "Proving Sanctity: Selection and Authentication of Saints in the Later Middle Ages," *Viator* 20 (1989), pp. 183–205.

33. Edited in Menestò, *Il processo*. For the surviving documentation, see "Introduzione," *ibid.*, pp. xlv–civ.

34. Although doctors regularly testified in canonization procedures beginning in the thirteenth century, their testimony almost always focused on authenticating miracles of healing performed by the saint and his or her relics, not corporeal signs of sanctity. Chiara was one of only two such cases Ziegler found in his study of this practice; the other was Pierre of Luxembourg (d. 1387), whose corpse did not undergo the changes natural to a dead body. See Ziegler, "Practitioners and Saints," esp. pp. 202–203.

35. The literature on this phenomenon is large and growing rapidly. See, in general, Roberto Rusconi, "Pietà, povertà e potere: Donne e religione nell'Umbria tardomedievale," in Daniel Bornstein and Roberto Rusconi (eds.), *Mistiche et devote nell'Italia tardomedievale* (Naples: Liguori, 1992), pp. 11–24, and many of the essays in Bornstein and Rusconi (eds.), *Women and Religion*, especially Clara Gennaro, "Clare, Agnes, and Their Earliest Followers: From the Poor Ladies of San Damiano to the Poor Clares" (1980), pp. 39–55; Mario Sensi,

"Anchoresses and Penitents in Thirteenth- and Fourteenth-Century Umbria" (1982), pp. 56–83; and Anna Benvenuti Papi, "Mendicant Friars and Female Pinzochere in Tuscany: From Social Marginality to Models of Sanctity" (1983), pp. 84–103. See also Giovanna Casagrande, "Movimenti religiosi umbri e Chiara da Montefalco," in Leonardi and Menestò (eds.), *S. Chiara da Montefalco*, pp. 53–70. The footnotes of these studies contain many other references.

36. André Vauchez, "La nascita del sospetto," trans. Monica Turi, in Gabriella Zarri (ed.), *Finzione e santità tra medioevo e età moderna* (Turin: Rosenberg and Sellier, 1991), pp. 41–42, and *Sainthood in the Later Middle Ages*, esp. pp. 348–86.

37. Mario Sensi, "La monacazione delle recluse nella valle Spoletina," in Leonardi and Menestò, *S. Chiara da Montefalco*, pp. 71–94, and Casagrande, "Movimenti religiosi," pp. 61–62.

38. See John Coakley, "Friars as Confidants of Holy Women in Medieval Dominican Hagiography," in Renate Blumenfeld-Kosinski and Timea Szell (eds.), *Images of Sainthood in Medieval Europe* (Ithaca, NY: Cornell University Press, 1991), pp. 222–46, and "Friars, Sanctity, and Gender: Mendicant Encounters with Saints, 1250–1325," in Clare A. Lees (ed.), *Medieval Masculinities: Regarding Men in the Middle Ages* (Minneapolis: University of Minnesota Press, 1994), pp. 91–110.

39. On Chiara's visions, see Chiara Frugoni, "Female Mystics, Visions, and Iconography," in Bornstein and Rusconi, *Women and Religion*, pp. 131, 134, 147.

40. Claudio Leonardi, "Committenze agiografiche nel Trecento," in Vincent Moleta (ed.), *Patronage and Public in the Trecento* (Florence: Olschki, 1986), pp. 37–58.

41. Nancy Caciola, *Discerning Spirits: Divine and Demonic Possession in the Middle Ages* (Ithaca, NY: Cornell University Press, 2003).

42. Vauchez, *Sainthood in the Later Middle Ages*, p. 523; see also p. 507 n.18. On the later history of these concerns, see the essays in Zarri, *Finzione e santità*; for a case of apparently feigned female sanctity, see Judith C. Brown, *Immodest Acts: The Life of a Lesbian Nun in Renaissance Italy* (New York: Oxford University Press, 1986).

43. Béranger, *Vita Sanctae Clarae*, p. 289; *Life of Saint Clare*, pp. 41–42.

44. Quoted in Menestò, *Il processo*, p. xxiv.

45. *Ibid.*, pp. 435–36. Chiara had earlier been diagnosed by Master Mercato of Gubbio as suffering from suffocation of the uterus resulting from amenorrhea due to the unusually tight closure of her vagina or hymen (*clausuram virginitatis*); out of modesty she refused the recommended treatment: masturbation by another woman (*ibid.*, p. 270).

46. On the growing importance of physical evidence in contemporary legal procedure, see Erich Holdefleiss, *Der Augenscheinsbeweis im mittelalterlichen deutschen Strafverfahren* (Stuttgart: Kohlhammer, 1933), pp. 2–10.

47. See especially Caroline Walker Bynum, "The Female Body and Religious Practice in the Later Middle Ages," *Fragmentation and Redemption: Essays on Gender and the Human Body in Medieval Religion* (New York: Zone Books, 1991), pp. 181–219, and *Holy Feast and Holy Fast: The Religious Significance of Food to Medieval Women* (Berkeley: University of California Press, 1987).

48. In her comprehensive comparison of the *vitae* of holy men and women who died during this period in Umbria and southern Tuscany, Catherine M. Mooney found that three — in fact, four, including Margherita of Cortona, whose embalming she seems to have overlooked — of the eleven women were embalmed, but none of the nine men; see Mooney, "Women's Visions, Men's Words: The Portrayal of Holy Women and Men in Fourteenth-Century Italian Hagiography," Ph.D. diss., Yale University, 1991, p. 218.

49. Amy Hollywood, *The Soul as Virgin Wife: Mechthild of Magdeburg, Marguerite Porete, and Meister Eckhart* (Notre Dame: University of Notre Dame Press, 1995), pp. 27–36; quotations on p. 35. See also Hollywood, "Inside Out: Beatrice of Nazareth and Her Hagiographer," in Catherine M. Mooney (ed.), *Gendered Voices: Medieval Saints and Their Interpreters* (Philadelphia: University of Pennsylvania Press, 1999), pp. 78–98.

50. Caciola, *Discerning Spirits*, esp. pt. 2. See also Caciola, "Mystics, Demoniacs, and the Physiology of Spirit Possession in Medieval Europe," *Comparative Studies in Society and History* 42 (2000), pp. 268–306, and Dyan Elliott, "The Physiology of Rapture and Female Spirituality," in Peter Biller and A.J. Minnis

(eds.), *Medieval Theology and the Natural Body* (Rochester, NY: York Medieval Press, 1997), pp. 141–73.

51. Caciola, *Discerning Spirits*, pp. 61–63. These themes form part of a broader set of ideas, which Jeffrey F. Hamburger has called the "new theology of the heart." This expressed itself in a variety of heart-based ideas, metaphors, and images, including the cult of the heart of Jesus, the emphasis on the wound in the side of the crucified Christ, and the metaphor of writing on the heart. See Hamburger, *Nuns as Artists: The Visual Culture of a Medieval Convent* (Berkeley: University of California Press, 1997), esp. ch. 4; Eric Jager, *The Book of the Heart* (Chicago: University of Chicago Press, 2000); and Marie-Anne Polo de Beaulieu, "La légende du coeur inscrit dans la littérature religieuse et didactique," *Le "Cuer" au Moyen Age: Réalité et Senefiance = Senefiance* 30 (1991), pp. 299–312.

52. Francis of Assisi, "Epistola I," *Opuscula sancti patris Francisci Assisiensis* (Quaracchi: Collegii S. Bonaventurae, 1904), pp. 93–94.

53. Ambrose, *Enarrationes in Psalmos* 47:10, cited in Hugo Rahner, "Die Gottesgeburt: Die Lehre der Kirchenväter von der Geburt Christi im Herzen des Gläubigen," *Zeitschrift für katholische Theologie* 59 (1935), p. 388.

54. Ambrose, *Commentaria in Lucam* 10.24, in *Corpus scriptorum ecclesiasticorum latinorum,* vol. 32, *Sancti Ambrosii opera*, pt. 2, ed. Karl Schenkl (Vienna: Tempsky, 1897), pp. 464–65.

55. See Rahner, "Die Gottesgeburt," pp. 390–91 and 396; Robert Linhardt, *Die Mystik des hl. Bernhard von Clairvaux* (Munich: Verlag Natur und Kultur, 1923), pp. 192–97; and, in general, Caroline Walker Bynum, "Jesus as Mother and Abbot as Mother: Some Themes in Twelfth-Century Cistercian Writing," *Jesus as Mother: Studies in the Spirituality of the High Middle Ages* (Berkeley: University of California Press, 1982), esp. pp. 114–24.

56. For this textual and visual tradition, see Amy Neff, "The Pain of *Compassio*: Mary's Labor at the Foot of the Cross," *Art Bulletin* 80 (1998), pp. 254–43, and Otto von Simpson, "*Compassio* and *Co-Redemptio* in Roger van der Weyden's *Descent from the Cross*," *Art Bulletin* 35 (1953), pp. 9–16. Some writers specified that this second childbirth involved Mary's heart rather than her uterus.

57. Bynum, "Jesus as Mother."

58. Béranger, *Vita Sanctae Clarae*, p. 454; *Life of Saint Clare*, p. 104 (translation emended slightly, for accuracy).

59. See Mooney, "*Imitatio Christi* or *Imitatio Mariae*? Clare of Assisi and Her Interpreters," in Mooney (ed.), *Gendered Voices*, pp. 52–77. Chiara's visions, even as recorded by Berengario, show her as identifying directly with Christ in his Passion and as experiencing his planting the cross in her heart exclusively in these terms.

60. On the nature and early history of this procedure, see Daniel Schäfer, *Geburt aus dem Tod: Der Kaiserschnitt an Verstorbenen in der abendländischen Kultur* (Hürtgenwald: Guido Pressler, 1999), chs. 1–3, and Renate Blumenfeld-Kosinski, *Not of Woman Born: Representations of Caesarean Birth in Medieval and Renaissance Culture* (Ithaca, NY: Cornell University Press, 1990), pp. 21–26. Both these studies should be used with care: Schäfer's is more reliable, although his Latin transcriptions should be checked against their sources; for important caveats regarding Blumenfeld-Kosinski's scholarship and interpretation, see the review by Monica H. Green in *Speculum* 67 (1992), pp. 380–81. For additional references regarding the medieval practice of this operation, see Green, "Bodies, Gender, Health, Disease: Recent Work on Medieval Women's Medicine," *Studies in Medieval and Renaissance History*, ser. 3, vol. 2 (2005), p. 19.

61. Bernard de Gordon, *Practica, seu Lilium medicinae* 7.15 (Venice: Giovanni and Gregorio de' Gregori, 1496), fol. 217v.

62. Louis-Fernand Flutre and Cornelius Sneyders de Vogel, *Li fet des Romains: Compilé ensemble de Saluste et de Suétoine et de Lucan; Texte du XIIIe siècle* (Paris: Droz, 1938), ch. 2, p. 8. On the Italian influence of this work, see Louis-Fernand Flutre, *Li fait des Romains dans les littératures française et italienne du XIIIe au XVe siècle* (Paris: Hachette, 1932). For more on the birth of Caesar, see Chapter Three and Chapter Five below.

63. It was described by Thomas Aquinas in *Summa theologiae* 3.68.11, and emphatically prescribed, in considerable detail, by a variety of ecclesiastical councils and synods. See Schäfer, *Geburt aus dem Tod*, pp. 32–33; Silvano Cavazza, "Double Death: Resurrection and Baptism in a Seventeenth-Century Rite,"

trans. Mary M. Gallucci, in Edward Muir and Guido Ruggiero (eds.), *History from Crime* (Baltimore: Johns Hopkins University Press, 1994), pp. 17–20; and Green, "Bodies, Gender, Health, Disease," pp. 19–20.

64. Giordano of Pisa (Giordano of Rivalto), *Prediche del Beato Fra Giordano da Rivalto, dell'Ordine dei predicatori recitate in Firenze dal MCCCIII al MCCCVI*, ed. Domenico Moreni (Florence: Magheri, 1831), sermon 1, vol. 1, pp. 5–6.

65. Menestò, *Il processo*, p. 85.

66. *Ibid.*, p. 87.

67. M.H. Laurent, "La plus ancienne légende de la B. Marguerite de Città di Castello," *Archivum fratrum praedicatorum* 10 (1940), p. 128.

68. Song of Solomon 8:6.

69. Katharine Park, "Impressed Images: Reproducing Wonders," in Jones and Galison (eds.), *Picturing Science*, pp. 254–71. On the broader meanings of sealing in medieval culture, see Brigitte Miriam Bedos-Rezak, "Medieval Identity: A Sign and a Concept," *American Historical Review* 105 (2000), pp. 1489–1533. On the use of the sealing metaphor in visionary piety, see, for example, Jeffrey F. Hamburger, "Seeing and Believing: The Suspicion of Sight and the Authentication of Vision in Late Medieval Art," in Alessandro Nova and Klaus Krüger (eds.), *Imagination und Wirklichkeit: Zum Verhältnis von mentalen und realen Bildern in der Kunst der Frühen Neuzeit* (Mainz: Von Zabern, 2000), pp. 58–59.

70. Albertus Magnus, *Book of Minerals*, trans. Dorothy Wyckoff (Oxford: Clarendon, 1967), 1.1.5, p. 22.

71. This idea appears in numerous ancient and contemporary sources, both Latin and vernacular; for example, an early fourteenth-century work by Francesco da Barberino, *Reggimento e costumi di donna*, ed. Giuseppe E. Sansone (Turin: Loescher-Chiantore, 1957), pt. 16, p. 226. See Jacqueline Marie Musacchio, "Imaginative Conceptions in Renaissance Italy," in Geraldine A. Johnson and Sara F. Matthews Grieco (eds.), *Picturing Women in Renaissance and Baroque Italy* (Cambridge: Cambridge University Press, 1997), pp. 42–60, and *The Art and Ritual of Childbirth in Renaissance Italy* (New Haven: Yale University Press, 1999), pp. 128–34; and, more generally, David Freedberg, *The Power of Images:*

Studies in the History and Theory of Response (Chicago: University of Chicago Press, 1989), pp. 2–4. See also Chapter Three below.

72. Park, "Impressed Images."

73. Jacobus de Voragine, "Sermo III de S. Francisco," *Sermones pulcherrimi variis scripturarum doctrinis referti de sanctis per anni totius circulum concurrentibus* (Paris: n.p., 1510), sermo 264, sig. [Ddvii v–viii r]: *Sanctus ergo Franciscus in visione sibi facta ymaginabatur seraphim crucifixum, et tam fortis ymaginatio extitit quod vulnera passionis in carne sua impressit.* See Davidson, "Miracles of Bodily Transformation," esp. pp. 116–18. To attribute the stigmata to the effects of Francis's imagination was in no sense to debunk them, given the central role played by the imagination in contemporary theological accounts of ecstatic vision. On the three levels of vision described by Augustine in his discussion of the rapture of Paul, see Suzannah Biernoff, *Sight and Embodiment in the Middle Ages* (New York: Palgrave Macmillan, 2002), pp. 25–26.

74. Béranger, *Vita Sanctae Clarae*, p. 169; *Life of Saint Clare*, p. 27. See also Elliott, "Physiology of Rapture and Female Spirituality," esp. pp. 144–47.

75. Laurent, "La plus ancienne légende," p. 122.

76. Park, "Impressed Images," pp. 262–63.

77. Jacobus de Voragine, *The Golden Legend: Readings on the Saints*, trans. William Granger Ryan (Princeton, NJ: Princeton University Press, 1993), vol. 1, pp. 140–43.

78. Hans Belting, *The Image and Its Public in the Middle Ages: Form and Function of Early Paintings of the Passion* trans. Mark Bartusis and Raymond Meyer (1981; New Rochelle, NY: Caratzas, 1990), pp. 58 and 80 (quotation); in general, see chs. 4, 6, 7. On these ideas as manifested in thirteenth- and fourteenth-century Umbrian and Tuscan painting, see, for example, Anne Derbes, *Picturing the Passion in Late Medieval Italy: Narrative Painting, Franciscan Ideologies, and the Levant* (Cambridge: Cambridge University Press, 1996); Rona Goffen, *Spirituality in Conflict: Saint Francis and Giotto's Bardi Chapel* (University Park: Pennsylvania State University Press, 1988); and Jeryldene Wood, "Perceptions of Holiness in Thirteenth-Century Italian Painting: Clare of Assisi," *Art History* 4 (1991), pp. 301–28. Caroline Walker Bynum complicates Belting's point in

"Seeing and Seeing Beyond: The Mass of St. Gregory in the Fifteenth Century," in Jeffrey F. Hamburger and Anne-Marie Bouché (eds.), *The Mind's Eye: Art and Theological Argument in the Middle Ages* (Princeton, NJ: Princeton University Press, 2005), pp. 208–40.

79. While Chiara's heart was publicly displayed, there is no indication that her body was placed on view uncovered before the Counter-Reformation period. Her opaque wooden coffin, preserved in the monastery, dates from the fifteenth century, and the glass coffin in which her corpse now rests, from the later sixteenth century. Margherita's embalmed body is currently on display in a glass casket beneath the altar of the Dominican church in Città di Castello (as in Chiara's case, a Counter-Reformation innovation), while two of the stones are preserved in a silver reliquary; the third was given to Ferdinando di Borbone, Duke of Parma, in the late eighteenth century and subsequently lost. See Ileana Tozzi, "Tra mistica e politica: L'esperienza femminile nel Terz'Ordine della penitenza di San Domenico," *Rassegna storica online* 1 n.s 4 (2003). She is currently the object of a renewed canonization campaign as the patron saint of the disabled, as literature available in the church makes clear.

80. Laurent, "La plus ancienne légende," p. 126. Cf. *Vita Beatae Margaritae*, chs. 7–8, pp. 26–28.

81. Laurent, "La plus ancienne légende," p. 127.

82. *Ibid.*

83. On the rivalry between Dominicans and Franciscans in the visual arts, see Derbes, *Picturing the Passion*, p. 201 n.121, and Dieter Blume, "Ordenskonkurrenz und Bildpolitik: Franziskanische Programme nach dem theoretischen Armutsstreit," in Hans Belting and Dieter Blume (eds.), *Malerei und Stadtkultur in der Dantezeit* (Munich: Hirmer, 1989), pp. 149–71.

84. Derbes, *Picturing the Passion*, p. 17; Joanna Cannon, "Simone Martini, the Dominicans, and the Early Sienese Polyptich," *Journal of the Warburg and Courtauld Institutes* 45 (1982), pp. 75–76.

85. Béranger, *Vita Sanctae Clarae*, p. 98; *Life of Saint Clare*, p. 22.

86. Béranger, *Vita Sanctae Clarae*, p. 96; *Life of Saint Clare*, p. 19. See also Menestò, *Il processo*, for example, pp. 270 and 330.

87. See Chapter Four and Chapter Five below, as well as Patricia Simons, "Anatomical Secrets: *Pudenda* and the *Pudica* Gesture," in Gisela Engel (ed.), *Das Geheimnis am Beginn der europäischen Moderne* (Frankfurt: Klostermann, 2002), pp. 302–27. For some of the complexities involved in analyzing narratives of female martyrdom without falling into anachronism, see Madeline H. Caviness, *Visualizing Women in the Middle Ages: Sight, Spectacle, and Scopic Economy* (Philadelphia: University of Pennsylvania Press, 2001), and Elizabeth A. Castelli, *Visions and Voyeurism: Holy Women and the Politics of Sight in Early Christianity*, ed. Christopher Ocker (Berkeley: Center for Hermeneutical Studies, 1995).

88. On the meanings of Christ's nakedness, see Derbes, *Picturing the Passion*, pp. 31–32. There was obviously special discomfort involved in the display of the female genitals, which are rarely if ever shown in religious images or even in medieval medical illustrations.

89. Against the anachronistic reading of violence in medieval devotional images and texts, see Caroline Walker Bynum, "Violent Imagery in Late Medieval Piety," *Bulletin of the German Historical Institute* 30 (2002), pp. 3–36.

90. Paravicini Bagliani, *Pope's Body*, pp. 122–32; Sallmann, *Naples et ses saints*, pp. 296–98.

91. Biernoff, *Sight and Embodiment*, p. 137. See also Belting, *Image and its Public*, esp. pp. 56–58, and Robert Scribner, "Ways of Seeing in the Age of Dürer," in Dagmar Eichberger and Charles Zika (eds.), *Dürer and His Culture* (Cambridge: Cambridge University Press, 1998), pp. 93–117.

92. See, in general, Caciola, *Discerning Spirits*, ch. 3.

93. On this altarpiece and the circumstances of its production, see Gaudenz Freuler, "Andrea di Bartolo, Fra Tommaso d'Antonio Caffarini, and Sienese Dominicans in Venice," *Art Bulletin* 69 (1987), pp. 572–76.

94. Jacopo Scalza, *Leggenda latina della B. Giovanna detta Vanna d'Orvieto*, ed. Vincenzo Marreddu (Orvieto: Sperandio Pompei, 1853), p. 134; Raymond of Capua, *The Life of Catherine of Siena*, trans. Conleth Kearnes (Wilmington, DE: Glazier, 1980), pp. 185–86.

95. Chiara's experience of this piercing may represent her visionary understanding of compunction (*compunctio cordis*), the deep and wounding remorse

for one's sins that was fundamental to the soul's progress: Joseph de Guibert, "La componction du coeur," *Revue d'ascétique et de mystique* 15 (1934), esp. pp. 225–36. See also Hamburger, *Nuns as Artists*, ch. 3.

96. Fernando Salmón and Montserrat Cabré, "Fascinating Women: The Evil Eye in Medical Scholasticism," in Roger French (ed.), *Medicine from the Black Death to the French Disease* (Aldershot: Ashgate, 1998), pp. 53–84.

97. *Ibid.*, p. 51. See also Ruth H. Cline, "Heart and Eyes," *Romance Philology* 25 (1971–1972), pp. 262–97, and Sebastian Neumeister, "Das Bild der Geliebten im Herzen," in Ingrid Kasten, Werner Paravicini, and René Pérennec (eds.), *Kultureller Austausch und Literaturgeschichte im Mittelalter/Transferts culturels et histoire littéraire au moyen age* (Sigmaringen: Jan Thorbecke, 1998), esp. pp. 320–30.

98. Biernoff, *Sight and Embodiment*, pp. 31–34 and 156, and Adrian Randolph, "Regarding Women in Sacred Space," in Johnson and Matthews Grieco (eds.), *Picturing Women*, esp. p. 39.

99. See Bynum, "Jesus as Mother."

CHAPTER TWO: THE SECRETS OF WOMEN

1. Salimbene de Adam, *Cronica*, ed. Giuseppe Scalia (Bari: Laterza, 1966), vol. 2, p. 894.

2. On the evidence for anatomies in medieval Byzantium and the Islamic world, see Introduction n.28.

3. *Chronicon parmense ab anno MXXXVIII usque ad annum MCCCXXXVIII*, ed. Giuliano Bonazzi (Città di Castello: Lapi, 1902), *ad annum* 1286: "*Item eo anno fuit magna mortalitas hominum et bestiarum in civitate et episcupatu Parme, et maxime de bestiis menutis.*" The epidemic is also recorded as having killed both birds and humans north of the Alps; see Alfonso Corradi, *Annali delle epidemie occorse in Italia dalle prime memorie fino al 1850*, vol. 1 (Bologna: Gamberini e Parmeggiani, 1865), p. 155. On the immediate nature of Salimbene's sources for events after 1284, when he began to record them almost daily, see Jacques Paul, "Salimbene, testimone e cronista," in Jacques Paul and Mariano D'Alatri, *Salimbene da Parma: Testimone e cronista* (Rome: Istituto Storico dei Cappuccini, 1992), pp. 21–27.

4. See Chapter One n.30. For dissection in connection with medical instruction at the University of Bologna, see Nancy G. Siraisi, *Taddeo Alderotti and His Pupils: Two Generations of Italian Medical Learning* (Princeton, NJ: Princeton University Press, 1981), pp. 110–14.

5. Taddeo Alderotti, *Expositio in Aphorismorum Ipocratis volumen, Expositiones in arduum Aphorismorum Ipocratis volumen, in divinum Pronosticorum Ipocratis librum* (Venice: Luca Antonio Giunta, 1527), fol. 140r: "*Ad hanc questionem non possum cum certitudine respondere, quia nec invenio eam determinatam ab auctoribus expresse, neque anothomiam vidi in muliere pregnante.*" On the dating of this commentary, see Siraisi, *Taddeo Alderotti*, p. 40.

6. Salimbene's disapproval of Frederick II's opening of a human body in connection with an experiment regarding human digestion, in contrast, certainly stemmed from the fact that Frederick's subjects had not been allowed to die of natural causes but had been killed to satisfy the emperor's "curiosity"; see Salimbene, *Cronica, ad annum* 1250, vol. 2, p. 515.

7. For examples of this practice in Florence, Perugia, and Orvieto in 1348, see Katharine Park, *Doctors and Medicine in Early Renaissance Florence* (Princeton, NJ: Princeton University Press, 1985), p. 97, and Vincenzo Busacchi, "Necroscopie trecentesche a scopo anatomo-patologico in Perugia," *Rivista di storia della medicina* 9 (1965), pp. 160–64. The Perugia autopsy of 1348, like the Cremona autopsy of 1286, revealed "a small abscess full of poison [*una bessica picola piena de veneno*] around the heart" of the plague victim, a coincidence that may be evidence for the continuity of the practice between the late thirteenth and mid-fourteenth centuries; see Busacchi, "Necroscopie," pp. 161–62.

8. *Consilia* were formal letters of medical advice; see Jole Agrimi and Chiara Crisciani, *Les consilia médicaux*, trans. Caroline Viola (Turnhout: Brepols, 1994), and Siraisi, *Taddeo Alderotti*, pp. 270–302. Although most *consilia* were addressed to individuals, epidemics might inspire works addressed to whole communities. One of the conditions under which Taddeo was retained as a municipal physician in Venice in 1293 was that he would supply a formal opinion in the case of an epidemic specifying "from what things the population had to abstain and what measures they needed to take"; quoted in Ugo Stefanutti, *Documentazioni crono-*

logiche per la storia della medicina, chirurgia e farmacia in Venezia dal 1258 al 1332 (Venice: Ongania, 1961), p. 22.

9. Helmut Puff, *Sodomy in Reformation Germany and Switzerland, 1400–1600* (Chicago: University of Chicago Press, 2003), p. 74. The trope of unspeakability was considerably more specific in its uses than the trope of secrecy, almost always signaling a reference to the sexual sin of sodomy.

10. On these two different traditions of secrecy, see William Eamon, *Science and the Secrets of Nature: Books of Secrets in Medieval and Early Modern Culture* (Princeton, NJ: Princeton University Press, 1994), chs. 1–2, and Pamela O. Long, *Openness, Secrecy, Authorship: Technical Arts and the Culture of Knowledge from Antiquity to the Renaissance* (Baltimore: Johns Hopkins University Press, 2001), chs. 2–3, esp. pp. 88–95 (on proprietary craft knowledge).

11. Karma Lochrie focuses on this particular idea of secrecy in *Covert Operations: The Medieval Uses of Secrecy* (Philadelphia: University of Pennsylvania Press, 1999), esp. the introduction and ch. 3. On the many possible connotations of secrecy, see Long, *Openness, Secrecy, Authorship*, p. 7.

12. See Puff, *Sodomy in Reformation Germany*, ch. 3, and Lochrie, *Covert Operations*, ch. 1.

13. Pseudo-Albertus Magnus, *De secretis mulierum et virorum* (Leipzig: Melchior Lotter, 1501); chapter divisions refer to this edition. This work has been translated by Helen Rodnite Lemay as *Women's Secrets: A Translation of Pseudo-Albertus Magnus's De secretis mulierum with Commentaries* (Albany: State University of New York Press, 1992). I have used her translation as a point of departure (but not her chapter divisions, which differ from those in the 1505 edition), checking it against the 1505 Latin text of the work itself and the commentary (incipit "Scribit philosophus") that she calls Commentary A. The colophon of the earliest known manuscript (in a private collection in Paris, dated ca. 1300) refers to the text as *De secretis mulierum*, while Commentary A (earliest manuscript 1353) refers to it as *Secreta mulierum et virorum*, the usage followed by the 1505 edition. There is no critical edition of the highly variable Latin text and commentaries (two of the latter of which are excerpted by Lemay), and there are significant differences between the manscript tradition, which does not

include chapter divisions, and the printed editions. On the work itself, see Margaret R. Schleissner, "*Secreta mulierum*," in Kurt Ruh (ed.), *Die deutsche Literatur des Mittelalters: Verfasserlexikon*, 2nd ed., vol. 8 (Berlin: Walter de Gruyter, 1992), pp. 986–93, and Monica H. Green, "'Traittié tout de mençonges': The *Secrés des dames*, 'Trotula,' and Attitudes toward Women's Medicine in Fourteenth- and Early-Fifteenth-Century France," in Marilynn Desmond (ed.), *Christine de Pizan and the Categories of Difference* (Minneapolis: University of Minnesota Press, 1998), pp. 146–78.

14. Pseudo-Albertus Magnus, *De secretis mulierum et virorum*, ch. 1, sig. a2v; cf. Lemay, *Women's Secrets*, p. 59.

15. For an extended discussion of the meanings of "secret" in medieval texts on generation, see Monica H. Green, "From 'Diseases of Women' to 'Secrets of Women': The Transformation of Gynecological Literature in the Later Middle Ages," *Journal of Medieval and Early Modern Studies* 30 (2000), pp. 7–14. Lochrie ignores the epistemological issues in *De secretis mulierum* in her otherwise useful study of secrecy in medieval texts; see Lochrie, *Covert Operations*, pp. 118–31.

16. Biblioteca Apostolica Vaticana, ms. Vat. Pal. Lat. 1264; quoted in Lynn Thorndike, "Further Consideration of the *Experimenta*, *Speculum Astronomiae*, and *De secretis mulierum* Ascribed to Albertus Magnus," *Speculum* 30 (1955), p. 430. This is the commentary Lemay calls Commentary A.

17. The definitive treatment of the meanings and epistemological implications of *experimenta* is Jole Agrimi and Chiara Crisciani, "Per una ricerca su *experimentum-experimenta*: Riflessione epistemologica e tradizione medica (secoli XIII–XV)," in Pietro Janni and Innocenzo Mazzini (eds.), *Presenza del lessico greco e latino nelle lingue contemporanee* (Macerata: Università degli Studi di Macerata, 1990), pp. 9–49. Other useful discussions appear in Eamon, *Science and the Secrets of Nature*, pp. 53–58, and Michael McVaugh, "Two Montpellier Recipe Collections," *Manuscripta* 20 (1976), pp. 175–80.

18. Pseudo-Albertus Magnus, *De secretis mulierum et virorum*, ch. 4, sig. e2r. Lemay mistranslates "*docte mulieres vel helene in hac arte*" as "harlots, and women learned in the art of midwifery" (*Women's Secrets*, p. 102), whereas *helene* is

probably a corruption of *lene*, procuresses (Monica H. Green, personal communication); the text clearly and consistently refers to midwives as *obstitrices*, a term that does not appear in this passage. This set of ideas is an old one and appears in Hippocratic texts; see Helen King, *Hippocrates' Woman: Reading the Female Body in Ancient Greece* (London: Routledge, 1998), pp. 136–39.

19. Pseudo-Albertus Magnus, *De secretis mulierum et virorum*, ch. 3, sig. c4r–v (text) and c5r (commentator), and ch. 4, sig. e2r; cf. Lemay, *Women's Secrets*, pp. 88 and 102.

20. Pseudo-Albertus Magnus, *De secretis mulierum et virorum*, ch. 3, sig. c5r; cf. Lemay, *Women's Secrets*, p. 88. See Agrimi and Crisciani, "Per una ricerca," and "Medici e 'vetulae' dal Duecento al Quattrocento: Problemi di una ricerca," in Paolo Rossi, Lucilla Borcelli, Chiaretta Poli, and Giancarlo Carabelli, *Cultura popolare e cultura dotta nel Seicento* (Milan: Angeli, 1983), pp. 147–50.

21. For more on this model of medical learning, see, for example, Siraisi, *Taddeo Alderotti*, pp. 120–25, and R.K. French, *Medicine Before Science: The Rational and Learned Doctor from the Middle Ages to the Enlightenment* (Cambridge: Cambridge University Press, 2003), esp. pp. 84–117.

22. See Agrimi and Crisciani, "Medici e 'vetulae'"; "Immagini e ruoli della 'vetula' tra sapere medico e antropologia religiosa (secoli XIII-XV)," in Agostino Paravicini Bagliani and André Vauchez (eds.), *Poteri carismatici e informali: Chiesa e società medioevali* (Palermo: Sellerio, 1992), pp. 224–61; and "Savoir médical et anthropologie religieuse: Les représentations et les fonctions de la *vetula* (XIIIe–XVe siècle)," *Annales: Economies, Sociétés, Civilisations* 48 (1993), pp. 1281–1308. As Sarah Kay shows, the Old Woman's speech in the *Roman de la rose* illustrates these themes in a literary mode: "Women's Body of Knowledge: Epistemology and Misogyny in the *Romance of the Rose*," in Sarah Kay and Miri Rubin (eds.), *Framing Medieval Bodies* (Manchester: Manchester University Press, 1994), esp. pp. 216–19. Cf. Guillaume de Lorris and Jean de Meun, *Le Roman de la rose*, ed. Daniel Poirion (Paris: Garnier-Flammarion, 1974), ll. 122802–805: "*N'onc ne fui d'Amors a escole / Ou l'en leüst la theorique, / Mes je sai bien par la pratique. / Experiment m'en ont fait sage, / Que j'ai hantés tout mon aage.*"

23. Lanfranco of Milan, *Practica...que dicitur ars completa totius cyrurgie* 2.5,

in Guy of Chauliac, *Cyrurgia Guidonis de Cauliaco et Cyrurgia Bruni, Theodorici, Rolandi, Lanfranci, Rogerii, Bertapalie* (Venice: Bonetus Locatellus, 1498), fol. 206va: "*Nullas enim in eo [use of medicines] ponemus, nisi illas quibus longo tempore sumus usi, et quas a reverendis doctoribus medicis, ac etiam mulieribus habuimus, que omnes sine dubio in casibus suis sunt experte.*"

24. Bruno of Longobucco, prologue to *Cyrurgia magna*, in Guy of Chauliac, *Cyrurgia Guidonis*, fol. 83ra: "*Sint [surgeons] etiam viri litterati, aut ab eo qui novit litteras ad minus artem adiscant, vix enim aliquem absque litteris hanc artem comprehendere puto, sed tempore presente nedum ydiote. Immo quod indecentius et horribilius iudicatur, viles femine et presumptuose artem hanc usurpaverunt et abutuntur ea, quia licet curant, ut refert Almansor, nec artem nec ingenium habent.*"

25. The authoritative study of this development is Siraisi, *Taddeo Alderotti*, esp. chs. 4–5. For developments specific to surgery, see Michael McVaugh, "Surgical Education in the Middle Ages," *Dynamis* 20 (2000), esp. pp. 286–93.

26. For Guglielmo's biography, derived mostly from autobiographical remarks in his writings, see Mario Tabanelli (ed.), *La chirurgia italiana nell'Alto Medioevo* (Florence: Olschki, 1965), vol. 2, pp. 449–506, and Jole Agrimi and Chiara Crisciani, "The Science and Practice of Medicine in the Thirteenth Century According to Guglielmo da Saliceto, Italian Surgeon," in Luis García-Ballester, Roger French, Jon Arrizabalaga, and Andrew Cunningham (eds.), *Practical Medicine from Salerno to the Black Death* (Cambridge: Cambridge University Press, 1994), pp. 63–64.

27. For manuscripts and printed editions of these works, see Tabanelli (eds.), *La chirurgia italiana*, vol. 2, pp. 513–21, and Agrimi and Crisciani, "Science and Practice of Medicine," p. 66.

28. Guglielmo of Saliceto, *Summa conservationis et curationis* (Venice: Marinus Saracenus, 1490), 1.178 and 1.180, sig. i5vb and i6va.

29. *Ibid.*, preface, sig. a2ra; see also h4vb. On this point, see Jole Agrimi and Chiara Crisciani, *Edocere medicos: Medicina scolastica nei secoli XIII–XV* (Naples: Guerini, 1988), pp. 185–86, and Nancy G. Siraisi, "How to Write a Latin Book on Surgery: Organizing Principles and Authorial Devices in Guglielmo da Saliceto and Dino del Garbo," in García-Ballester, French, Arrizabalaga, and

Cunningham (eds.), *Practical Medicine*, p. 94, and *Taddeo Alderotti*, pp. 14–23.

30. Teodorico Borgognoni of Lucca, preface to *Cyrurgia*, in Guy of Chauliac, *Cyrurgia Guidonis*, fol. 106ra: "*Quia vero modico valde tempore fui cum domino Hugone predicto, neque videre neque comprehendere neque discere ad plenum potui expertissimas curas suas.*"

31. *Ibid.* See Agrimi and Crisciani, *Edocere medicos*, p. 187, and "Science and Practice of Medicine," p. 64; and Siraisi, *Taddeo Alderotti*, pp. 14–15. On the tensions around the development of proprietary attitudes toward craft knowledge in thirteenth-century Europe, see Long, *Openness, Secrecy, Authorship*, ch. 3, esp. pp. 88–93, and Eamon, *Science and the Secrets of Nature*, pp. 27–37.

32. Taddeo Alderotti, *Expositio in Johannitii Isagogarum libellum*, ch. 2, in *Expositiones*, fol. 343va: "*et per hoc separatur practica que est scientia a practica usuali quam faciunt vetule.*"

33. On the duties of *medici condotti*, see Vivian Nutton, "Continuity or Rediscovery? The City Physician in Classical Antiquity and Medieval Italy," in Andrew W. Russell (ed.), *The Town and State Physician in Europe from the Middle Ages to the Enlightenment* (Wolfenbüttel: Herzog August Bibliothek, 1981), pp. 9–46, and Park, *Doctors and Medicine*, pp. 87–96.

34. On medical licensing and the varieties of female medical practice in late medieval and Renaissance Italy, see Katharine Park, "Medicine and Magic: The Healing Arts," in Judith C. Brown and Robert C. Davis (eds.), *Gender and Society in Renaissance Italy* (London: Longman, 1998), pp. 129–49. Important overviews appear in Monica H. Green, "Women's Medical Practice and Health Care in Medieval Europe," in Judith M. Bennett (ed.), *Sisters and Workers in the Middle Ages* (Chicago: University of Chicago Press, 1989), pp. 39–78, and "Documenting Medieval Women's Medical Practice," in García-Ballester, French, Arrizabalaga, and Cunningham (eds.), *Practical Medicine*, pp. 322–52.

35. On Mondino and on the composition of this work, which may well have been compiled in several stages, see Piero P. Giorgi and Gian Franco Pasini's introduction to Mondino de' Liuzzi, *Anothomia di Mondino de' Liuzzi da Bologna, XIV secolo*, ed. Piero P. Giorgi and Gian Franco Pasini (Bologna: Istituto per la Storia dell'Università di Bologna, 1992), pp. 88–91; Siraisi, *Taddeo Alderotti*, pp.

66–68 and 111–14; and G. Martinotti, "L'insegnamento dell'anatomia a Bologna prima del secolo XIX," *Studi e memorie per la storia dell'Università di Bologna* 2 (1911), pp. 25–32.

36. Mondino de Liuzzi, prologue to *Anothomia di Mondino*, p. 98. This edition is based on a single fourteenth-century manuscript from Bologna, probably contemporary with Mondino, and is the best Latin text available. It is unlikely that he obtained both the female cadavers he dissected in January and March 1316 from the judicial authorities, given the infrequency of, especially, female executions; it may be that, like late fifteenth-century anatomists, he drew on the dead from local hospitals.

37. Document transcribed in Mary Niven Alston, "The Attitude of the Church Towards Dissection before 1500," *Bulletin of the History of Medicine* 16 (1944), pp. 233–35. As Alston points out, the indictment was primarily concerned with their vandalism of the cemetery and violation of a grave, not the dissection itself.

38. There is no evidence that Guglielmo ever took part in either an autopsy or a dissection of a human being (as opposed to animals). One of his cases, involving a man who had been wounded in the neck by an arrow, is sometimes cited in this connection, but his description makes clear that this involved only a surface inspection; cf. Francesco Puccinotti, *Storia della medicina*, vol. 1, *Medicina antica* (Livorno: Wagner, 1850), pt. 1, p. 357 n.1.

39. On the epistemology of Taddeo and his pupils, see Siraisi, *Taddeo Alderotti*, ch. 5; see also, in general, Agrimi and Crisciani, *Edocere medicos*, esp. chs. 1 and 5.

40. *De juvamentis membrorum* was an anonymous Latin translation of an abbreviated Arabic version of Galen's *De usu partium* (*On the Usefulness of the Parts of the Body*), which included only the first ten books of the latter. Galen's work had also been translated into Latin in its entirety from the Greek by Niccolò of Reggio in the early fourteenth century. *De juvamentis membrorum* was much more influential than Niccolò's translation, however, and was printed in Galen, *Galieni Pergamensis medicorum omnium principis Opera*, ed. Diomedes Bonardus (Venice: For Filippo Pinzio da Caneto, 1490), vol. 1, sig. ee1r–gg1r.

The first edition of *De usu partium*, on the other hand, was not published until 1528. See Roger French, "*De juvamentis membrorum* and the Reception of Galenic Physiological Anatomy," *Isis* 70 (1979), pp. 96–109, and, in general, Richard J. Durling, "A Chronological Census of Renaissance Editions and Translations of Galen," *Journal of the Warburg and Courtauld Institutes* 24 (1961), pp. 230–305. As for *De interioribus*, there were in fact two separate Latin versions of Galen's *De locis affectis* that bore that title. One of these, by an anonymous translator, was translated from the Arabic, while the other, from the Greek, was the work of Burgundio of Pisa. The text of the latter is edited in Richard J. Durling and Fridolf Kudlien (gen. eds.), *Galenus Latinus*, vol. 1, *Burgundio of Pisa's Translation of Galen's...De interioribus*, ed. Richard J. Durling (Stuttgart: Steiner, 1992). The former, which circulated more widely, is printed in Galen, *Galieni Pergamensis ... Opera*, vol. 2, sig. l4r–o4v. Galen's great work on anatomical dissection, *De anatomicis administrationibus*, was not known until much later; the first Latin translation, by Demetrios Chalkokondyles, was published (by Jacopo Berengario of Carpi) in 1529. Other, shorter anatomical works by Galen, notably *De anatomia matricis* (*On the Anatomy of the Uterus*), which was also translated by Niccolò of Reggio, had very limited diffusion and influence until the late fifteenth century, when it, too, was printed in Galen, *Galieni Pergamensis... Opera*, vol. 1, sig. zz2v–3v.

41. On Bartolomeo's life and works, see Siraisi, *Taddeo Alderotti*, pp. 45–49. On his forensic autopsies, see Ladislao Münster, "Alcuni episodi sconosciuti o poco noti sulla vita e sull'attività di Bartolomeo da Varignana," *Castalia: Rivista di storia della medicina* 10 (1954), pp. 207–15, and Joseph Shatzmiller, "The Jurisprudence of the Dead Body: Medical Practition [*sic*] at the Service of Civic and Legal Authorities," *Micrologus* 7 (1999), p. 228.

42. Bartolomeo of Varignana, Commentary on *De interioribus*, Vatican: Biblioteca Apostolica Vaticana, ms. Vat. Lat. 4452, fol. 84r. On this commentary, see Nancy G. Siraisi, "Taddeo Alderotti and Bartolomeo da Varignana on the Nature of Medical Learning," *Isis* 68 (1977), pp. 33–39, and *Taddeo Alderotti*, p. 124. I am grateful to Siraisi for lending me a microfilm of the manuscript. On the importance of anatomy in the learned physician's claim to epistemological

authority based on the knowledge of causes, see French, *Medicine Before Science*, pp. 113–14.

43. Bartolomeo of Varignana, Commentary on *De interioribus*, fol. 84v.

44. See n.24 above and King, *Hippocrates' Woman*, p. 135.

45. On these developments, see Park, "Medicine and Magic," pp. 130–38, and Monica H. Green, *Making Women's Medicine Masculine: The Rise of Male Authority in Pre-Modern Gynecology* (Oxford: Oxford University Press, forthcoming 2007).

46. See Green, "From 'Diseases of Women' to 'Secrets of Women.'" Green summarizes this tradition in "Secrets of Women," in Margaret Schaus (ed.), *Women and Gender in Medieval Europe: An Encyclopedia* (New York: Routledge, 2006).

47. Green, "From 'Diseases of Women' to 'Secrets of Women,'" pp. 12–14 (quotation on p. 14).

48. *Ibid.*, p. 28. See also Green, "'Traittié tout de mençonges,'" esp. pp. 163–67.

49. Good general studies of women's bodies as portrayed in medieval learned literature include Danielle Jacquart, "La morphologie du corps féminin selon les médecins de la fin du Moyen Age," *Micrologus* 1 (1993), pp. 81–98; Danielle Jacquart and Claude Thomasset, *Sexuality and Medicine in the Middle Ages*, trans. Matthew Adamson (1985; Princeton, NJ: Princeton University Press, 1988), esp. chs. 1, 2, 5; Joan Cadden, *Meanings of Sex Difference in the Middle Ages: Medicine, Science, and Culture* (Cambridge: Cambridge University Press, 1993), passim; and Charles T. Wood, "The Doctors' Dilemma: Sin, Salvation, and the Menstrual Cycle in Medieval Thought," *Speculum* 56 (1981), pp. 710–27.

50. The Italian text survives in three known manuscripts, all preserved in Florence: Biblioteca Nazionale Centrale di Firenze, ms. Palat. 557, pp. 214–35 (early fifteenth century); Biblioteca Riccardiana, ms. 2228, fol. 71r–74r (fifteenth century); and Biblioteca Riccardiana, ms. 2350, fol. 79r–88r (fifteenth century) — references from Monica H. Green (personal communication). My quotations come from ms. Palat. 557. The Italian text corresponds closely to the version of the French text in Manuscript B (Paris: Bibliothèque Nationale, ms. fr. 631) in *Ce sont les secrés des dames, deffendus à révéler*, ed. Al[exandre] C[olson]

and Ch[arles]-Ed[ouard] C[azin] (Paris: Edouard Rouveyre, 1880). (The French editors' decision to publish the text under their initials shows that the association of this material with secrecy was still alive and well in the late nineteenth century.) The only organizational departures from the French text are the division of the Italian text into twenty-nine numbered chapters, some shuffling of the material in chs. 10–13 (on the physical effects of the planets, especially the moon), and the addition of a thirtieth, wholly unrelated chapter on phlebotomy. For discussions of the French text, see Dinora Corsi, "'Les secrés des dames': Tradition, traductions," *Médiévales* 14 (1988), pp. 47–57, and, especially, Green, "'Traittié tout de mençonges,'" pp. 150–53.

51. *I segreti delle femine*, chs. 28–29, ms. Palat. 557, p. 232.

52. *Ibid.*, chs. 13 and 29, pp. 221 and 233.

53. This difference is already signaled in the title by the use of the respectful *donne*, rather than the pejorative *femine*, to refer to women. I have consulted the following manuscripts of *Le segrete cose delle donne*: Florence, Biblioteca Laurenziana, Plut. 73, 51, fol. 52r–56v (fourteenth century); Biblioteca Laurenziana, ms. Redi 172, vol. 1, fol. 73r–82v (fourteenth century); Florence, Biblioteca Riccardiana, ms. 2165, fol. 70r–76r (1433); Biblioteca Riccardiana, ms. 2500, fol. 60r–64r (fifteenth century); Biblioteca Riccardiana, ms. 2175, fol. 44r–45v (second half of the fifteenth century). For more complete manuscript information, see Monica H. Green, "A Handlist of the Latin and Vernacular Manuscripts of the So-Called *Trotula* Texts: Part II: The Vernacular Texts and Latin Re-Writings," *Scriptorium* 51 (1997), pp. 100–101, and "Medieval Gynecological Literature: A Handlist," *Women's Healthcare in the Medieval West: Texts and Contexts* (Burlington, VT: Ashgate, 2000), Appendix p. 30. An imperfect transcription of ms. Redi 172 appears in Giuseppe Manuzzi (ed.), *Il libro delle segrete cose delle donne* (Florence: Tipografia del Vocabolario, 1863 [actually, post-1874]), pp. 1–20. My own quotations come from this text, corrected to conform to the orthography of the manuscript.

54. Green, "Handlist of the Latin and Vernacular Manuscripts," p. 100. See, in general, Green, "'Traittié tout de mençonges,'" pp. 158–67, and "The Development of the *Trotula*," *Women's Healthcare*, ch. 5.

55. Prologue, *Le segrete cose delle donne*, ms. Redi 172, vol. 1, fol. 73r; cf. *Liber de sinthomatibus mulierum*, [2], in Monica H. Green (ed. and trans.), *The Trotula: A Medieval Compendium of Women's Medicine* (Philadelphia: University of Pennsylvania Press, 2001), pp. 104–106.

56. Prologue, *I segreti delle femine*, ms. Palat. 557, p. 214: "*Una donna mi preghò per diricta cortesia ch'io iscrivessi alcuna cosa profictievole. E come che di ciò io sia di poco ingiengno e none usato di scrivere nè di dectare, nondimeno io l'otrapederò per lo suo amore, però che'l suo amore m'a partito il core e'l senno, in tale modo per modo che io non desidero se non di fare cosa che le piaccia e che agrado le sia, però ch'io porto in nel mio cuore il suo amore sanza ripetimento. Et simile io spero avere meglio o alcuno conforto da llei. Et però ch'ell'è sì dengnia e sì prefecta donna che di tucti i beni è sanza pari al mondo d'onore e di bellezza, io vi raconterò per la preghiera* [misreading of French *première*] *una materia che io truovo in uno libro il quale ci divisa i segreti delle femine. Io priego questa amorosa donna che quand'ela leggierà il libro ch'ella non si crucci con meco nè di meno m'ami, che io qui racconti loro sagreti il più brieve ch'io posso.*"

57. *Ibid.*, ch. 16, p. 223.

58. Pseudo-Albertus Magnus, *De secretis mulierum et virorum*, ch. 3, sig. c5r, and ch. 4, sig. e2v.

59. *I segreti delle femine*, ch. 13, ms. Palat. 557, p. 221: "*Et sappiate ancora che gli huomini ànno ricevute molte gravi malactie nel membro con alcune femmine però ch'elle mettono il membro dell'uomo in uno bucho dentro alla porta dal lato alla diricta via, e facendo così magagnono l'uomo alcuna volta malamente*"; *ibid.*, ch. 16, p. 223: "*Tucta volta elli* [miscarriage] *adiviene per cruccio e spesso per ira e per troppa fatica e però che le malvagie femmine fanno quand'elle dubitano d'essere grosse, però che quando elle si discredono elle s'affaticono elleno in danzare, trescando, mangiando e bevendo e saltando e usando con lei* [sic] *humini, che troppo affaticano la matricie, si lo distrugono per fare l'antica arte. Sì che tucto l'anullano, chè per così fare si destruggie e mangangnasi la porta della donna.*" On the powerful meanings of *doctrina*, which had academic associations, see Agrimi and Crisciani, *Edocere medicos*, ch. 2.

60. Pseudo-Albertus Magnus, *De secretis mulierum et virorum*, ch. 4, sig. e4r.

61. *I segreti delle femine*, ch. 17, ms. Palat. 557, p. 223: "*Ma e' si truova poche savie donne che levano i fanciulli che perfectamente il sappino et anche ben l'aiutino et però si perdono molti fanciulli*." This switch comes from the insertion into the French *Secrés* of a passage on midwives from Thomas of Cantimpré's *Liber de natura rerum*; for details, see Ann Ellis Hanson and Monica H. Green, "Soranus of Ephesus: *Methodicorum princeps*," in Wolfgang Haase and Hildegard Temporini (eds.), *Aufstieg und Niedergang der Römischen Welt*, pt. 2, vol. 37.2 (Berlin: Walter de Gruyter, 1994), p. 1058.

62. The midwife entrusted with the birth in the Latin treatise on which this section of the work is based, the *Book of Women's Symptoms* (*Liber de sinthomatibus mulierum*), as well as in *I segreti delle femine*, has completely disappeared, leaving the *fisici* as the sole experts on the topic. See *Le segrete cose delle donne*, ch. 6, ms. Redi 172, vol. 1, fol. 78v–79r; cf. *Liber de sinthomatibus mulierum*, [115]-[122], in Green, *Trotula*, pp. 104–106.

63. On the scant evidence of a female readership for medical books of any sort in this period; see Monica H. Green, "The Possibilities of Literacy and the Limits of Reading: Women and the Gendering of Medical Literacy," *Women's Healthcare*, ch. 7. Unlike *I segreti delle femine*, where the other treatises in the manuscripts suggest a lay owner or commissioner, at least three and probably four of the five known manuscripts containing *Le segrete cose delle donne* were made by or for medical men. Two of these (BRF, ms. Ricc. 2500 and ms. Ricc. 2175) were most likely physicians; their copies contain many Latin medical recipes and *consilia*. For skeleton descriptions of the contents of these manuscripts, see Mahmoud Salem Elsheikh, *Medicina e farmacologia nei manoscritti della Biblioteca Riccardiana di Firenze* (Rome: Vecchiarelli, 1990), pp. 51 and 61, and Green, "Handlist of the Latin and Vernacular Manuscripts," pp. 100–101. The third (BRF, ms. Ricc. 2165), dated 1437, probably belonged to a surgeon; in addition to the vernacular medical works with which *I segreti delle donne* typically circulated, this manuscript included a variety of Italian medical recipes and *consilia* and the surgical summa of one Master Francesco, translated into Italian, "*acciò che ciascheduno che vuole attendere a imparare a ffare l'arte della cierosia, sanza che egli sappia gramaticha, la può bene fare e adoperare senza timore nè dif-*

fetto, pure che egli sappia bene ciò che si contiene in questo libro e che sia stato con buono medicho a pratichare nella predetta arte" (fol. 87v). List of contents in Elsheikh, *Medicina e farmacologia*, p. 46. The evidence concerning the fourth manuscript (BLF, ms. Redi 172, vol. 1) is less conclusive, but the fact that it contains a table of contents for a vernacular treatise on "all kinds of fever" (fol. 35r–37v) suggests that it may also have belonged to a medical practitioner, albeit not a university-educated physician.

64. For the 1241 case, see Jacqueline Murray, "On the Origins and Role of 'Wise Women' in Causes for Annulment on the Grounds of Male Impotence," *Journal of Medieval History* 16 (1990), pp. 247–48 n.3. For Bartolomeo of Varignana's involvement in two such cases, see Shatzmiller, "Jurisprudence of the Dead Body," p. 226. For a later case from Pistoia, see Alberto Chiappelli, "Di un singolare procedimento medico-legale tenuto in Pistoia nell'anno 1375 per supposizione d'infante," *Rivista di storia critica delle scienze mediche e naturali* 10 (1919), pp. 129–35. Gentile of Foligno (d. 1348) was asked by the eminent jurist Cino of Pistoia for a general opinion on the length of pregancy, doubtless for use in one or more cases of this sort; see Roger French, *Canonical Medicine: Gentile da Foligno and Scholasticism* (Leiden: Brill, 2001), p. 276.

65. See Green, "Women's Medical Practice," and Park, "Medicine and Magic," p. 131.

66. See Jacqueline Marie Musacchio, *The Art and Ritual of Childbirth in Renaissance Italy* (New Haven: Yale University Press, 1999), ch. 1, esp. pp. 32–33, on familial reactions. Public concerns focused on the need to provide an adequate labor supply and to discourage nonreproductive sex in the form of sodomy, which was seen as distracting men from reproductive, marital sex. On the important theme of fertility in late medieval medical and natural philosophical literature, see, in general, Cadden, *Meanings of Sex Difference*, ch. 5.

67. Details in Park, "Medicine and Magic," pp. 129–31; see also Iris Origo, *The Merchant of Prato, Francesco di Marco Datini* (London: Cape, 1957), pp. 159–62.

68. As Monica H. Green points out, we still know very little about the development of midwifery as a specialized area of expertise and an occupational

identity in Christian Europe. It may have emerged as late as the thirteenth century; see Green, "Bodies, Gender, Health, Disease," *Studies in Medieval and Renaissance History*, ser. 3, vol. 2 (2005), pp. 15–16.

69. Ms. Ricc. 2500, fol. 68r–71v.

70. The relatively strong presence of Italian physicians in the areas of gynecology and obstetrics, and their interest in this type of practice, may explain why midwives appear not to have been licensed or regulated in the fifteenth century in Italy, unlike many parts of northern Europe; such licensing implies a degree of occupational recognition and autonomy that fifteenth-century Italian midwives may not have enjoyed. On the very different situation in northern Europe, see Sibylla Flügge, *Hebammen und heilkundige Frauen: Recht und Rechtswirklichkeit im 15. und 16. Jahrhundert* (Frankfurt: Stroemfeld, 1998), chs. 5–11.

71. Quoted in Fausto M. de' Reguardati, *Benedetto de' Reguardati da Norcia: "Medicus tota Italia celeberrimus"* (Trieste: Lint, 1977), p. 50 n.8; on Reguardati's reputation for special expertise in the area of obstetrics and gynecology, see pp. 114–28. My thanks to Monica Azzolini for this reference.

72. On this general phenomenon, see Green, *Making Women's Medicine Masculine*; for a census of the surviving works, see the relevant entries in Green, "Medieval Gynecological Literature."

73. Guglielmo of Saliceto, *Summa conservationis et curationis* 1.163–80, sig. h8rb–i6vb. For a sense of some of the contents, see Helen Rodnite Lemay, "William of Saliceto on Human Sexuality," *Viator* 12 (1981), pp. 165–81.

74. Guglielmo of Saliceto, *Summa conservationis et curationis* 1.178, sig. i5vb: "*Hec autem narrationes de panniculis et umbilico et generatione cordis et epatis et cerebri ex vesicis dictis superius et superfluitatibus que retinentur cum fetu usque ad horam fetus phisice sunt, et verificate per anothomiam et narrationem obstetricum et mulierum filiantium, et que partus multos suis temporibus sustinuerint cum aborsu, et absque aborsu michi non videtur quod ista per medicum modis aliis possint verificari.*" Animals, especially pigs, had been used in anatomical teaching at the medical center of Salerno since at least the mid-twelfth century; see R.K. French, *Dissection and Vivisection in the European Renaissance* (Aldershot: Ashgate, 1999), pp. 14–16; Nancy G. Siraisi, *Medieval and Early Renaissance Medicine: An Introduction*

to *Knowledge and Practice* (Chicago: University of Chicago Press, 1990), p. 86; and, for English translations of the principal texts, George W. Corner, *Anatomical Texts of the Earlier Middle Ages: A Study in the Transmission of Culture* (Washington: Carnegie Institution of Washington, 1927).

75. Antonio Guaineri, *Commentariolus de egritudinibus matricis*, ch. 35, *Opus preclarum ad praxim non mediocriter necessarium cum Joannis Falconis nonnullis non inutiliter adjunctis* (Lyon: Constantinus Fradin, 1525), fol. 163v. This work was a revision and expansion of the original *De propriis mulierum aegritudinibus, seu De matricibus* ([Padua: Stendal], 1474). See, in general, Helen Rodnite Lemay, "Anthonius Guainerius and Medieval Gynecology," in Julius Kirshner and Suzanne F. Wemple (eds.), *Women of the Medieval World: Essays in Honor of John H. Mundy* (Oxford: Blackwell, 1985), pp. 317–36, and "Women and the Literature of Obstetrics and Gynecology," in Joel T. Rosenthal (ed.), *Medieval Women and the Sources of Medieval History* (Athens: University of Georgia Press, 1990), pp. 189–209.

76. Guaineri, *Commentariolus*, ch. 6, fol. 138r.

77. Michele Savonarola, *Il trattato ginecologico-pediatrico in volgare: Ad mulieres ferrarienses de regimine pregnantium et noviter natorum usque ad septennium* (Milan: Società Italiana di Ostetricia e Ginecologia, 1952), bk. 2, ch. 2, p. 104.

78. *Ibid.* 1.5, p. 55; 2.3, p. 132.

79. *Ibid.* 2.2, p. 109.

80. *Ibid.* 2.2, p. 122.

81. *Ibid.* 2.2, p. 116.

82. *Ibid.* 2.2, pp. 116–24.

83. On the increased visibility of women's genitals, see, for example, the explicit statements in Savonarola, *Il trattato* 1.5, p. 61, and Antonio Benivieni, *De abditis nonnullis ac mirandis morborum et sanationum causis*, 28–29, ed. Giorgio Weber (Florence: Olschki, 1994), pp. 89–91, as well as Chapter Three below.

84. Preoccupied with reproduction, medical writers from this period ignored women's external genitals; the clitoris, known to Greek medical writers, was not "rediscovered" until the middle of the sixteenth century: Jacquart and

Thomasset, *Sexuality and Medicine in the Middle Ages*, pp. 44–46; Katharine Park, "The Rediscovery of the Clitoris: French Medicine and the *Tribade*, 1570-1620," in David Hillman and Carla Mazzio (eds.), *The Body in Parts: Fantasies of Corporeality in Early Modern Europe* (New York: Routledge, 1997), pp. 171–93.

85. Mondino de' Liuzzi, *Anothomia di Mondino*, pp. 236–38 and 250–52. Mondino used these physical connections to argue against the idea that the uterus could move around within the body. See, in general, Cadden, *Meanings of Sex Difference*, chs. 4–5.

86. Guaineri, *Commentariolus*, ch. 1, fol. 134r.

87. See also Gianna Pomata, "Menstruating Men: Similarity and Difference of the Sexes in Early Modern Medicine," in Valeria Finucci and Kevin Brownlee (eds.), *Generation and Regeneration: Tropes of Reproduction in Literature and History from Antiquity to Early Modern Europe* (Durham: Duke University Press, 2001), pp. 109–52.

88. Claude Thomasset, "Le corps féminin ou le regard empêché," *Micrologus* 1 (1993), p. 102.

89. See n.74 above.

90. [Ps.-]Galen, *De spermate*, in Galen, *Galieni Pergamensis... Opera*, vol. 1, sig. xx4r–yy2v. On this doctrine, see Robert Reisert, *Der siebenkammerige Uterus: Studien zur mittelalterlichen Wirkungsgeschichte und Entfaltung eines embryologischen Gebärmuttermodells* (Hannover: Wellm, 1986), esp. pp. 74–75; Fridolf Kudlien, "The Seven Cells of the Uterus: The Doctrine and Its Roots," *Bulletin of the History of Medicine* 39 (1965), pp. 415–23; and Cadden, *Meanings of Sex Difference*, pp. 198–99 and (for other medieval theories of sex determination) pp. 130–34 and 195–201. Although Galen's *On the Anatomy of the Uterus* did in fact describe the placenta and umbilical cord on the basis of animal dissections, it had relatively little circulation until it was published in 1490 as part of Galen's collected works: Galen, *Galieni Pergamensis... Opera*, vol. 1, sig. zz2v–3v. On this translation, see Durling, "Chronological Census," pp. 233 and 292, and Chapter Four n.17.

91. Mondino's commentary has been edited: Mondino de' Liuzzi, *Expositio super capitulum de generatione embrionis Canonis Avicennae cum quibusdam quaes-*

tionibus, ed. Romana Martorelli Vico (ca. 1320; Rome: Istituto Storico Italiano per il Medio Evo, 1993). The commentaries of Dino and Tommaso, as well as that of Tommaso's contemporary Jacopo of Forlì, can be found in *Expositio Jacobi supra capitulum de generatione embrionis cum questionibus eiusdem,* ed. Bassianus Politus (Venice: Bonetus Locatellus, 1502). Dino's commentary on Hippocrates' *De natura fetus* also appears in this volume.

92. Mondino de' Liuzzi, *Expositio super capitulum de generatione embrionis,* [lectio 6], p. 75: "*cum anothomizatur fetus non natus.*" See also [lectio 8], p. 105.

93. Tommaso del Garbo, *Expositio super capitulo de generatione embryonis... ,* in Jacopo of Forlì, *Expositio supra capitulum de generatione embrionis*, fol. 39v.

94. Mondino de' Liuzzi, *Anothomia di Mondino*, pp. 240–43.

95. Although Donatello's relief *The Heart of the Miser* was almost certainly based on his observation of a dissection, the heart is not represented in an anatomical or medical context; on this work, see Chapter Five below.

96. *Fasiculo de medicina*, attr. Johannes de Ketham, ed. and trans. Sebastiano Manilio (Venice: Giovanni and Gregorio de' Gregori, 1494), sig. d1r. This work has been reprinted in facsimile in *The Fasciculo di medicina, Venice 1493*, ed. Charles Singer (Florence: Lier, 1925), vol. 2, and Tiziana Pesenti, *Fasiculo de medicina in volgare* (Tresivo: Antilia, 2001), vol. 1. For an excellent introduction to medical and editorial aspects of the work, as well as a basic bibliography, see Pesenti, *Fasiculo de medicina in volgare*, vol. 2, *Il "Fasciculus medicinae," ovvero, Le metamorfosi del libro umanistico*, and "Editoria medica tra Quattro e Cinquecento: L'*Articella* e il *Fasciculus medicine*," in Ezio Riondato (ed.), *Trattati scientifici nel Veneto fra il XV e XVI secolo* (Vicenza: Pozza, 1985), pp. 1–28. See also Charles Singer, introduction to *Fasiculo di medicina*, vol. 1, and Karl Sudhoff, introduction to the facsimile of the earlier Latin edition of the collection, *The Fasciculus medicinae of Johannes de Ketham, Alemanus: Facsimile of the First (Venetian) Edition of 1491*, ed. and trans. Charles Singer (Milan: Lier, 1924). On the unconvincing attribution to Johannes de Ketham, see Pesenti, *Fasiculo de medicina in volgare*, vol. 2, pp. 49–53.

97. Pesenti details the differences between the 1491 and 1494 editions in *Fasiculo de medicina in volgare*, vol. 2, ch. 3.

98. On the practice of anatomy in hospitals in this period, including the presence of artists such as Leonardo, see Monica Azzolini, "Leonardo da Vinci's Anatomical Studies in Milan: A Re-examination of Sites and Sources," in Jean A. Givens, Karen M. Reeds, and Alain Touwaide (eds.), *Visualizing Medieval Medicine and Natural History, 1200–1550* (Burlington, VT: Ashgate, 2006), pp. 152–54 and 161–67, and Domenico Laurenza, *Leonardo nella Roma di Leone X (c. 1513–16): Gli studi anatomici, la vita, l'arte* (Vinci: Biblioteca Leonardiana, 2004), pp. 29–30.

99. For the German manuscript antecedents of this figure, see Pesenti, *Fasiculo de medicina in volgare*, vol. 2, pp. 17–18; various of these are illustrated in the plates appended to *Fasciculus medicinae*, ed. Singer.

100. On the use of this phrase in contemporary Italian writing on art, see Joanna Woods-Marsden, "*Ritratto al naturale*: Questions of Realism and Idealism in Early Renaissance Portraits," *Art Journal* 46 (1987), pp. 209–16. Although the claim to have drawn something from nature was not always accurate, comparison with modern anatomical images of the dissected uterus and vagina confirm that this woodcut was ultimately based on the drawing of a dissection. As Pesenti argues, it seems also to reflect the textual tradition of Mondino's *Anatomy*, which was added to the 1494 edition; Pesenti, *Fasiculo de medicina in volgare*, vol. 2, pp. 143–44.

101. The best general history of anatomical illustration in the late Middle Ages and the Renaissance is still Robert Herrlinger, *History of Medical Illustration from Antiquity to A.D. 1600* (1967; London: Pitman Medical, 1970). See also Loris Premuda, *Storia dell'iconografia anatomica* (Milan: Martello, 1957); K.B. Roberts and J.D.W. Tomlinson, *The Fabric of the Body: European Traditions of Anatomical Illustration* (Oxford: Clarendon, 1992), chs. 1–7; and Mimi Cazort, Monique Kornell, and K.B. Roberts, *The Ingenious Machine of Nature: Four Centuries of Art and Anatomy* (Ottawa: National Gallery of Canada, 1996). The last three should be used with care.

102. The (now) nine-figure series included five diagrams of the arteries, veins, bones, nerves, and muscles superimposed on male bodies, as well as four of individual organs: the male genitals, the thoracic viscera, the female genitals,

and the brain and eyes. Three of the eleven known manuscripts that contain illustrated copies of this work are definitely Italian: Pisa, Biblioteca Universitaria di Pisa, ms. 735 (first half of the fourteenth century); Milan, Biblioteca Ambrosiana, ms. Trivulziano 836 (mid-fourteenth century); and Oxford, Bodleian Library, Cod. e Museo 19 (mid-fourteenth century). On this series and its sources, see Ynez Violé O'Neill, "The Fünfbilderserie — A Bridge to the Unknown," *Bulletin of the History of Medicine* 51 (1977), pp. 538–49; Carlo Maccagni, "Frammento di un codice di medicina del secolo XIV (manoscritto N. 735, Già Codice Roncioni N. 99) della Biblioteca Universitaria di Pisa," *Physis* 11 (1969), pp. 311–78; and Luigi Belloni, "Gli schemi anatomici trecenteschi (serie dei cinque sistemi e occhio) del Codice Trivulziano 836," *Rivista di storia delle scienze mediche e naturali* 41 (1950), pp. 193–207. The figure on the bottom right-hand corner of fol. 2r of the Pisa manuscript is not, *pace* O'Neill and Maccagni, a hermaphroditic image combining elements of the male and female genitals but a straightforward drawing of the former, complicated by a scribal error in one of the captions (l. 108 in Maccagni, "Frammento di un codice di medicina," p. 348), where *uuluam* (vulva) is substituted for *uirgam* (penis).

103. For the history of the Muscio series, see Hanson and Green, "Soranus of Ephesus," pp. 1023–24 and 1072–73 (list of manuscripts).

104. Chantilly, Musée de Condé, ms. 344, fol. 257r–73v. (Guido refers to Philip as Philip VII.) On this manuscript, see Ernest Wickersheimer, "L'Anatomie' de Guido de Vigevano, médecin de la reine Jeanne de Bourgogne (1345)," *Archiv für Geschichte der Medizin* 7 (1913), pp. 1–25, with figures, and *Anatomies de Mondino dei Liuzzi et de Guido de Vigevano* (Paris: Droz, 1926). The style of the cadaver is very close to depictions by contemporary Tuscan artists. See, for example, Richard Offner, *A Critical and Historical Corpus of Florentine Painting*, pt. 3, vol. 5 (New York: College of Fine Art, New York University, 1947), pls. XVI and LI3. I have not seen the image in the manuscript copy of Guglielmo of Saliceto's *Chirurgia* in Venice, Biblioteca Nazionale Marciana, reproduced in Premuda, *Storia dell'iconografia anatomica*, pl. 18. Judging by the style of the painting, which shows an unambiguously dissected body, it seems unlikely that it dates from the fourteenth century, as Premuda claims (pp. 34–35).

105. Transcribed in Wickersheimer, "L'Anatomie' de Guido de Vigevano," p. 5.

106. *Ibid.*, p. 11.

107. See Green, "Introduction," *Trotula*, pp. 22–34.

108. On the lack of any universally acknowledged papal prohibition, see Introduction; German and French writers tended to recognize the existence of such a prohibition in the fourteenth and early fifteenth centuries, while Italian writers did not.

109. See Karl Sudhoff, "Abbildungen zur Anatomie des Maître Henri de Mondeville (ca. 1260 bis ca. 1320)," in *Ein Beitrag zur Geschichte der Anatomie im Mittelalter speziell der anatomischen Graphik: Nach Handschriften des 9. bis 15. Jahrhunderts* (Leipzig: Barth, 1908), pp. 82–89 and pl. 24.

110. On the production of early printed book illustrations in Venice, the personnel involved in that production, and their division of labor, see David Landau and Peter Parshall, *The Renaissance Print, 1470–1550* (New Haven: Yale University Press, 1994), p. 36, and Marino Zorzi, "Stampa, illustrazione libraria e le origini dell'incisione figurativa a Venezia," in Mauro Lucco (ed.), *La pittura nel Veneto: Il Quattrocento* (Milan: Electa, 1990), vol. 2, pp. 686–702. On Manilio's role in producing the new image of the female figure in the 1494 edition, see Pesenti, *Fasiculo de medicina in volgare*, vol. 2, p. 143. I am grateful to Karen Rosoff Encarnación and Lisa Pon for their help in researching and thinking through these aspects of the *Fasiculo* illustration.

111. These mistakes were quickly corrected in the second Latin edition of 1495, which used the new woodcut but inserted the proper caption numbers: *Fasciculus medicine* (Venice: Giovanni and Gregorio de' Gregori, 1495).

112. On the novel and complicated representational problems that confronted anatomical illustrators working from dissected bodies, see Martin Kemp, "'The Mark of Truth': Looking and Learning in Some Anatomical Illustrations from the Renaissance and Eighteenth Century," in W.F. Bynum and Roy Porter (eds.), *Medicine and the Five Senses* (Cambridge: Cambridge University Press, 1993), esp. pp. 88–89.

113. *Fasiculo de medicina*, sig. d3r–e2r: *Le problema overo interrogationi delli*

membri genitali, cioè de la matrice e testiculi overo secreti della donna. (The ovaries were referred to as the female "testicles.") For a detailed summary of this text, extracted from a longer treatise on human physiology known as *Omnes homines*, which dated from the late thirteenth or early fourteenth century, see Margaret Schleissner, "Sexuality and Reproduction in the Late Medieval 'Problemata Aristotelis,'" in Josef Domes, Werner Gerabek, Bernhard D. Haage, Christoph Weisser, and Volker Zimmermann (eds.), *Licht der Natur: Medizin in Fachliteratur und Dichtung* (Göppingen: Kümmerle, 1994), pp. 383–98, and Cadden, *Meanings of Sex Difference*, pp. 89–104.

114. For an explication of the scene in this image, see Jerome J. Bylebyl, "Interpreting the *Fasciculo* Anatomy Scene," *Journal of the History of Medicine and Allied Sciences* 45 (1990), pp. 285–316, and Pesenti, *Fasiculo de medicina in volgare,* vol. 2, pp. 114–36.

115. On the limited literacy of even patrician women in this period, see Christiane Klapisch-Zuber, "Les clefs florentines de Barbe-Bleue: L'apprentissage de la lecture," *La maison et le nom: Stratégies et rituel dans l'Italie de la Renaissance* (Paris: Editions de l'Ecole des Hautes Etudes en Sciences Sociales, 1990), pp. 309–30. On the lack of evidence for women as readers of medical texts, see Green, "Possibilities of Literacy."

116. See Patricia Simons, "Anatomical Secrets: *Pudenda* and the *Pudica* Gesture," in Gisela Engel (ed.), *Das Geheimnis am Beginn der europäischen Moderne* (Frankfurt: Klostermann, 2002), pp. 302–27.

117. See Cornelius O'Boyle, "Gesturing in the Early Universities," *Dynamis* 20 (2000), esp. pp. 268–75.

CHAPTER THREE: THE MOTHER'S PART

1. These events are noted in Filippo's *ricordanza*, in Florence, Archivio di Stato di Firenze (hereafter ASF): Carte Strozziane, 5, 22, fol. 97r. For background on the Strozzi family in the fifteenth century in relation to women, marriage, and lineage, see Ann Crabb, *The Strozzi of Florence: Widowhood and Family Solidarity in the Renaissance* (Ann Arbor: University of Michigan Press, 2000), and Lorenzo Fabbri, *Alleanza matrimoniale e patriziato nella Firenze del '400:*

Studio sulla famiglia Strozzi (Florence: Olschki, 1991). Crabb discusses the marriage of Fiametta and Filippo on pp. 206–11.

2. ASF, Carte Strozziane, 5, 22, fol. 97r: "*A dì 23 detto piaque a ddio chiamare a sse la benedetta anima della Fiametta mia donna e a dì 24 fu sopellita in Sancta Maria Novella nella nostra uxata sepoltura. Il male suo fu che nel parto di sopra non purghò bene, e niente di meno, anchora che avesse auto in detto parto qualche dì un po' di febre e alchune doglie, di 4 dì era stata bene ed era levata e stavasi per chasa. E'l dì sulle 21 hora li chominciò una doglia grande intorno al quore. Giròssi sul letucio e di qui si fecie portare ne' letto, ramarichandosi sempre grandemente del chuore. E per molti ripari vi si facessino per donne e per medici nulla govò. Chè circha a hora 23 finy.*"

3. Christiane Klapisch-Zuber, "Le dernier enfant: Fécondité et vieillissement chez les Florentines (XIVe–XVe siècle)," in Jean-Pierre Bardet, François Lebrun, and René Le Mée (eds.), *Mesurer et comprendre: Mélanges offerts à Jacques Dupâquier* (Paris: Presses Universitaires de France, 1993), pp. 277–90, and "Les femmes et la mort à la fin du Moyen Age," in Stéphane Toussaint (ed.), *Ilaria del Carretto e il suo monumento: La donna nell'arte, la cultura et la società del '400* (Lucca: S. Marco, 1995), pp. 219–20.

4. This is the only Master Lodovico included by Benedetto Dei in his list of physicians active in Florence in 1470; see the transcription in Giuseppina Carla Romby, *Descrizioni e rappresentazioni della città di Firenze nel XV secolo* (Florence: Libreria Editrice Fiorentina, 1976), p. 68.

5. ASF, Carte Strozziane, 5, 22, fol. 97r: "*Fecy aprire il chorpo e infra li altri vi fu a vederllo Maestro Lodovico, e dissemi poi aver trovato la matricie piena di sangue putrafatto, e che questo la fecie perire. E apresso che avea il feghato molto ghuasto e simile il polmone, e che già il polmone si era cominciato a picchare alle reny, e che se non periva di questo male sarebbe chaduta nel tixiccho. Dettemi la sua perdire grande passione perchè ogni dì mi contentavo de molto di ley per molte buone parti che in ley rengniavono.*"

6. For the contours of private practice in this period, see Katharine Park, *Doctors and Medicine in Early Renaissance Florence* (Princeton, NJ: Princeton University Press, 1985), pp. 109–17.

7. ASF, Corporazioni Religiose Soppresse dal Governo Francese 95, 212, fol. 171r, quoted in n.81 below.

8. Florence, Biblioteca Nazionale Centrale di Firenze (hereafter BNCF): Fondo Nazionale, 2, 2, 357, fol. 58r, quoted in n.58 below, and ASF, Carte Strozziane 2, 22, fol. 64v, quoted in n.59 below.

9. Figures from list of *ricordanze* and other family memoirs in Christiane Klapisch-Zuber, *La maison et le nom: Stratégies et rituels dans l'Italie de la Renaissance* (Paris: Editions de l'Ecole des Hautes Etudes en Sciences Sociales, 1990), pp. 244–48. My figure of fifty-eight includes all of the *ricordanze* listed by Klapisch-Zuber that cover any of the years between 1470, the earliest point at which I would expect to find a private autopsy of this sort, and 1520, the limit of the purview of this chapter, supplemented by the *ricordanza* of Cambio di Manno Petrucci, cited in n.8. I assume that Klapisch-Zuber, whose research on women and death in these texts was exhaustive, picked up all the relevant autopsies in "Les femmes et la mort." My thanks to her for these references.

10. Carlo Malagola (ed.), *Statuti delle università e dei collegi dello studio bolognese* (Bologna: Zanichelli, 1888), pp. 289–90 (statute of 1405); Alessandro Gherardi (ed.), *Statuti della Università et studio fiorentino dell' anno MCCCLXXXVII, seguiti da un'appendice di documenti dal MCCCXX al MCCCCLXXII* (Florence: Cellini, 1881), p. 74; and Tiziana Pesenti, *Fasiculo di medicina in volgare* (Padua: Università degli Studi, 2001), vol. 1, pp. 122–27 (on the statutes of the University of Padua from 1465). On the situation in Milan and at the University of Pavia, see Monica Azzolini, "Leonardo da Vinci's Anatomical Studies in Milan: A Re-Examination of Sites and Sources," in Jean A. Givens, Karen M. Reeds, and Alain Touwaide, *Visualizing Medieval Medicine and Natural History, 1200–1550* (Burlington, VT: Ashgate, 2006), pp. 158–63.

11. London, Wellcome Library for the History and Understanding of Medicine, ms. 5265: "*e questo sera utile non solamente a la universitade ymo a tuto il mondo.*" My thanks to Vivian Nutton for this reference. On problems concerning the supply of cadavers for dissection, see Katharine Park, "The Criminal and the Saintly Body: Autopsy and Dissection in Renaissance Italy," *Renaissance Quarterly* 47 (1994), pp. 6–14.

12. Nancy G. Siraisi, *Medieval and Early Renaissance Medicine: An Introduction to Knowledge and Practice* (Chicago: University of Chicago Press, 1990), p. 89.

13. On contemporary funerary ritual and its importance in patrician identity, see Sharon T. Strocchia, *Death and Ritual in Renaissance Florence* (Baltimore: Johns Hopkins University Press, 1992), ch. 6, and Francis William Kent, *Household and Lineage in Renaissance Florence: The Family Life of the Capponi, Ginori, and Rucellai* (Princeton, NJ: Princeton University Press, 1977), pp. 258–70.

14. Antonio Benivieni, *De abditis nonnullis ac mirandis morborum et sanationum causis* (Florence: Filippo Giunta, 1507); ed. and (English) trans. Charles Singer (Springfield, IL: Thomas, 1954); ed. and (Italian) trans. Giorgio Weber (Florence: Olschki, 1994). Weber's edition includes not only the complete Latin text of 1507 but also almost one hundred cases omitted by Girolamo, based on the transcription of a manuscript now lost. For cases included in the printed edition, I will note the chapter numbers, which are shared by the Singer translation and the Weber edition; for the unpublished cases, I will use page references to the Weber edition. On Benivieni's life and works, which include a regimen of health for Lorenzo di Piero de' Medici, see Ugo Stefanutti, "Benivieni, Antonio," in *Dizionario biografico degli italiani*, vol. 8 (Rome: Istituto della Enciclopedia Italiana, 1966), pp. 543–45 and the bibliography therein. The best discussion of *De abditis* is Nancy G. Siraisi, "'Remarkable' Diseases, 'Remarkable' Cures, and Personal Experience in Renaissance Medical Texts," *Medicine and the Italian Universities, 1250–1600* (Leiden: Brill, 2001), esp. pp. 231–44.

15. *Benivieni,* De abiditis, chs. 83 and 89 (criminal); pp. 168–69 (forensic autopsy, where it is not clear whether the corpse was actually opened).

16. *Ibid.*, chs. 34 and 76.

17. *Ibid.*, pp. 226–27.

18. *Ibid.*, pp. 223–24. *Mancinus* may be a Latinization of Macigni, the lineage of Fiametta's mother-in-law, Alessandra.

19. *Ibid.*, ch. 32.

20. *Ibid.*, ch. 36 (Antonio Bruni). Benivieni mostly describes the opening of the body using passive constructions, as here, making it difficult to determine who was doing the cutting. In one case, however (the autopsy of an otherwise

unidentified young man called Giuliano), an assistant — *minister*, probably a barber or surgeon — did the cutting, while in another (the autopsy of the son of Piero Adimari) the context suggests that Benivieni may have performed the operation himself: "*nos cum annuente patre emortuum corpus incidissemus.*" See *ibid.*, chs. 33 and 37.

21. See Park, *Doctors and Medicine*, p. 115; collaboration provided a certain amount of professional protection if the patient died. The physicians who participated in the anatomy were not necessarily the patient's regular doctors. Thus the physician who attended Fiametta during her pregnancy was "Maestro Moise" (almost certainly Mosè di Giuseppe Spagnolo), who was paid two florins "*per curare la Fiametta nel suo parto*" (ASF, Carte Strozziane, 5, 22, fol. 54r; see below, n.42); nonetheless, Filippo Strozzi does not mention him as present among the physicians at the autopsy, referring instead to Master Lodovico Toscanelli.

22. For example, Benivieni, *De abiditis*, chs. 3 (a "certain noblewoman") and 79 (daughter of Ruggiero Corbinelli).

23. *Ibid.*, ch. 79. For a similar statement, see ch. 94 (Monna Diamante).

24. See Roger French, *Canonical Medicine: Gentile da Foligno and Scholasticism* (Leiden: Brill, 2001), ch. 3, esp. pp. 130–40, and Gianna Pomata, *Contracting a Cure: Patients, Healers, and the Law in Early Modern Bologna*, trans. Gianna Pomata (Baltimore: Johns Hopkins University Press, 1998), ch. 5 (the body understood in terms of obstruction and flow).

25. For Torni's *consilium* describing the autopsy of the adolescent son of a Florentine wool merchant in 1496, see Bernardo Torni, *Opuscoli filosofici e medici*, ed. Marina Messina Montelli (Florence: Nuova Italia, 1982), pp. 39–44; English translation (not based on this critical edition) in Lynn Thorndike, *Science and Thought in the Fifteenth Century: Studies in the History of Medicine and Surgery, Natural and Mathematical Science, Philosophy and Politics* (1929; New York: Hafner, 1967), pp. 126–31. On Cardano, see Anthony Grafton and Nancy G. Siraisi, "Between the Election and My Hopes: Girolamo Cardano and Medical Astrology," in William R. Newman and Anthony Grafton (eds.), *Secrets of Nature: Astrology and Alchemy in Early Modern Europe* (Cambridge, MA: MIT Press, 2001), pp. 84–85.

26. On Lorenzo de' Medici's death and funeral, see Strocchia, *Death and Ritual*, pp. 215–16. On his autopsy, see Antonio Benivieni, *De regimine sanitatis ad Laurentium Medicem*, ed. Luigi Belloni (Turin: Società Italiana di Patologia, 1951), p. 15 n.4.

27. Guy of Chauliac, *La grande chirurgie*, Bibliothèque Universitaire de Montpellier, ms. fr. 184, fol. 14v; cf. Guy of Chauliac, *Inventarium sive chirurgia magna*, ed. Michael R. McVaugh (Leiden: Brill, 1997). The painting appears at the beginning of the first book of the treatise, which addresses anatomy; for a brief discussion of it, see Gerhard Wolf-Heidegger and Anna Maria Cetto, *Die anatomische Sektion in bildlicher Darstellung* (Basel: Karger, 1967), p. 131.

28. Other examples can be found in Wolf-Heidegger and Cetto, *Die anatomische Sektion*, pp. 418–21.

29. On the idealization of death in childbirth as the hallmark of the "good wife," see Klapisch-Zuber, "Les femmes et la mort," pp. 120–21 (quotation on p. 121). On women, family, and marriage in Renaissance Florence, see the essays of Klapisch-Zuber, many of them collected in *La maison et le nom*; and *Women, Family, and Ritual in Renaissance Italy*, trans. Lydia Cochrane (Chicago: University of Chicago Press, 1985). The situation was complicated in practice, as Klapisch-Zuber and others have emphasized; see, for example, Thomas Kuehn, "Some Ambiguities of Female Inheritance Ideology in the Renaissance," *Continuity and Change* 2 (1987), pp. 11–36.

30. Although Filippo Strozzi was wealthy enough to accommodate wet nurses in his own household, most babies were sent out to rural wet nurses within two weeks after birth, after an initial period of nursing by their mothers; see Louis Haas, *The Renaissance Man and His Children: Childbirth and Early Childhood in Florence 1300–1600* (Houndmills: Macmillan, 1998), pp. 113–15, and, in general, Christiane Klapisch-Zuber, "Blood Parents and Milk Parents: Wet Nursing in Florence, 1300–1530," *Women, Family, and Ritual*, pp. 132–64.

31. Several days after Fiametta's death, Filippo paid Maria a fee of 1 lira, 7 soldi, 4 danari "for having looked after [*ghuardato*] Fiametta during the birth"; see Carte Strozziane 5, 22, fol. 54r. On birth practices in general, particularly in Florence, see Haas, *Renaissance Man*, ch. 2, and Jacqueline Marie Musacchio, *The*

Art and Ritual of Childbirth in Renaissance Italy (New Haven: Yale University Press, 1999), chs. 1–2 and passim. *Guardadonne* seem to have stayed with the new mother for three to five weeks.

32. Musacchio, *Art and Ritual*, p. 52.

33. For some sample payments, see Haas, *Renaissance Man*, pp. 41–42.

34. See, for example, Sibylla Flügge, *Hebammen und heilkundige Frauen: Recht und Rechtswirklichkeit im 15. und 16. Jahrhundert* (Frankfurt: Stroemfeld, 1998), chs. 5–11.

35. Monica H. Green, "Bodies, Gender, Health, Disease: Recent Work on Medieval Women's Medicine," *Studies in Medieval and Renaissance History* ser. 3, vol. 2 (2005), pp. 15–16.

36. Michele Savonarola, *Il trattato ginecologico-pediatrico in volgare: Ad mulieres ferrarienses de regimine pregnantium et noviter natorum usque ad septennium*, ed. Luigi Belloni (Milan: Società Italiana di Ostetricia e Ginecologia, 1952), exists only in two manuscripts; see Belloni's introduction, pp. i–xii. See, in general, Monica H. Green, "The Possibilities of Literacy and the Limits of Reading: Women and the Gendering of Medical Literacy," *Women's Healthcare in the Medieval West: Texts and Contexts* (Burlington, VT: Ashgate, 2000), ch. 7, and, on the relatively restricted literacy of Florentine patrician women, Christiane Klapisch-Zuber, "Les clefs florentines de Barbe-Bleue: L'apprentissage de la lecture," *La maison et le nom*, pp. 309–30. I know of only one treatise on these matters dedicated to a particular woman, the *Enneas muliebris* (Ferrara: L. de Rubeis, 1502) of the physician Ludovico Bonacioli, addressed to Lucrezia Borgia, the duchess of Ferrara, whom Bonacioli attended during several of her deliveries; see Luigi Samoggia, "Lodovico Bonacciolo medico ostetrico di Lucrezia Borgia in Ferrara," *Atti della Accademia dei Fisiocratici di Siena, Sezione medico-fisica* n.s. 13 (1974), fasc. 1, pp. 513–31.

37. For the role of a female empiric in treating the infertility of one very wealthy woman, see Katharine Park, "Medicine and Magic: The Healing Arts," in Judith C. Brown and Robert C. Davis, *Gender and Society in Renaissance Italy* (London: Longman, 1998), pp. 129–31.

38. Examples of royal women who used physicians in their deliveries

include Lucrezia Borgia and Bianca Maria Visconti, the duchess of Milan and wife of Francesco Sforza. For the former, see Samoggia, "Lodovico Bonacciolo," and for the latter, see Fausto M. de' Reguardati, *Benedetto de' Reguardati da Norcia: "Medicus tota Italia celeberrimus"* (Trieste: Lint, 1977), pp. 114–28. Physicians would normally have been discouraged from taking active part in deliveries, not only by contemporary ideas of female modesty and decorum (*onestà*) but also because manual intervention was considered beneath the dignity of the physician.

39. See Katharine Park, "Medicine and Magic"; Haas, *Renaissance Man*, p. 33; Reguardati, *Benedetto de' Reguardati*, p. 114; Musacchio, *Art and Ritual*, pp. 139–40; and, in general, Joan Cadden, *Meanings of Sex Difference in the Middle Ages: Medicine, Science, and Culture* (Cambridge: Cambridge University Press, 1993), pp. 230–32 and 249–53.

40. ASF, Carte Strozziane, 5, 22, fol. 97r and 54r.

41. See Umberto Cassuto, *Gli ebrei a Firenze nell'età del Rinascimento* (Florence: Galletti e Cocci, 1918), pp. 90, 184, 186; and Park, *Doctors and Medicine*, pp. 72–75.

42. ASF, Carte Strozziane, fol. 54v. Master Mosè's fee was substantially larger than the one or two lire typically paid to midwives in this period; there were roughly four lire to the florin.

43. Letters of Filippo di Filippo Strozzi to his brother Lorenzo (19.i.1525, 20.i.1525, 25.i.1525, 3.ii.1525), ASF, Carte Strozziane, 3, 108, fol. 77r–78r, 79r, 88r–v. Clarice was married to Filippo Strozzi's son Filippo by his second marriage.

44. On the nature and early history of this procedure, which I also mentioned in Chapter One, in relation to the opening of Chiara of Montefalco, see Daniel Schäfer, *Geburt aus dem Tod: Der Kaiserschnitt an Verstorbenen in der abendländischen Kultur* (Hürtgenwald: Guido Pressler, 1999), chs. 1–3, and Renate Blumenfeld-Kosinski, *Not of Woman Born: Representations of Caesarean Birth in Medieval and Renaissance Culture* (Ithaca, NY: Cornell University Press, 1990), pp. 21–26. Blumenfeld-Kosinski's general argument concerning the relations between physicians and midwives is without foundation.

45. Gaspare Nadi, *Diario bolognese*, ed. Corrado Ricci and A. Bacchi della

Lega (Bologna: Romagnoli dall'Acqua, 1886), p. 52. It is impossible to tell from his title and description (*medego*) if Master Giovanni was a surgeon or a physician. For the early fourteenth-century Pisan case described by the preacher Giordano of Rivalto, which involved the presence of four doctors, see Chapter One n.64.

46. See Jacopo Berengario da Carpi, *Carpi Commentaria cum amplissimis additionibus super Anatomia Mundini*... (Bologna: Girolamo de' Benedetti, 1521), fol. 211v–212r, for an early sixteenth-century case from Bologna described in Introduction n.8, and E. Tardito, "Tagli cesarei di altri tempi," *Minerva medica* 59/supplement to n. 94 (24 November 1968), pp. 5079–5106, for a case from Vercelli in 1546.

47. ASF, Mediceo avanti il Principato, 35, 746 (inventory) = 35, 747 (microfilm, spool 59), letter of 24.i.1477: "*Carissimo mio Lorenzo. I' son' tanto oppresso da passione e dolore per l'acerbissimo e inopinato chaso della mia dolcissima sposa che io medesimo non so dove mi sia. La quale chome aria inteso ieri chome piacque a dio a hore xxii sopra parto passò di questa presente vita, e lla creatura sparata lei gli chavamo di chorpo morta, che m'è stato anchora doppio dolore. Son' certissimo che per la tua solita pietà avendomi chompassione m'arai per ischusato s'io non ti scrivo a longho.*"

48. For the case for identifying this relief with Francesca's tomb, see Musacchio, *Art and Ritual*, pp. 29–31; the father strongly resembles Giovanni Tornabuoni as depicted in contemporary paintings. In addition to the iconographic problems, however, it is difficult to explain the appearance of the Strozzi arms on the relief and to account for its presence in Florence; I thank Jonathan Katz Nelson for his advice on these issues. The right-hand section of the relief recalls ancient Greek tombstones of women who died in childbirth, although it is not clear how well known these were in late fifteenth-century Italy. On these tombstones, see Ursula Vedder, "Frauentod-Kriegertod im Spiegel der attischen Grabkunst des 4. Jhr. v. Chr.," *Mitteilungen des Deutschen Archäologischen Instituts* 103 (1988), pp. 161–91, and Nancy Demand, *Birth, Death, and Motherhood in Classical Greece* (Baltimore: Johns Hopkins University Press, 1994), ch. 7.

49. For example, Musacchio, *Art and Ritual*, figures 3 and 33.

50. Angelo Poliziano, *Prose volgari inedite e poesie latine e greche edite e inedite*, ed. Isidoro Del Lungo (Florence: Barbèra, 1867), p. 62. For an extended account of this episode, see Gaetano Pieraccini, *La stirpe de' Medici di Cafaggiolo: Saggio di ricerche sulla trasmissione ereditaria dei caratteri biologici* (Florence: Vallechi, 1924–1925), vol. 2, pp. 134–36.

51. ASF, Mediceo avanti il Principato, 30, 274 (Clarice to Lorenzo, 7.ix.1478): "*È in questo punto giunto qui Maestro Stefano, et per la gratia di dio m'ha trovata assai meglo al mio parere, et moto m'a confortata. Lui vi scriverà più a pieno quello ne intende.*"

52. *Ibid.*, 31, 283 (Clarice to Lorenzo, 8.ix.1478): "*Maestro Stephano m'à fatto lui intendere come ancora quelli altri medici che non porto pericolo.*" Clarice was staying in Pistoia at the time. Master Stefano's letters: *Ibid.*, 31, 282 (7.ix.1478) and 31, 284 (8.ix.1478).

53. ASF, Carte Strozziane, 3, 108, fol. 79r (25.i.1525/6).

54. Christiane Klapisch-Zuber, "Les femmes et la mort," p. 219, and "Le dernier enfant," pp. 282–83. See, in general, Musacchio, *Art and Ritual*, pp. 24–31.

55. The degree to which pregnancy and childbirth were medicalized was an obvious function of wealth and social status. Roughly half the population of Florence was composed of laborers of restricted and unreliable means, and while we know nothing about the childbirth practices of laborers and the poor in the city, they certainly could not have afforded the services of a physician, while inhabitants of the countryside would have had no access to one in any case.

56. See, for example, n.52 above (Clarice) and the *ricordanze* of Tommaso di Francesco Giovanni (ASF, Carte Strozziane 2, 16bis, entry of 8.vii.1452), Tribaldo d'Amerigo de' Rossi (BNCF, Fondo Nazionale 2, 2, 357, fol. 58r), and Cambio di Manno Petrucci (ASF, Carte Strozziane 2, 22, fol. 64v), the last two quoted in nn.58–59 below.

57. See n.82 below.

58. *Ricordanza* of Tribaldo de' Rossi, BNCF, Fondo Nazionale 2, 2, 357, fol. 58r: "*Richordo chome piachue a meser domenedio, avendo l'Alesandra dona di Piero Ripetti male auto già due mesi incircha, tirò a se la sua benedetta anima a dì 16 di*

marzo 1492 a ore [lacuna] *di notte, e'l dì medesimo a ore 24 la sopelirono in Sa'
Lorenzo di Firenze nela sepoltura di detto Piero Ripetti. Aveva ghuasto la matricie e'l
feghato e polmoni, chè dicho' le done la fecie isparare. Era istata a marito 2 anni e 9
mesi meno 2 dì, cioè mesi 33 meno 2 dì apunto."*

59. *Ricordanza* of Cambio and Giovanni di Manno Petrucci, ASF, Carte
Strozziane 2, 22, fol. 64v: "*Richordo chome ogi questo dì d'achosto a ore 13 piaque
a Dio trare a se la benedetta anima della Gratia mia dona e figliuola di Lionardo
d'Andrea B[uondelmonti?], e fe'lla sopelire in Santa Maria Novella cho' 16 chopie di
frati di detta Santa Maria Novella e 8 chopie di frati di Santo Branchatio. . . . Erano
alla sua morte l'Alesandra dona fu di Lorenzo d'Alesandro Buondelmonti sua sorella
et la Luchrezia dona fu di Marcho di Tolosino de' Medici, e fecio'la sparare. E'l male
suo era della matricie e del* [illegible: mich . . . sa? milza?] *biancha.*"

60. Alessandro Valori, "L'onore femminile attraverso l'epistolario di Margherita e Francesco Datini da Prato," *Giornale storico della letteratura italiana* 175
(1998), pp. 53–83.

61. *Ibid.* In a letter from 1469, for example, Fiametta Adimari reassured
her husband, who was nervous about her attending the wedding of one of his
kinswomen in his absence, noting that she had returned to her house with other
women; see Crabb, *Strozzi of Florence*, p. 208.

62. Cadden, *Meanings of Sex Difference*, p. 249; Haas, *Renaissance Man*,
p. 139.

63. Cadden, *Meanings of Sex Difference*, pp. 91–92.

64. *Ibid.*, pp. 109–30. See also Jane Fair Bestor, "Ideas about Procreation
and their Influence on Ancient and Medieval Views of Kinship," in David
I. Kertzer and Richard P. Saller (eds.), *The Family in Italy from Antiquity to the
Present* (New Haven: Yale University Press, 1991), p. 151 and passim; Gianna
Pomata, "Legami di sangue, legami di seme: Consanguineità e agnazione nel
diritto romano," *Quaderni storici* 86 (1994), pp. 299–334; and Giulia Calvi, *Il
contratto morale: Madri e figli nella Toscana moderna* (Rome: Laterza, 1994), esp.
pp. 30–31, 63–64, and 108.

65. See, in general, Cadden, *Meanings of Sex Difference*, pp. 23–25 and
130–34.

66. *Ibid.*, pp. 130–34 and 195–201. See, for example, Mondino de' Liuzzi, *Expositio super capitulum de generatione embrionis Canonis Avicennae cum quibusdam quaestionibus*, ed. Romana Martorelli Vico (Rome: Istituto Storico Italiano per il Medio Evo, 1993), [lectio 12], pp. 172–73.

67. ASF, Signori e Collegi, Deliberazioni, ordinaria autorità, 42, fol. 5v–6r: the law in question, governing women's clothing, was aimed "*ad refrenandam barbaram et indomitam feminarum bestialitatem, que non memores sue nature fragilitatis et quod viris subdite sint, . . . ipsos viros cogunt mellifluis venenis ipsis subicere, immemoresque viros portant qui ab ipsis hominibus procreantur, ipse que tamquam sacculum semen naturale perfectum ipsorum virorum retinent, ut homines fiant.*"

68. Leon Battista Alberti, *The Family in Renaissance Florence*, bk. 1, trans. Renée Neu Watkins (Columbia: University of South Carolina Press, 1969), p. 47; Italian text in Leon Battista Alberti, *I libri della famiglia*, ed. Ruggiero Romano and Alberto Tenenti, new edition by Francesco Furlan (Turin: Einaudi, 1994), pp. 36–37.

69. Marsilio Ficino, "Epistola ad fratres vulgaris," *Supplementum Ficinianum: Marsilii Ficini florentini philosophi platonici opuscula inedita et dispersa*, ed. Paul Oskar Kristeller (Florence: Olschki, 1937), vol. 2, pp. 113 and 115. On contemporary understandings of father-son relationships, ideal and real, see Kent, *Household and Lineage*, pp. 45–49.

70. Janet Adelman, *Suffocating Mothers: Fantasies of Maternal Origin in Shakespeare's Plays, Hamlet to the Tempest* (New York: Routledge, 1992), pp. 211–13.

71. Christiane Klapisch-Zuber, "La mère cruelle: Maternité, veuvage et dot," *La maison et le nom*, pp. 249–50. On the naturalization of patrilineal ideas of the family, see Bestor, "Ideas about Procreation," and Pomata, "Legami di sangue."

72. See the reference in Chapter One nn.71–72.

73. See Jacqueline Marie Musacchio, "Imaginative Conceptions in Renaissance Italy," in Geraldine A. Johnson and Sara F. Matthews Grieco (eds.), *Picturing Women in Renaissance and Baroque Italy* (Cambridge: Cambridge University Press, 1997), pp. 42–60, and *Art and Ritual*, pp. 128–34.

74. Mondino de' Liuzzi, *Expositio super capitulum de generatione embrionis*, [lectio 12], p. 176. By the fourteenth century, this theory had replaced the older theory known as pangenesis, which held that the male seed was not generated solely in the testicles but also sprang from other parts of the body, such as the legs or eyes; this meant that it could transmit defects from throughout the body, such as lameness or blindness. On pangenesis, see Cadden, *Meanings of Sex Difference*, pp. 91–92. Late fifteenth-century medical writers who, following Galen, argued for the existence of real female seed, pointed to illnesses inherited through the female line as evidence of their position; for example, Gabriele de' Zerbis, "Anathomia matricis pregnantis: Et est sermo de anathomia et generatione embrionis," *Liber anathomie corporis humani et singulorum membrorum illius* (Venice: Bonetum Locatellum for Octavianus Scotus, 1502), fol. 1r (treatise separately foliated).

75. For example, Avicenna, *Liber canonis* 4.3.3.1 (1507; Hildesheim: Olms, 1964), fol. 442v: "*Et quandoque accidit propter hereditatem et propter complexionem embrionis ex qua creatus est in se propter complexionem que est ei, aut acquisitam in matrice propter dispositionem que est ei, sicut si accidat ei ut sit conceptio in dispositione menstruorum.*" On conception during menstruation, see Ottavia Niccoli, "'Menstruum quasi monstruum': Monstrous Births and Menstrual Taboo in the Sixteenth Century," trans. Mary M. Gallucci, in Edward Muir and Guido Ruggiero (eds.), *Sex and Gender in Historical Perspective* (Baltimore: Johns Hopkins University Press, 1990), pp. 1–25.

76. Paolo da Certaldo, *Libro di buoni costumi*, ed. Alfredo Schiaffini (Florence: Le Monnier, 1945), pp. 83–84.

77. The following narrative is taken from Adalberto Pazzini, "La medicina alla corte dei Gonzaga a Mantova," in *Mantova e i Gonzaga nella civiltà del Rinascimento* (Mantua: Mondadori, 1977), pp. 320–23; Luca Beltrami, "L'annullamento del contratto di matrimonio fra Galeazzo M. Sforza e Dorotea Gonzaga (1463)," *Archivio storico lombardo*, ser. 2, 6 (1889), pp. 126–32; and Gregory Lubkin, *A Renaissance Court: Milan under Galeazzo Maria Sforza* (Berkeley: University of California Press, 1994), pp. 45–46.

78. Quoted in Pazzini, "La medicina alla corte dei Gonzaga," p. 323. This

seems to have been in large part a pretext, as Francesco Sforza by this time had his eye on a more prestigious marriage alliance with the House of Savoy.

79. See n.2 above.

80. See nn.58 and 59 above.

81. ASF, Corporazioni Religiose Soppresse dal Governo Francese 95, 212, fol. 171r: "*Richordo chome questo dì da sera circha a ore 1 1/2 di notte morì la Bartolomea mi' donna ch'era d'eta d'anni 42 incircha. E morì in Firenze in chasa mia in chamera terena. E morì di mal di matrice che gli fe' frusso di ventre che gli durò mesi 18 in circha, e mai i medici trovarono rimedio. Lasciòmi 6 figliuoli nati di lei ed aveva [?] 4 femine e 2 maschi. E pperchè mi prechò la facessi sparare per vedere se'l mal suo fussi di tisicho, per porre la medicina alle figliuole o ad altri. E chosì fe'. E si trovò il suo [?] male di natura ch'era la matrice inchallito in modo chon rasoio no' si poteva tagliare. Lascòmi d'incharicho maritasse una fanciulla e l'anno gli facessi dire qualche messa e altro non voleva da mme in vita. À bene fatti beni a sua satisfazione. E domandòmi perdonanza di quanto aveva dato per dio sanza domandarmene. Era stata mecho ani 27 e mesi 3 incircha, senpre in buona parte e charità ch'era una donna venerabile e di gran governo e senpre visuta chon stimo di dio.*"

82. See n.5 above. Despite the references to *tisico* (consumption) in both of these cases, I have found no mention of this illness in fifteenth-century discussions of constitutional disease; on the pervasiveness of what modern writers would call tuberculosis on Renaissance Florence, see Ann G. Carmichael, "The Health Status of Florentines in the Fifteenth Century," in Marcel Tetel, Ronald G. Witt, and Rona Goffen (eds.), *Life and Death in Fifteenth-Century Florence* (Durham, NC: Duke University Press, 1989), pp. 28–45.

83. Torni, *Opuscoli*, p. 40.

84. *Ibid.*, pp. 40 (quotations) and 43–44.

85. On the historiography of imperial Rome in late medieval and Renaissance Italy, see Arturo Graf, *Roma nella memoria e nelle immaginazioni del medio evo* (Turin: Loescher, 1923; repr. Sala Bolognese: Arnaldo Forni, 1987); he describes the basic sources on pp. 197–98. See also Louis-Fernand Flutre, *Li fait des Romains dans les littératures française et italienne du XIIIe au XVe siècle* (Paris: Hachette, 1932). For a list of the many late fifteenth-century manuscripts relat-

ing to this theme in Florentine libraries — many of them owned by Florentine residents or citizens — see Ernesto Giacomo Parodi, "Le storie di Cesare nella letteratura italiana dei primi secoli," *Studi di filologia romanza* 4 (1889), p. 414.

86. BNCF, ms. Palat. 471, fol. 74r: "*Giulio Cesare fu el primo imperadore di Roma et fu figliuolo d'un nobile humo chiamato Chatullo. Gli poeti e gli autori dichano che Cesare diciese d'Enea troiano et fu troiano, e ffu cholui e 'l quale ebbe solo tucti gli onori del mondo. . . . E Cesare fu decto perchè morì la madre prima che llui nasciesse, non de chè fu tagliata la madre e llui ne fu tracto vivo del chorpo.*" This manuscript, which was copied in 1479 for the Florentine Antonio di Lionardo de' Nobili (see explicit), belongs to the most common family of texts of this work, which Flutre calls Family B; see Flutre, *Li fait des Romains dans les littératures*, p. 274. The text varies little from manuscript to manuscript, except for orthography. I have also consulted the versions in BNCF, Fondo Nazionale 2, 1, 363, fol. 63v–75v (late fifteenth century) and Fondo Nazionale 2, 4, 280, fol. 79r–90r (1455). The source of this story is probably Isidore of Seville, *Etymologiae* 9.3.12.

87. Cf. Louis-Fernand Flutre and K. Sneyders de Vogel, *Li fet des Romains: Compilé ensemble de Saluste et de Suétoine et de Lucan*, ch. 2 (Paris: Droz, 1938), p. 8: "*Gaius Juilles Cesar fu tant en vantre sa mere que il covint le ventre tranchier et ouvrir ainz que il en poïst oissir; et trova l'en que il avoit mout granz cheveux. Por ce fu il apelez Cesar par sornon, car cist mosz Cesar puet senefier ou cheveleüre ou tranchement.*" For the circulation of *Li fet des Romains* in Italy, see Flutre, *Li fait des Romains dans les littératures*, and, in general, see Graf, *Roma*, ch. 8.

88. On the iconographic tradition of this scene, see Louis-Fernand Flutre, "La naissance de César," *Aesculape* 24 (1934), pp. 244–50; Friedrich von Zglinicki, *Geburt: Eine Kulturgeschichte in Bildern* (Braunschweig: Westermann, 1983), ch. 5; and Blumenthal-Kosinski, *Not of Woman Born*, ch. 2.

89. BNCF, ms. Palat. 471, fol. 74v–75r: "*Nerone inperadore doppo la morte de Caudio [sic] regniò xiiii anni e sei mesi e xxx dì, e fu pessimo huomo. . . . Fecie uccidere sua madre per vedere el luogho dove lui era estato, cioè la fecie sparare, e ancho lui per vedere chome furono grandi e fuochi di Troia fecie mectere fuocho da quatro parti di Roma. Poi chomandò a' filosafi che llo facessono ingravidare acciò che lui parturisse*

un figliuolo che somigliasse lui, perciò dubitava che la moglie no' gli parturisse figliuoli legitimi. E chosì fecieno e filosafi perciò che gli missono una rochia [sic: in other mss., *ranocchia*] *in chorpo. Al'utimo a un mezzo miglio fuori di Roma lui estesso s'uccise e lla sua charne ne fu portata da lupi e da chani.*" The story of Nero's opening of Agrippina seems to have been a medieval invention; see Ernest Langlois, *Origines et sources du Roman de la Rose* (Paris: Thorin, 1891), pp. 128–29, and "La traduction de Boèce par Jean de Meun," *Romania* 42 (1913), p. 347. The principal ancient sources on which it was based were Suetonius, *Lives of the Caesars* 6.34, which describes Nero's handling of Agrippina's limbs and his call for drink; Tacitus, *Annals* 14.9, which indicates that he had her stabbed in the womb, but not that he had her body opened; Dio Cassius, *Roman History* 61.14 (ditto); and the brief reference in Boethius, *Consolation of Philosophy* 2.m6. Although I know of no Italian illustrations of this scene, it was pictured in a number of French manuscripts of both the *Roman de la rose* and Laurent de Premierfait's French version of Boccaccio's *De casibus virorum illustrium*; see Wolf-Heidegger and Cetto, *Die anatomische Sektion*, pp. 131–42, and Chapter Five below.

90. Pliny, *Natural History* 7.9.47. On the heroic associations of "Caesarean" birth, see Schäfer, *Geburt aus dem Tod*, pp. 19–21, and Nadia Maria Filippini, *La nascita straordinaria: Tra madre e figlio, la rivoluzione del taglio cesareo (sec. XVIII–XIX)* (Milan: Angeli, 1995), pp. 21–22. On the lack of evidence for its practice in the Roman world, see Danielle Gourevitch, "Chirurgie obstétricale dans le monde romain: Césarienne et embryotomie," in Véronique Dasen (ed.), *Naissance et petite enfance dans l'Antiquité* (Fribourg: Academic Press, 2004), pp. 240–44.

91. Ovid, *Metamorphoses* 2.612–32.

92. Shakespeare, *Macbeth* 4.1.80 and 5.8.15–16. The phrase "none of woman born" echoes the Latin surnames often given to children who were born in this way: *nonnatus* or *ingenitus*, that is, unborn.

93. Adelman, *Suffocating Mothers*, p. 143. This is in concord with ancient and medieval theories of birthing, which interpreted it as the result of fetal rather than maternal activity and labor; see Ann Ellis Hanson, "A Long-Lived 'Quick Birther' (*okytokion*)," in Dasen (ed.), *Naissance et petite enfance*, pp. 269–80.

94. Adelman, *Suffocating Mothers*, p. 146.

95. Jacobus de Voragine, "Life of St. Peter Apostle," *The Golden Legend: Readings on the Saints*, trans. William Granger Ryan (Princeton, NJ: Princeton University Press, 1993), vol. 1, p. 347. Latin text in Graf, *Roma*, vol. 1, pp. 339–40.

96. *Ibid.*

97. See, for example, Cristelle L. Baskins, *Cassone Painting, Humanism, and Gender in Early Modern Italy* (Cambridge: Cambridge University Press, 1998); Jacqueline Musacchio, "The Rape of the Sabine Women on Quattrocento Marriage Panels," in Trevor Dean and K.J.P. Lowe (eds.), *Marriage in Italy, 1300–1650* (Cambridge: Cambridge University Press, 1998), pp. 66–82; and Ellen Callmann, "The Growing Threat to Marital Bliss as Seen in Fifteenth-Century Florentine Paintings," *Studies in Iconography* 5 (1979), esp. pp. 84–86.

98. See, for example, Kuehn, "Some Ambiguities of Female Inheritance Ideology," and Kent, "La famiglia patrizia fiorentina nel Quattrocento."

CHAPTER FOUR: THE EVIDENCE OF THE SENSES

1. [*Leggenda anonima di Elena Duglioli*], Bologna, Biblioteca Comunale dell' Archiginnasio di Bologna (BCAB), ms. B4314, fol. 166r. The two most important studies of Elena are Gabriella Zarri, "L'altra Cecilia: Elena Duglioli Dall'Olio (1472–1520)," *Le sante vive: Cultura e religiosità femminile nella prima età moderna* (Turin: Rosenberg and Sellier, 1990), pp. 165–96, and Gianna Pomata, "A Christian Utopia of the Renaissance: Elena Duglioli's Spiritual and Physical Motherhood (ca. 1510–1520)," in Kaspar von Greyerz, Hans Medick, and Patrice Veit (eds.), *Von der dargestellten Person zum erinnerten Ich: Europäische Selbstzeugnisse als historische Quellen (1500–1850)* (Cologne: Böhlau, 2001), pp. 323–53. For a summary of the manuscript sources for Elena's life and miracles, see Pomata, "Christian Utopia," pp. 324–35 n.3.

2. [*Leggenda anonima*], fol. 171r: "*adoravanlo et basciavanlo come preciosa reliquia, spargendole in tutta quella mattina adosso odorifere herbe.*"

3. See Gabriella Zarri, "Storia di una committenza," in *L'Estasi di Santa Cecilia di Raffaello da Urbina nella Pinacoteca Nazionale di Bologna* (Bologna: ALFA, 1983), pp. 21–42.

4. Pietro Ritta of Lucca, *Narrativa della vita e morte della Beata Elena Duglioli dall'Oglio...*, BCAB, ms. Gozzadini 292, fol. [2r]: "*come che in quelle virginee mammille habbi a perseverare detto latte per insino al fine del mondo.*" Regarding Elena's heart, see Ritta, *Narrativa*, fol. [3v] and the anonymous *Del cuore di Helena vergine dal signore realmento tolto con molte cose circa quello sopra ogni natura accadute...*, in BCAB, ms. Gozzadini 292. (The manuscript is unfoliated.) The *Narrativa* is much shorter than Elena's anonymous *vita*, although the relationship between the two is close, indicating that they, along with *Del cuore*, all came out of the close circle of her immediate disciples. Like a number of Elena's followers, Ritta was a regular canon. On the issues raised by Elena's claims of virginal lactation, see Pomata, "Christian Utopia."

5. Ritta, *Narrativa*, fol. 177v–78r.

6. These and subsequent *experientie* on Elena's body are described in detail in the [*Leggenda anonima*], fol. 176v–93v, and in a more abbreviated fashion in *Del cuore*, fol. [22r]–[24v].

7. On Damiano, see Vittorio Putti, *Berengario da Carpi: Saggio biografico e bibliografico, seguito dalla traduzione del "De fractura calvae sive cranei"* (Bologna: Cappelli, 1937), p. 112.

8. Zarri, "L'altra Cecilia," p. 169.

9. [*Leggenda anonima*], fol. 179r: "*fu nel loco del cor ritrovata una cosa dal cor tanto dissimile, che homo pratico di tal humana parte non mai l'haveria per cor ricognoscuto. Et li astanti medici dissero quella esser una molta strana cosa, et che non mai più il simile a quello audito haveano.*"

10. *Ibid.*, fol. 178v.

11. Béranger de Saint-Affrique, *Vita Sanctae Clarae de Cruce: Ordinis eremitarum S. Augustini*, ed. Alfonso Semenza, *Analecta augustiniana* 17–18 (1939–1941), p. 406; translated as Berengario di Sant'Africano, *Life of Saint Clare of Montefalco*, trans. Matthew J. O'Connell, ed. John E. Rotelle (Villanova, PA: Augustinian Press, 1998), p. 89. According to Béranger, these men "issued the opinion that [the objects] could not have been naturally formed, but only by divine power."

12. M.H. Laurent, "La plus ancienne légende de la B. Marguerite de Città di Castello," *Archivum fratrum praedicatorum* 10 (1940), [ch. 26], p. 127.

13. Nancy Caciola, *Discerning Spirits: Divine and Demonic Possession in the Middle Ages* (Ithaca, NY: Cornell University Press, 2003), ch. 6; see, in general, Gabriella Zarri, "'Vera' santità, 'simulata' santità: Ipotesi e riscontri," and André Vauchez, "La nascita del sospetto," in Zarri (ed.), *Finzione e santità tra medioevo ed età moderna* (Turin: Rosenberg & Sellier, 1991), pp. 9–36 and 39–51.

14. On the elaboration of the authority of university medicine in the late Middle Ages, see R.K. French, *Medicine before Science: The Rational and Learned Doctor from the Middle Ages to the Enlightenment* (Cambridge: Cambridge University Press, 2003), as well as, in general, Nancy G. Siraisi, *Taddeo Alderotti and His Pupils: Two Generations of Italian Medical Learning* (Princeton, NJ: Princeton University Press, 1981), and *Medieval and Early Renaissance Medicine: An Introduction to Knowledge and Practice* (Chicago: University of Chicago Press, 1990).

15. Vivian Nutton, "The Diffusion of Ancient Medicine in the Renaissance," *Medicina nei secoli* 14 (2002), pp. 468–69.

16. See Chapter Two above, and in general R.K. French, *Dissection and Vivisection in the European Renaissance* (Aldershot: Ashgate, 1999), ch. 3; Andrea Carlino, *Books of the Body: Anatomical Ritual and Renaissance Learning*, trans. John Tedeschi and Anne C. Tedeschi (1994; Chicago: University of Chicago Press, 1999), chs. 3–4; and Siraisi, *Medieval and Early Renaissance Medicine*, ch. 4.

17. Galen, *Galieni Pergamensis medicorum omnium principis Opera*, ed. Diomedes Bonardus (Venice: For Filippo Pinzio da Caneto, 1490). For details regarding the translation and circulation of all these texts, see Richard J. Durling, "A Chronological Census of Renaissance Editions and Translations of Galen," *Journal of the Warburg and Courtauld Institutes* 24 (1961), pp. 230–305. On Galen's teachings concerning generation, female anatomy, and female physiology in these works, see Rebecca Flemming, *Medicine and the Making of Roman Women: Gender, Nature, and Authority from Celsus to Galen* (Oxford: Oxford University Press, 2000), ch. 6. *On the Anatomy of the Uterus* had been translated by Niccolò of Reggio in the early fourteenth century but had had no discernable influence on anatomical writing. For an English version (based not on this but on a sixteenth-century Latin translation), see "On the Anatomy of the Uterus," trans. Charles Mayo Goss, *Anatomical Record* 144 (1962), pp. 77–83.

18. See Chapter Three above.

19. The most detailed description of these works, exclusive of Vesalius' *Tabulae anatomicae sex*, *Fabrica*, and *Epitome*, appears in L.R. Lind, *Studies in Pre-Vesalian Anatomy: Biography, Translations, Documents* (Philadelphia: American Philosophical Society, 1975), which contains bibliographies regarding each writer. See also French, *Dissection and Vivisection*, chs. 3–5, and Carlino, *Books of the Body*, pp. 191–225. I discuss the work of Vesalius in Chapter Five.

20. Alessandro Benedetti, *Historia corporis humani, sive Anatomice*, 5.35, ed. and trans. (into Italian) Giovanna Ferrari (Florence: Giunti, 1998), p. 350: "*Hortor omnes, tum tyrones tum veteranos medicos vel chirurgos, ad frequens huiusmodi theatrum, quod singulis saltem annis celebrandum sit, quoniam in eo vera videmus, aperta contemplamur, ut opera naturae tanquam viventia nostris subiaceant oculis.*" On Benedetti's life and career, see Giovanna Ferrari, *L'esperienza del passato: Alessandro Benedetti filologo e medico umanista* (Florence: Olschki, 1996), ch. 1. An English translation of the *Anatomice* appears in Lind, *Studies in Pre-Vesalian Anatomy*, pp. 81–137, but I have used my own translations here.

21. Benedetti, *Anatomice*, 5.35, p. 350. See Ferrari's remarks on this passage and its sources in *L'esperienza del passato*, pp. 154–55.

22. Girolamo Manfredi, proemium to *Anothomia*, trans. A. Mildred Westland, in Charles Singer, "A Study in Early Renaissance Anatomy, with a New Text: The *Anothomia* of Hieronymo Manfredi (1490)," in Charles Singer (ed.), *Studies in the History and Method of Science* (Oxford: Clarendon, 1917), p. 131. This work, based heavily on Mondino's *Anatomy*, was never published and exists in a single manuscript in the Bodleian Library.

23. See Ferrari, *L'esperienza del passato*, pp. 169–71.

24. See Giovanna Ferrari, "Public Anatomy Lessons and the Carnival: The Anatomy Theatre of Bologna," *Past and Present* 117 (1987), pp. 50–106, and Cynthia Klestinec, "A History of Anatomy Theaters in Sixteenth-Century Padua," *Journal of the History of Medicine and Allied Sciences* 59 (2004), pp. 404–409.

25. See Berengario's numerous dismissive remarks concerning the intellectual utility of public dissections in Jacopo Berengario da Carpi, *Carpi commentaria cum amplissimis additionibus super Anatomia Mundini* . . . (Bologna: Girolamo

de' Benedetti, 1521), fol. 119v, 438r, 479v, 516r. Ippolito of Montereale's account is transcribed in Dorothy M. Schullian, "An Anatomical Demonstration by Giovanni Lorenzo of Sassoferrato, 19 November 1519," in *Miscellanea di scritti di bibliografia ed erudizione in memoria di Luigi Ferrari* (Florence: Olschki, 1952), pp. 487–94. On dissection in hospitals, see Monica Azzolini, "Leonardo da Vinci's Anatomical Studies in Milan: A Re-Examination of Sites and Sources," in Jean A. Givens, Karen Reeds, and Alain Touwaide (eds.), *Visualizing Medieval Medicine and Natural History, 1200–1550* (Burlington, VT: Ashgate, 2006), pp. 152–54 and 161–67; on anatomies on private patients, see Chapter Three.

26. Berengario, preface to *Commentaria*, fol. 6v, and preface to *Isagogae breves perlucidae ac uberimae in anatomiam humani corporis* (Bologna: Girolamo de' Benedetti, 1522), fol. 2r; translated in *A Short Introduction to Anatomy*, trans. L.R. Lind, with anatomical notes by Paul G. Roofe (Chicago: University of Chicago Press, 1959), p. 35. The best introduction to Berengario's work is R.K. French, "Berengario da Carpi and the Use of Commentary in Anatomical Teaching," in Andrew Wear, R.K. French, and Iain M. Lonie (eds.), *The Medical Renaissance of the Sixteenth Century* (Cambridge: Cambridge University Press, 1985), pp. 42–74.

27. Berengario, *Commentaria*, fol. 466r and 259v–60r.

28. See, for example, *ibid.*, fol. 153r, 168v–176v, and 184v–239r.

29. See Chapter Three above and, in general, Ian Maclean, *The Renaissance Notion of Woman: A Study in the Fortunes of Scholasticism and Medical Science in European Intellectual Life* (Cambridge: Cambridge University Press, 1980), pp. 35–39, and Joan Cadden, *Meanings of Sex Difference in the Middle Ages: Medicine, Science, and Culture* (Cambridge: Cambridge University Press, 1993), pp. 117–30.

30. Berengario, preface to *Commentaria*, fol. 4r. For a sense of what was in fact Galen's much larger oeuvre on anatomy, see Vivian Nutton, Introduction to Andreas Vesalius, *De humani corporis fabrica*, trans. Daniel Garrison and Malcolm Hast, http://vesalius.northwestern.edu.

31. Note that the uterus and vagina (which included the external genitals and was usually referred to as the *collum* or "neck" of the the uterus) were typically treated as a single organ.

32. Sebastiano Bontempi, "Vita beatae Columbae Reatinae," in *Acta sanctorum quotquot toto orbe coluntur*, vol. 5, *Maii* (Antwerp: Michael Cnobarus, 1685; facs. repr. Brussels: Culture et Civilisation, 1968), pp. 319*–98*, no. 216. On Colomba's life and Bontempi's *vita*, see Roberto Rusconi, "Colomba da Rieti: La signoria dei Baglioni e la 'seconda Caterina,'" in Enrico Menestò and Robert Rusconi, *Umbria sacra e civile* (Turin: Nuova Eri Editioni Rai, 1989), pp. 211–26 and the bibliography on p. 235, and the essays in Giovanna Casagrande and Enrico Menestò (eds.), *Una santa, una città* (Spoleto: Centro Italiano di Studi sull'Alto Medioevo, 1991), esp. Gabriella Zarri, "Colomba da Rieti e i movimenti religiosi femminili del suo tempo," pp. 89–108; Giovanna Casagrande, "Terziarie domenicane a Perugia," pp. 109–60; and Enrico Menestò, "La 'leggenda' della beata Colomba e il suo biografo," pp. 161–76. For more extensive background on the late fifteenth- and early sixteenth-century spiritual and political movement to which Colomba belonged, see Gabriella Zarri, "Le sante vive," *Le sante vive*, ch. 3. On Bontempi and his relationship to Colomba, see John Coakley, "Friars as Confidants of Holy Women in Medieval Dominican Hagiography," in Renate Blumenfeld-Kosinski and Timea Szell (eds.), *Images of Sainthood in Medieval Europe* (Ithaca, NY: Cornell University Press, 1991), pp. 24–43.

33. Bontempi, "Vita beatae Columbae," no. 217.

34. *Ibid.*, no. 120. Bontempi's detailed description of their investigation occupies the entire remainder of his *vita*, nos. 120–28.

35. *Ibid.*, no. 120. Bontempi gives no indication of the identity of these men, with the exception of "a certain expert physician from Spain" (no. 122).

36. *Ibid.*, nos. 121–22. On physiological deficiencies that were thought to produce symptoms that resembled visionary behavior, see Dyan Elliott, "The Physiology of Rapture and Female Spirituality," in Peter Biller and A.J. Minnis (eds.), *Medieval Theology and the Natural Body* (Rochester, NY: York Medieval Press, 1997), esp. pp. 157–58, and Nancy Caciola, "Mystics, Demoniacs, and the Physiology of Spirit Possession in Medieval Europe," *Comparative Studies in Society and History* 42 (2000), pp. 268–306.

37. Bontempi, "Vita beatae Columbae," no. 120: "*Comitabantur eos et cives*

ipsi, sensu astuti et perspicui intellectu, apud quos vigent gymnasia doctrinarum, assiduusque est conflictus periculo litterario, ac juge spectaculum examinis rigorosi."

38. See Paola Zambelli, "L'immaginazione e il suo potere (da al-Kindi, al-Farabi e Avicenna al Medioevo latino e al Rinascimento," in Albert Zimmermann and Ingrid Craemer-Ruegenberg (eds.), *Orientalische Kultur und europäisches Mittelalter* (Berlin: Walter de Gruyter, 1985), pp. 188–206, and Armand Maurer, "Between Reason and Faith: Siger of Brabant and Pomponazzi on the Magic Arts," *Mediaeval Studies* 18 (1956), pp. 1–18.

39. See, for example, French, *Medicine before Science*, pp. 102–105.

40. See, in general, Lorraine Daston and Katharine Park, *Wonders and the Order of Nature, 1150–1750* (New York: Zone Books, 1998), pp. 145–46 and 159–64, and Stuart Clark, *Thinking with Demons: The Idea of Witchcraft in Early Modern Europe* (Oxford: Clarendon, 1997), chs. 14–15.

41. Pietro Pomponazzi, *De naturalium effectuum causis; sive, De incantationibus* (1567; Hildesheim: Olms, 1970). (The work initially circulated only in manuscript and was first printed in 1556.) See Martin L. Pine, *Pietro Pomponazzi: Radical Philosopher of the Renaissance* (Padua: Antenore, 1986), ch. 3.

42. On the considerable local resistance to Elena's cult from both Dominicans, who were allied with the rival political party of the recently deposed Bentivoglio family, and monks promoting the cult of Caterina Vigri, who saw Elena as an undesirable rival, see Zarri, "L'altra Cecilia," pp. 183–84.

43. [*Leggenda anonima*], fol. 187v: "*Fra li medici era schisma et contrarietà di sententie."*

44. *Ibid.*, fol. 187v–188r: "*Secondo, il Leone et il chirugico diceano tal incorruttione esser causata naturalmente dalla complexione et dispositione di tal parte, laquale per esser nervosa e piena di arterie è più tarda al risolversi che il resto dil corpo, qual dicevano posser esser sta' aiutata dalla myrra et aloes et allume che per filial riverentia da alcuni pietosi figlioli nell'exenterato corpo con stoppa era sta' posta. Tertio dissero che quella candidezza era la carne mamillare, qual più candida per natura è dil resto, et che ivi non era nè esser posseva latte, essendo dil tutto cessate le cause di quello, che sono li menstrui et calor naturale, il cui segno anchor era il non vedersi fluire."* On the relationship between menstrual blood and milk, see

335

Gianna Pomata, "La meravigliosa armonia: Il rapporto fra seni ed utero dall' anatomia vascolare all'endocrinologia," in Giovanna Fiume (ed.), *Madri: Storia di un ruolo sociale* (Venice: Marsilio, 1995), pp. 45–81.

45. Cf. Aristotle, *Meteorology* 4.1.379a3–b10.

46. Pietro Pomponazzi, *Dubitationes in quartum Meteorologicorum Aristotelis librum* (Venice: Francesco de' Franceschi da Siena, 1563), 16–24, fol. 9r–18r; Ludovico Boccadiferro, *Lectiones in quartum meteororum Aristotelis librum* (Venice: Francesco de' Franceschi da Siena, 1563), pp. 21–68. My thanks to Craig Martin for help with obtaining, dating, and interpreting these texts.

47. [*Leggenda anonima*], fol. 188v: "*Primo che fra le parti dil corpo facillime alla corruttione le mamelle della donna sono da esser connumerate, et che facilillimamente si corrompono, sì per causa del superabundante humore, qual etiamdio per la tenerezza et mollitie di quella carne glandosa, et di ciò non esser da dubitare. Secondo disse che non essendovi humana fraude tal duratione da esser attribuita a supernatural vertude. Tertio che defender si posseva in quella sustantia candida esser latte commixto congelato.*" On the controversy over *De immortalitate animae*, see Pine, *Pietro Pomponazzi*, pp. 124–38 and ch. 3.

48. *Ibid.*, fol. 188v–89r: "*che ivi in quella carne bianca glandosa era latte congelato, attribuendolo però come artifice naturale fuor del miraculo al calor naturale ivi conservato, dicendo che ridiculo era il dire non vi esser calore, per cui mezzo il latte, che in morte vi era, nè credibile tutto fluisse, come nel proprio loco fu conservato, et asserire ivi non esser latte, per non vedersi fluire, non se conveniva ad un dotto. Imperochè essendo ivi coagulato al modo che vi era, etiamdio con la forza dil foco non si potrebe risolvere, questo protestò dil tutto esser vero, et esser parato in ogni loco tal verità deffendere.*"

49. This man may be the same as Virgilio Gerardus, A.D., M.D. (Bol.), one of the authors of the laudatory poems appended to Berengario's *Commentaria*.

50. [*Leggenda anonima*], fol. 192r–v: "*Li medici, che sempre nimicano li miraculi, et all'opra di natura confugono, concesseron fin alhora il petto esser incorrotto, ma per cause naturali, che non sapevano, dicendo però che era molto alterato dal stato dil primo experimento, et che havea qualche grado di corruttione, tanto però pochi, che già doi mesi passati ne dovea haverne in molto magior numero. Dissero anchora, il*

humor ritrovatovi et mollitie indicar la propinqua resolutione, ma ben molto stupire della tanto longa duratione, et così scrissero al sommo Pontefice, et furono in animo di mandargli un peccio di quella mamillar sostantia, acciò dalli romani medici fosse dato il giudicio sopra l'oraculo dil conservato latte."

51. [*Leggenda anonima*], fol. 178v (Chiara); Bontempi, "Vita beatae Columbae," no. 120 (Margherita).

52. Bontempi, "Vita beatae Columbae," no. 217: "My heart has been made like melting wax in the middle of my belly." Cf. Psalm 22: 14–18 (Psalm 21 in Bontempi's numeration). On the heart as the traditional site of divine inspiration, see Caciola, *Discerning Spirits*, and Chapter One above.

53. Lind, *Studies in Pre-Vesalian Anatomy*, p. 70; possibly inspired by Colomba, Benedetti later wrote a book on "prodigious" fasts. The discussion of the wound in Christ's side appears in Berengario, *Commentaria*, fol. 336v–37r.

54. On the extraordinary development of medical semiology in this period, see Ian Maclean, *Logic, Signs, and Nature in the Renaissance: The Case of Learned Medicine* (Cambridge: Cambridge University Press, 2002), esp. ch. 8.

55. See Nancy G. Siraisi, "La comunicazione del sapere anatomico ai confini tra diritto e agiografia: Due casi del secolo XVI," in Massimo Galuzzi, Gianni Micheli, and Maria Teresa Monti (eds.), *Le forme della comunicazione scientifica* (Milan: FrancoAngeli, 1998), pp. 419–38; Jean-Michel Sallmann, *Naples et ses saints à l'âge baroque: 1540–1750* (Paris: Presses Universitaires de France, 1994), pp. 308–10; and Katharine Park, "Holy Autopsy: Saintly Bodies and Medical Expertise, 1300–1600," in Julia Hairston and Walter Stephens (eds.), *The Body in Early Modern Italy* (Baltimore, MD: Johns Hopkins University Press, forthcoming 2007).

56. Giovanni Incisa della Rocchetta and Nello Vian (eds.), with the assistance of P. Carlo Gasbarri, *Il primo processo per San Filippo Neri nel Codice Vaticano Latino 3798 e in altri esemplari dell'Archivio dell'Oratorio di Roma* (Vatican City: Biblioteca Apostolica Vaticana, 1957), vol. 2, p. 227.

57. I have found no reference to this discovery in either the canonization records in Celestino Piana, "I processi di canonizzazione su la vita di S. Bernardino da Siena," *Archivum Franciscanum Historicum* 44 (1951), pp. 87–160

and 383–435, or the early lives collected in *Acta sanctorum*, vol. 5, *Maii*, pp. 257*–318*. It is possible that the reference appears in an early *vita* and was censored by the Bollandist editors of the latter, as Gianna Pomata has suggested (personal communication). On the fate of Bernardino's viscera and the knife used to embalm him, see Martino Bertagna, *L'Osservanza di Siena: Studi storici* (Siena: Osservanza, 1964), vol. 1, pp. 81–82, and vol. 2, pp. 68–69. My thanks to Machtelt Israels for this reference.

58. After teaching in Bologna in the 1470s and 1480s, Zerbi eventually finished his academic career at the University of Padua. On these two scholars, see French, *Dissection and Vivisection*, ch. 3; Lind, *Studies in Pre-Vesalian Anatomy*, pp. 141–64; and the authoritative studies on Berengario: Putti, *Berengario da Carpi*, and French, "Berengario da Carpi."

59. Zerbi supplemented his discussion of the nonpregnant uterus in the three books of his *Book of Anatomy* with a separate treatise on the anatomy of the pregnant uterus: Gabriele de' Zerbi, *Liber anathomie corporis humani et singulorum membrorum illius* (Venice: Bonatus Locatellus for Octavianus Scotus, 1502), fol. 40r–47r, together with the fourteen separately numbered folios on the pregnant uterus that appear after the third book. (In contrast, Zerbi devoted only twelve folios to the brain and ten to the heart.) Berengario, *Commentaria*, fol. 181r–287v. Berengario's *Isagogae breves*, meant for beginners, does not discuss the pregnant uterus, which dramatically abbreviates the treatment: Berengario, *Short Introduction*, pp. 76–83.

60. Berengario, *Commentaria*, fol. 195r.

61. *Ibid.*, fol. 6v.

62. On Benedetti, see n.20 above. On the controversial nature of illustration in the descriptive sciences of this period, see Ferrari, *L'esperienza del passato*, pp. 155–56 and 322–23, and Sachiko Kusukawa, "Illustrating Nature," in Marina Frasca-Spada and Nick Jardine (eds.), *Books and the Sciences in History* (Cambridge: Cambridge University Press, 2000), esp. pp. 105–108.

63. On Leonardo's work on anatomy, see Kenneth D. Keele, *Leonardo da Vinci's Elements of the Science of Man* (New York: Academic Press, 1983), and Domenico Laurenza, *De figura umana: Fisiognomica, anatomia e arte in Leonardo*

(Florence: Olschki, 2001). On Leonardo's discussion of the female reproductive system and the embryo, see Keele, *Leonardo da Vinci's Elements*, pp. 348–62, and Laurenza, *De figura umana*, pp. 95–126.

64. On Berengario's interest in the visual arts, see Putti, *Berengario da Carpi*, pp. 83 and 116. On the woodcuts of the *Commentaries*, most of which were reused in the first edition of the *Short Introduction*, see Putti, *Berengario da Carpi*, pp. 131–92; French, "Berengario da Carpi," pp. 61–62, and *Dissection and Vivisection*, pp. 101–102; and Ludwig Choulant, *Geschichte und Bibliographie der anatomischen Abbildung nach ihrer beziehung auf anatomische Wissenschaft und bildende Kunst* (Leipzig: Weigel, 1852), pp. 136–42. Regarding the identities of the designers of these images, see n.93 below.

65. Berengario, *Commentaria*, fol. 226v. On the meanings of the woman's gesture, see Maurizio Rippa Bonati, "*Manuum munus*: Per una iconografia del 'toccar con mano,'" in A. Olivieri and M. Rinaldi (eds.), *All'incrocio dei saperi: La mano* (Padua: CLUEB, 2004), pp. 325–36.

66. Jole Agrimi and Chiara Crisciani, *Edocere medicos: Medicina scolastica nei secoli XIII–XV* (Naples: Guerini, 1988), esp. ch. 3. On the text-based nature of much anatomical writing, see Carlino, *Books of the Body*, ch. 5.

67. Zerbi, *Liber anathomie* 2.1, fol. 43r. This determination belongs to one of Zerbi's many *additiones*, more advanced reflections that alternated with his basic exposition (the *textus*), which was intended to be read during the dissection. It is not clear whether the *additiones* were communicated orally. Cf. proemium to *ibid.*, fol. 2r: "*Introductiones anathomice legende dum cadaver humanum dissectatur et anathomizatur intelligantur omnes que in textu adducentur exceptis additionibus.*"

68. *Ibid.*, fol. 43v–44r. *De spermate* was one of the *loci classici* for the seven cells of the uterus in medieval medical texts; see Robert Reisert, *Der sieben kammerige Uterus: Studien zur mittelalterlichen Wirkungsgeschichte und Entfaltung eines embryologischen Gebärmuttermodells* (Hanover: Wellm, 1986), pp. 41–44. It is no longer attributed to Galen and should not be confused with the apparently authentic *De semine*.

69. Berengario, *Commentaria*, fol. 218r–v and 184v–85r. Cf. Mondino de'

Liuzzi, *Anothomia di Mondino de' Liuzzi da Bologna, XIV secolo*, eds. Piero P. Giorgi and Gian Franco Pasini (Bologna: Istituto per la Storia dell'Università di Bologna, 1992), pp. 246–47. Berengario was unaware that Mondino had corrected himself in a subsequent work, written in 1320, which survives in a single manuscript; see Mondino de' Liuzzi, *Expositio super capitulum de generatione embrionis Canonis Avicennae cum quibusdam quaestionibus*, ed. Romana Martorelli Vico (Rome: Istituto Storico Italiano per il Medio Evo, 1993), [lectio 12], p. 143.

70. Berengario, *Commentaria*, fol. 204r and 194r–v.

71. *Ibid.*, fol. 208v–210v. Cf. Galen, *On the Usefulness of the Parts of the Body*, trans. Margaret Tallmadge May (Ithaca, NY: Cornell University Press, 1968), vol. 2, pp. 628–30.

72. Berengario, *Commentaria*, fol. 211r.

73. Thomas Laqueur, *Making Sex: Body and Gender from the Greeks to Freud* (Cambridge, MA: Harvard University Press, 1990), chs. 2–4.

74. Zerbi, *Liber anathomie*, fol. 47r. On sixteenth-century humanist medicine, and Galenism in particular, see Siraisi, *Medieval and Early Renaissance Medicine*, pp. 189–92, and Nutton, "Diffusion of Ancient Medicine" and the literature cited therein.

75. See Katharine Park, "Itineraries of the 'One-Sex Body': A History of an Idea," forthcoming.

76. Mondino de' Liuzzi, *Anathomia*, pp. 228–29. Mondino explains this with a long list of anatomical and physiological differences and similarities; the most fundamental difference is the inability of the female testicles to create real seed. Niccolò of Reggio translated *On the Use of Parts* in 1317, after Mondino had completed his treatise. It was known, for example, by Guy of Chauliac, but it was not fully assimilated before the late fifteenth century. See French, *Dissection and Vivisection*, pp. 67 and 82–83, and Guy of Chauliac, *Inventarium sive Chirurgia magna*, ed. Michael R. McVaugh (Leiden: Brill, 1997), vol. 2, pp. 47–49.

77. See MacLean, *Renaissance Notion of Woman*, pp. 32–34; Park, "Itineraries of the 'One-Sex Body'"; and Michael Stolberg, "A Woman Down to Her Bones: The Anatomy of Sexual Difference in the Sixteenth and Early Seventeenth Centuries," *Isis* 94 (2003), pp. 274–99. (Laqueur's and Londa Schie-

binger's rejoinders to Stolberg in the same issue are unconvincing.) For a more general critique of Laqueur's argument, see Katharine Park and Robert A. Nye, "Destiny Is Anatomy" (essay-review of Laqueur, *Making Sex*), *New Republic*, 18 February 1991, pp. 53–57.

78. Berengario, *Commentaria*, fol. 209r–210v; Zerbi, *Liber anathomie*, 2.1, fol. 47r.

79. On these general issues, see Chapter Three and, for more detail, Cadden, *Meanings of Sex Difference*, ch. 1 and pp. 119–30.

80. Berengario, *Commentaria*, fol. 232v; see also fol. 199r–200v and in general 227r–39v. Cf. Zerbi, *Liber anathomie* 2.1, fol. 47r, and *Anatomia matricis pregnantis*, fol. 1r–v.

81. Berengario, *Commentaria*, fol. 239r–v: "*magis adhaerens sensui.*"

82. Zerbi, proemium to *Liber anathomie*, fol. 1r.

83. On the first woman: Berengario, *Commentaria*, fol. 191v. According to Berengario, this woman had had sex four days before she was hanged, so that, in addition to the fetus, her uterus contained unmistakably male semen — an important finding, since this allowed for the possibility of superfetation, the conception of an additional child. On the same woman, see also fol. 222v. On the second woman: *ibid.*, fol. 424v. This anatomy allowed Berengario to demonstrate, in contradiction to the erroneous opinion of Galen "and the entire body of physicians," that the fetal membrane he called the *elancoydea* (Galen's *allantoeides*) envelops the entire fetus.

84. *Ibid.*, fol. 211v–12r. See Introduction at n.8.

85. Berengario, *Commentaria*, fol. 211v and 225r. See also Berengario, *Short Introduction*, ch. 1, p. 82.

86. Berengario, *Commentaria*, fol. 213v.

87. For a mid-sixteenth-century procedure performed by a barber, see E. Tardito, "Tagli cesarei di altri tempi," *Minerva medica* 59/supplement to n. 94 (24 November 1968), pp. 5079–5106.

88. Berengario, *Commentaria*, fol. 248r. Note that Mondino had already begun to study fetal development by dissecting miscarried fetuses; see Chapter Two n.92.

89. This belief had already been questioned two and a half centuries earlier by Guglielmo of Saliceto, who also lamented his dependence on "the testimony of midwives and ... women who have experienced many births and miscarriages." See Guglielmo of Saliceto, *Summa conservationis et curationis* (Venice: [Marinus Saracenus], 1490), 1.178, sig. i5vb, and Chapter Two n.75.

90. Berengario, *Commentaria*, fol. 263r.

91. *Ibid.*, fol. 262v.

92. Berengario had all but two of these reproduced in the first edition of his *Short Introduction* (1522), adding a new diagram of the uterus, and he made further adjustments in the second edition of 1523. Additional bibliography on the illustrations in n.64 above.

93. On the designers of the woodcuts, see Marzia Faietti and Daniela Scaglietti Kelescian, *Amico Aspertini* (Modena: Artioli, 1995), pp. 339–41, and Alberto Serra-Zanetti, *L'arte della stampa in Bologna nel primo ventennio del Cinquecento* (Bologna: A spese del Comune, 1959), pp. 51–55, *pace* Mimi Cazort, Monique Kornell, and K.B. Roberts, *The Ingenious Machine of Nature: Four Centuries of Art and Anatomy* (Ottawa: National Gallery of Canada, 1996), p. 38. On Girolamo Benedetti, see Serra-Zanetti, *L'arte della stampa*, pp. 80–103. I am grateful to Jonathan Katz Nelson, Lisa Pon, Erika Naginski, and Kathleen Weil-Garris Brandt for help in researching and thinking about Berengario's illustrations.

94. For example, the figures in Berengario, *Commentaria*, fol. 87r and 520v.

95. See A. Hyatt Mayor, *Artists and Anatomists* (New York: Metropolitan Museum of Art, 1984), pp. 92–93.

96. For the text of the caption, see the translation of n.65.

97. The caption of figure 4.2 reads, "You have in this figure the uterus with its horns on the sides, beneath which are the testicles in their natural place attached to the seminal vessels, which terminate at the uterus, as you see. They have their origin above, around the region of the kidneys, ... and this uterus is illustrated large, as if it were pregnant. In front of it is the bladder with the urethral ducts; its neck terminates in the vagina a little above the cleft which is called the vulva.... These things can be seen better in the anatomy of a pregnant

woman, although you can also see them in one who is not pregnant" (Berengario, *Commentaria*, fol. 225v). Similarly, the caption of figure 4.3 calls attention to the relationship among the seminal vessels, testicles, and uterine horns, and that between the cervix and the uterus (*ibid.*, fol. 226r). The "horns," which also appear in the female figure of the *Fasiculo de medicina* (figure 2.1), are probably Galen's reading of the supporting ligaments.

98. Bontempi, "Vita beatae Columbae," and Ritta, *Narrativa*, and Letter to Pope Leo X (12.x.1520), in BCAB, ms. Gozzadini 292 (unfoliated).

99. Coakley, "Friars as Confidants of Holy Women," p. 245. See also Coakley, "Friars, Sanctity, and Gender: Mendicant Encounters with Saints," in Clare A. Lees (ed.), *Medieval Masculinities: Regarding Men in the Middle Ages* (Minneapolis: University of Minnesota Press, 1994), pp. 91–110, and, in general, Catherine Mooney, "Women's Visions, Men's Words: The Portrayal of Holy Women and Men in Fourteenth-Century Italian Hagiography," Ph.D. diss. Yale University, 1991, ch. 7.

100. At Elena's death in 1520, Ritta claimed to have been her confessor for at least fourteen years: Ritta, *Narrativa*, fol. [6r].

101. [*Leggenda anonima*], fol. 47r: "*fra celesti chori, come un'altra Cecilia fra canti e suoni alli superni refferendo.*"

102. *Del cuore*, fol. 33r.

103. See Zarri, "Storia," and "L'altra Cecilia," pp. 165–68 and 175–78, and, in general, *L'Estasi di Santa Cecilia di Raffaello da Urbino*. On the four saints surrounding Cecilia in the painting, see Zarri, "L'altra Cecilia," n.66. Thomas Connolly focuses on musicology and musical iconography in *Mourning into Joy: Music, Raphael, and Saint Cecilia* (New Haven: Yale University Press, 1994).

104. On the Augustinian model of spiritual and (external) sensory experience that shaped the understanding of perception in this period, see Suzannah Biernoff, *Sight and Embodiment in the Middle Ages* (Houndmills: Palgrave Macmillan, 2002), pp. 25–26, and the last section of Chapter One above.

105. On the related iconographic tradition of David's discarded harp, see Connolly, *Mourning into Joy*, ch. 4.

106. On the tradition of *I modi*, see Bette Talvacchia, *Taking Positions: On the*

Erotic in Renaissance Culture (Princeton, NJ: Princeton University Press, 1999); on the eroticism of the pudic gesture, see Patricia Simons, "Anatomical Secrets: *Pudenda* and the *Pudica* Gesture," in Gisela Engel (ed.), *Das Geheimnis am Beginn der europäischen Moderne* (Frankfurt: Klostermann, 2002), pp. 302–27.

107. Charles Estienne, *De dissectione partium corporis humani* (Paris: Simon de Colines, 1545). See Talvacchia, *Taking Positions*, chs. 7–8.

108. See Phyllis Pray Bober and Ruth Rubinstein, with Susan Woodford, *Renaissance Artists and Antique Sculpture: A Handbook of Sources* (London: Miller, 1986), pp. 152–53 and 166–67.

Chapter Five: The Empire of Anatomy

1. Andreas Vesalius, *De humani corporis fabrica* (Basel: Joannes Oporinus, 1543). The authoritative treatment of Vesalius's works, career, and the prepara-tion of the *Fabrica* is still Charles D. O'Malley, *Andreas Vesalius of Brussels, 1514–1564* (Berkeley: University of California Press, 1964); see esp. chs. 6–8. For a compact and up-to-date introduction to Vesalius, see Vivian Nutton, Introduction, to Daniel H. Garrison and Malcolm Hast's online translation of the *Fabrica* (in preparation), at http://vesalius.northwestern.edu, and R.K. French, *Dissection and Vivisection in the European Renaissance* (Aldershot: Ash-gate, 1999), ch. 5.

2. On the *Fabrica* as a novel enterprise in the history of anatomical publica-tion and the title page as a reflection of new sixteenth-century ideas regarding the importance of dissection, see Andrea Carlino, *Books of the Body: Anatomical Ritual and Renaissance Learning*, trans. John Tedeschi and Anne C. Tedeschi (Chicago: Chicago University Press, 1999), ch. 1, esp. pp. 39–53, and Andrew Cunningham, *The Anatomical Renaissance: The Resurrection of the Anatomical Projects of the Ancients* (Aldershot: Scolar Press, 1997), ch. 4. O'Malley is wrong to identify my figure 5.2 as based on the body of another woman, the mistress of a Paduan monk. Vesalius clearly states that figures 24 and 27 of Book V of the *Fabrica* (my figures 5.1 and 5.2) are based on the anatomy of the executed crim-inal; see Vesalius, *Fabrica* 5.15, p. 539. For Berengario's illustrations, see Chapter Four above.

344

3. On the illustrations of the *Fabrica* and the artists involved in producing them, see David Rosand and Michelangelo Muraro, *Titian and the Venetian Woodcut* (Washington, DC: International Exhibitions Foundation, 1976), esp. sect. 6; Michelangelo Muraro, "Tiziano e le anatomie del Vesalio," in *Tiziano e Venezia: Convegno internazionale di studi, Venezia, 1976* (Vicenza: Pozza, 1980), pp. 307–16; and Martin Kemp, "A Drawing for the *Fabrica* and Some Thoughts upon the Vesalius Muscle-Men," *Medical History* 14 (1970), pp. 277–88.

4. Jacopo Berengario of Carpi, *Carpi Commentaria cum amplissimis additionibus super Anatomia Mundini* ... (Bologna: Girolamo de' Benedetti, 1521), fol. 194r–v.

5. Vesalius, *Fabrica* 5.15, p. 539. Here and throughout, unless otherwise indicated, I have used my own translations, which I have checked against Garrison and Hast's translation in progress (see n.1); I am grateful to them for access to a draft. The podestà was an official, appointed by the republic of Venice, who presided over the judicial and police systems of subject cities and oversaw executions.

6. Vesalius, *De humani corporis fabrica* (Basel: Oporinus, 1555), 5.15, p. 663.

7. *Statuta dominorum artistarum Achademiae Patavinae* (1498), 2.28; cited in Monica Azzolini, "Leonardo in Context: Medical Ideas and Practices in Renaissance Milan," Ph.D. diss., University of Cambridge, 2001, p. 156. On the prevalence of this kind of provision, see Katharine Park, "The Criminal and the Saintly Body: Autopsy and Dissection in Renaissance Italy," *Renaissance Quarterly* 47 (1994), p. 12. Exceptions were occasionally made to this rule; see n.20.

8. On the administration of criminal justice in Padua, Venice, and the Venetian territories, see Domenico Zorzi, "Sull'amministrazione della giustizia penale nell'età delle riforme: Il reato di omicidio nella Padova di fine Settecento," in Luigi Berlinguer and Floriana Colao (eds.), *Crimine, giustizia e società veneta in età moderna* (Milan: Giuffrè, 1989), esp. 273–282; Guido Ruggiero, *Violence in Early Renaissance Venice* (New Brunswick, NJ: Rutgers University Press, 1980), chs. 3 and 11 (on homicide); and Claudio Povolo, "Aspetti e problemi dell'amministrazione della giustizia penale nella Repubblica di Venezia, secoli

XVI–XVII," in Gaetano Cozzi (ed.), *Stato, società e giustizia nella Repubblica Veneta (sec. XV–XVIII)* (Rome: Jouvence, 1980), vol. 1, pp. 153–258.

9. Giuseppina de Sandre Gasparini, "La confraternità di S. Giovanni Evangelista della Morte in Padova e una 'riforma' ispirata dal vescovo Pietro Barozzi (1502)," in *Miscellanea Gilles Gérard Meersseman* (Padua: Antenore, 1970), vol. 1, pp. 765–815. On companies of this sort, see in general Adriano Prosperi, "Il sangue e l'anima: Ricerche sulle compagnie di giustizia in Italia," *Quaderni storici* 51 (1982), pp. 959–99; Vincenzo Paglia, *La morte confortata: Rite della paura e mentalità religiosa a Roma nell'età moderna* (Rome: Edizioni di Storia e Letteratura, 1982); Carlino, *Books of the Body*, pp. 99–116; and Samuel Y. Edgerton Jr., *Pictures and Punishment: Art and Criminal Prosecution During the Florentine Renaissance* (Ithaca, NY: Cornell University Press, 1985), esp. chs. 4–5.

10. Reformed statutes of 1502, 46–48, transcribed in de Sandre Gasparini, "La confraternità," pp. 808–11. Although the confraternity had female members, who took part in devotional and charitable activities, there is no evidence that they participated in the events surrounding executions, even those of female criminals.

11. Reformed statutes of 1502, 46, p. 808: "*e qualche volta el se ha trovato, e colui che l'à visto ne può render testimonianza, che a uno, el quale a disnar havea manzato quasi un capon intriego, essendo impichado do o 3 hore dapoi e concesso ali medici per far anathomia, no se li ha travado nì in el stomago nì in le intestine chossa alchuna, e fo iudicado che la grande angonia havesse risolta tuta quella materia chomo risove el sol l'aqua.*"

12. *Ibid.*, 47, p. 809. Such meditations were often facilitated by painted images; for examples, see Edgerton, *Pictures and Punishment*, ch. 5.

13. Reformed statutes of 1502, 47, p. 809. The statutes use the male pronoun throughout to refer to the condemned.

14. *Ibid.*, p. 810. On the events around the hanging of criminals and the confraternity's consignment of the corpse for dissection, see Carlino, *Books of the Body*, pp. 101–108.

15. The 1502 statutes are meticulous about the burial of severed body parts, requiring the brothers also to attend judicial mutilations and to bury the ampu-

tated hand or eye; see reformed statutes of 1502, 48, in de Sandre Gasparini, "Confraternità," p. 811.

16. On the development of this practice, see O'Malley, *Andreas Vesalius*, ch. 1; Carlino, *Books of the Body*, pp. 170–94; and Giovanna Ferrari, "Public Anatomy Lessons and the Carnival: The Anatomy Theatre of Bologna," *Past and Present* 117 (1987), pp. 50–106.

17. Transcribed in Azzolini, "Leonardo in Context," p. 156 n.73.

18. For a critique of the idea of "taboos" concerning corpse pollution, see Introduction. The second point, which has more to recommend it, looks back to Michel Foucault's brief discussion of execution rituals at the beginning of *Discipline and Punish: The Birth of the Prison*, trans. Alan Sheridan (1975; New York: Pantheon, 1977).

19. Alfonso Corradi, "Degli esperimenti tossicologici in *animae nobili* nel Cinquecento," *Rendiconti del Reale Istituto Lombardo di Scienze e Lettere*, ser. 2, 19 (1886), pp. 361–63. There is a detailed description of such an experiment in ASF, Carte Strozziane, ser. 1, 97, fol. 1r–7v. On (animal) vivisection in sixteenth-century anatomy, see French, *Dissection and Vivisection*, ch. 6.

20. Azzolini, "Leonardo in Context," ch. 5, and (on the infrequency of public dissections) Nutton, "Introduction," p. 11. Cynthia Klestinec (personal communication) has found one reference to a Paduan man who was executed for killing his wife and cutting her into pieces; the fact that he was dissected in Padua suggests that the particularly heinous nature of his crime might have convinced the officials to suspend the prohibition against dissecting locals. I know of no other cases that would suggest a punitive intent.

21. Carlino, *Books of the Body*, p. 80.

22. See Chapter Three n.11.

23. Vesalius, *Fabrica* 5.15, pp. 538–39. On the famous incident regarding the skeleton, see *ibid.* 1.39, pp. 161–62, translated in O'Malley, *Andreas Vesalius*, p. 64; on the heart of the man who had been quartered, see *ibid.* 6.9, p. 584. The willingness of Paduan students to engage in graverobbing was of evident concern to the Venetian governors of Padua — witness the legislation passed in February 1549/50, which imposed stiff fines for the "impious and inhuman"

practice of "taking dead bodies from their tombs, . . . as do certain foolhardy and wicked people in our city of Padua, to conduct private anatomies and make fats [presumably for magical and medicinal purposes] and sell the bones." Venice, Archivio di Stato, Senato Terra, 36, fol. 213r-v: "*Essendo in vero cosa impia et inhumana et da tutte le leggi divine et humane improbata il cavare delli corpi morti delle loro sepolture, come pare, che si facciano licito di fare alcun temerarii et scelerati nella città nostra di Padova, per far anatomie particolari et per far grassi et vendere li ossi.*" I owe this reference and transcription to Michelle A. Laughran (personal communication).

24. See, in general, Samuel W. Lambert, "The Initial Letters of the Anatomical Treatise, *De humani corporis fabrica*, of Vesalius," in Samuel W. Lambert, Willy Wiegand, and William M. Ivins Jr., *Three Vesalian Essays to Accompany the Icones anatomicae of 1934* (New York: Macmillan, 1952), pp. 3–24.

25. Mondino de' Liuzzi, *Expositio super capitulum de generatione embrionis Canonis Avicennae cum quibusdam quaestionibus*, ed. Romana Martorelli Vico (Rome: Istituto Storico Italiano per il Medio Evo, 1993), p. 157.

26. Baldasar Heseler, *Vesalius' First Public Anatomy at Bologna, 1540: An Eyewitness Report*, ed. and trans. Ruben Eriksson (Uppsala: Almqvist and Wiksells, 1959), pp. 198 and 180; see also p. 200. Vesalius performed and commented on the dissections that accompanied Corti's lectures.

27. Michael Sappol, *A Traffic of Dead Bodies: Anatomy and Embodied Social Identity in Nineteenth-Century America* (Princeton, NJ: Princeton University Press, 2002), pp. 80–90 (quotation on p. 80).

28. See Pamela H. Smith's discussion of "artisanal epistemology" in *The Body of the Artisan: Art and Experience in the Scientific Revolution* (Chicago: University of Chicago Press, 2004), esp. chs. 2–3.

29. Galen, *Claudii Galeni Pergameni, secundum Hippocratem medicorum facile principis opus De usu partium corporis humani*, trans. Niccolò of Reggio (Paris: Simon de Colines, 1528), bk. 14, p. 409.

30. Vesalius, *Fabrica* 5.15, pp. 530, 533, 535–36, 537. On the last point, see also 5.17, pp. 540–41.

31. *Ibid.* 5.15, p. 538; cf. Berengario, *Commentaria*, fol. 194r-v.

32. See O'Malley, *Andreas Vesalius*, pp. 201 and 436–37 n.7. Berengario's dissection of a pregnant criminal, described in *Commentaria*, fol. 191v and 222v, was highly unusual.

33. Vesalius, *Fabrica* 5.17, p. 540.

34. *Ibid.*, figure 30 of book 5 (Vesalius' numeration); see O'Malley, *Andreas Vesalius*, p. 174.

35. On Vesalius' career after the publication of the *Fabrica*, see O'Malley, *Andreas Vesalius*, chs. 11–12. On the differences between the first and second editions, see *ibid.*, pp. 273–79; Nancy G. Siraisi, "Vesalius and Human Diversity in *De humani corporis fabrica*," *Journal of the Warburg and Courtauld Institutes* 57 (1994), pp. 60–88; and Nutton, "Introduction," p. 5.

36. O'Malley, *Andreas Vesalius*, p. 173.

37. For other examples, including Berengario's title page woodcut, see Carlino, *Books of the Body*, ch. 1.

38. See Jonathan Sawday, *The Body Emblazoned: Dissection and the Human Body in Renaissance Culture* (London: Routledge, 1995), ch. 7, esp. pp. 70–72.

39. For polemical effect, Vesalius substantially understated Galen's engagement with human bodies, a point his critics and Galen's defenders were quick to point out; see Vivian Nutton, "André Vésale et l'anatomie parisienne," *Cahiers de l'Association Internationale des Etudes Françaises* 55 (2003), pp. 240–41.

40. Vesalius, *Fabrica* 5.15, p. 532.

41. Jacobus de Voragine, *The Golden Legend: Readings on the Saints*, trans. William Granger Ryan (Princeton, NJ: Princeton University Press, 1993), vol. 1, pp. 142–43. On the earlier history of the motif of the inscribed heart, see Eric Jager, *The Book of the Heart* (Chicago: University of Chicago Press, 2000); Marie-Anne Polo de Beaulieu, "La légende du coeur inscrit dans la littérature religieuse et didactique," in *Le "Cuer" au Moyen Age: Réalité et Senefiance = Senefiance* 30 (1991), pp. 299–312; and Chapter One above. On Botticelli's painting of the scene, see Ronald Lightbown, *Sandro Botticelli* (London: Paul Elek, 1978), vol. 2, pp. 67–69.

42. *Fior di virtù historiale* 8 (Florence: Jacopo di Carlo?, 1491); translated in *The Florentine Fior di virtù of 1491*, trans. Nicholas Fersen (Washington, D.C.: Library of Congress, 1953), n.p.

43. On this pictorial tradition in fifteenth- and sixteenth-century Italy, see Gerhard Wolf-Heidegger and Anna Maria Cetto, *Die anatomische Sektion in bildlicher Darstellung* (Basel: Karger, 1967), pp. 142–47 and 187–88, and Conrad de Mandach, *Saint Antoine de Padoue et l'art italien* (Paris: Laurens, 1899), pp. 224–30 and 289–94.

44. On the making of the altarpiece, see H.W. Janson, *The Sculpture of Donatello* (Princeton, NJ: Princeton University Press, 1963), pp. 162–87, and John White, "Donatello," in Giovanni Lorenzoni (ed.), *Le sculture del Santo di Padova* (Vicenza: Pozza, 1984), pp. 51–94. For a recent discussion, which focuses on the altar in relation to its contemporary viewers, see Geraldine A. Johnson, "Approaching the Altar: Donatello's Sculpture in the Santo," *Renaissance Quarterly* 52 (1999), pp. 626–66.

45. On the local cult of Saint Anthony, see Bernardino Bordin, "La devozione popolare a S. Antonio di Padova: Documenti e testimonianze," in Antonino Poppi (ed.), *Liturgia, pietà et ministeri al Santo* (Verona: Pozza, 1978), vol. 2, esp. pp. 108–14.

46. Sicco Ricci Polentone, *Vita* 1.35; edited in Vergilio Gamboso, "La *Sancti Antonii confessoris de Padua vita* di Sicco Ricci Polentone (ca. 1435)," *Il Santo* 11 (1971), p. 240. On Polentone and his sources, see *ibid.*, pp. 199–221 and 240–41 n.35. Cf. Luke 12:34.

47. For additional images of this sort, see Carlino, *Books of the Body*, ch. 1.

48. Michele Savonarola, a graduate of the medical faculty at Padua and then professor of medicine at the University of Ferrara, might have facilitated Donatello's attendance. (It was Savonarola who pressed Polentone to write Anthony's life in the first place.) On Florentine artists' interest in anatomy, see Bernard Schultz, *Art and Anatomy in Renaissance Italy* (Ann Arbor, MI: UMI Research Press, 1985), chs. 2–3, and Katharine Park, "Masaccio's Skeleton: Art and Anatomy in Early Renaissance Italy," in Rona Goffen (ed.), *Masaccio's Trinity* (Cambridge: Cambridge University Press, 1998), pp. 132–35.

49. Parallels between the two images have been discussed by Walter Artelt, "Das Titelbild zur 'Fabrica' Vesals und seine kunstgeschichtlichen Voraussetzungen," *Centaurus* 1 (1950), pp. 70–73, and Andrea Carlino, "Marsia, Sant'Antonio

ed altri indizi: Il corpo punito e la dissezione tra Quattro e Cinquecento," in Jean Céard, Marie-Madeleine Fontaine, and Jean-Claude Margolin (eds.), *Le corps à la Renaissance* (Paris: Aux Amateurs de Livres, 1990), pp. 129–38.

50. This figure has been interpreted in many ways — perhaps most convincingly as a figure of surface anatomy (in counterpoint to the skeleton in the center of the scene).

51. See Vesalius, preface to *Fabrica*, sig. 2r–3r. The best treatment of this theme is Carlino, *Books of the Body*, pp. 39–53.

52. See the references in n.9; Park, "Criminal and Saintly Body"; and Mitchell B. Merback, *The Thief, the Cross, and the Wheel: Pain and the Spectacle of Punishment in Medieval and Renaissance Europe* (Chicago: University of Chicago Press, 1999), which has a northern European focus.

53. See Carlino, "Marsia," pp. 134–36.

54. See Merback, *Thief, Cross, and Wheel*, pp. 272–78, and Edgerton, *Pictures and Punishment*, pp. 152–54.

55. See Amy Neff, "The Pain of *Compassio*: Mary's Labor at the Foot of the Cross," *Art Bulletin* 80 (1998), pp. 254–73; Otto Von Simpson, "*Compassio* and *Co-Redemptio* in Roger van der Weyden's *Descent from the Cross*," *Art Bulletin* 35 (1953), pp. 9–16; and Chapter One above.

56. See Park, "Criminal and Saintly Body," pp. 22–29.

57. See Park, "Masaccio's Skeleton," pp. 119–20; Loris Premuda, *Storia dell' iconografia anatomica* (Milan: Martello, 1957), for example, figures 2–3. The skeleton gained currency as a personification of death (as opposed to a representation of a dead person) only after the middle of the fifteenth century; see Alberto Tenenti, *Il senso della morte e l'amore della vita nel Rinascimento (Francia et Italia)* (Turin: Einaudi, 1957), p. 438.

58. See in general Leo Steinberg, *The Sexuality of Christ in Renaissance Art and in Modern Oblivion*, rev. ed. (Chicago: University of Chicago Press, 1996).

59. Carlino, "Marsia," p. 129.

60. See Edgerton, *Pictures and Punishment*, pp. 213–19, and William S. Heckscher, *Rembrandt's Anatomy of Dr. Nicolaas Tulp, an Iconological Study* (New York: New York University Press, 1958), pp. 117–21.

61. See Marie Tanner, *The Last Descendant of Aeneas: The Hapsburgs and the Mythic Image of the Emperor* (New Haven: Yale University Press, 1993), esp. pp. 67–69 and chs. 7–8.

62. Many of the manuscript and woodcut images of the death of Agrippina are reproduced in Wolf-Heidegger and Cetto, *Die anatomische Sektion*, figures 9/I–28a; explanatory text on pp. 231–42. See also F.W. Bourdillon, *The Early Editions of the Roman de la rose* (London: Bibliographical Society at the Chiswick Press, 1906). On images of Caesarean birth from the fourteenth through the sixteenth centuries, see Louis-Fernand Flutre, "La naissance de César," *Aesculape* 24 (1934), pp. 244–50; Friedrich von Zglinicki, *Geburt: Eine Kulturgeschichte in Bildern* (Braunschweig: Westermann, 1983), ch. 5; and Renate Blumenfeld-Kosinski, *Not of Woman Born: Representations of Caesarean Birth in Medieval and Renaissance Culture* (Ithaca, NY: 1990), ch. 2.

63. See Wolf-Heidegger and Cetto, *Die anatomische Sektion*, pp. 138–39, and, in general, Stephen K. Wright, *The Vengeance of Our Lord: Medieval Dramatizations of the Destruction of Jerusalem* (Toronto: Pontifical Institute of Mediaeval Studies, 1989), esp. pp. 186–88. The source in this case seems to have been the French translation of Boccaccio's *On the Fates of Famous Men* (see Chapter Three above).

64. Margaret D. Carroll, "The Erotics of Absolutism: Rubens and the Mystification of Sexual Violence," *Representations* 25 (1989), pp. 3–30. See also Jonathan Goldberg, "Fatherly Authority: The Politics of Stuart Family Images," in Margaret W. Ferguson, Maureen Quilligan, and Nancy J. Vickers (eds.), *Rewriting the Renaissance: The Discourses of Sexual Difference in Early Modern Europe* (Chicago: University of Chicago Press, 1986), pp. 3–32. This ideology of rape should be differentiated from its modern counterpart, in that it focused on the forcible abduction of women for purposes of procreation.

65. See Tanner, *Last Descendant of Aeneas*, ch. 7.

66. See Rona Goffen, *Titian's Women* (New Haven: Yale University Press, 1997), pp. 214–43.

67. Girolamo Cardano, *Encomium Neronis*, in *Opera omnia...*, ed. Charles Spon (Lyon: Huguetan and Ravaud, 1663), vol. 1, pp. 179–22. See the analyses of

this work in Alfonso Ingegno, "La storia come teatro della violenza: l' *Encomium Neronis*," *Saggio sulla filosofia di Cardano* (Florence: Nuova Italia, 1980), pp. 184–208, and Nancy G. Siraisi, "Girolamo Cardano and the Art of Medical Narrative," *Journal of the History of Ideas* 52 (1991), pp. 583–84. On Cardano's relationship to Vesalius, see Nancy G. Siraisi, *The Clock and the Mirror: Girolamo Cardano and Renaissance Medicine* (Princeton, NJ: Princeton University Press, 1997), passim.

68. See Ingegno, "La storia come teatro di violenza," p. 195. On the themes of gendered violence in Machiavelli, see Hanna Fenichel Pitkin, *Fortune Is a Woman: Gender and Politics in the Thought of Niccolò Machiavelli* (Berkeley: University of California Press, 1984).

69. Cardano, *Encomium*, p. 199. On the longstanding debate concerning the relative heinousness of matricide and patricide in the context of theories of generation, see Jane Bestor, "Ideas about Procreation and Their Influence on Ancient and Medieval Views of Kinship," in David I. Kertzer and Richard P. Saller (eds.), *The Family in Italy from Antiquity to the Present* (New Haven: Yale University Press, 1991), pp. 162–65.

70. Vesalius, preface to *Fabrica*, fol. 4r; translated by Garrison and Hast. On Vesalius' family background and career aspirations, see O'Malley, *Andreas Vesalius*, ch. 1 and pp. 187–91.

71. The relationship between the title page and images of Nero's opening of Agrippina was first pointed out by Sander L. Gilman, *Sexuality: An Illustrated History; Representing the Sexual in Medicine and Culture from the Middle Ages to the Age of AIDS* (New York: Wiley, 1989), p. 71, and Jerome J. Bylebyl, "Interpreting the *Fasciculo* Anatomy Scene," *Journal of the History of Medicine and Allied Sciences* 45 (1990), pp. 303–304. Girolamo Fabrici, who taught anatomy at Padua in the late sixteenth and early seventeenth centuries, invoked Nero's anatomy of Agrippina at the beginning of his treatise *The Formed Fetus* (1604): "It is the first beginnings of human life that I am setting before you.... Could one tell or invent a tale more magnificent, more mysterious, or more wonderful than this? They say that the emperor Nero himself (fascinated perhaps by the wonder of this theme) wished to look into the corpse of his dead mother and

gaze upon that first domicile of man, from which he himself had issued." See Girolamo Fabrici of Aquapendente, *De formato foetu*, translated in *The Embryological Treatises of Hieronymus Fabricius of Aquapendente*, ed. and trans. Howard B. Adelmann (Ithaca, NY: Cornell University Press, 1942), vol. 1, p. 237.

72. See O'Malley, *Andreas Vesalius*, p. 143.

73. See Mario Biagioli, "Galileo the Emblem Maker," *Isis* 81 (1990), pp. 230–58; Volker R. Remmert, "In the Sign of Galileo: Pictorial Representation in the Seventeenth-Century Copernican Debate," *Endeavour* 27 (2003), pp. 26–31, and the literature cited therein.

74. For an introduction to the theory and practice of early modern emblematics, which is the focus of an enormous body of scholarly literature, see Giancarlo Innocenti, *L'immagine significante: Studio sull'emblematica cinquecentesca* (Padua: Liviana, 1981), and Daniel S. Russell, *Emblematic Structures in Renaissance French Culture* (Toronto: University of Toronto Press, 1995), esp. ch. 5 (on emblems in court culture). On emblematic title pages, see Marion Kintzinger, *Chronos und Historia: Studien zur Titelblattikonigraphie historiographischer Werke vom 16. bis zum 18. Jahrhundert* (Wiesbaden: Harrassowitz, 1995); Margery Corbett and Ronald Lightbown, *The Comely Frontispiece: The Emblematic Title-Page in England, 1550–1660* (London: Routledge and Kegan Paul, 1979); and Karl Josef Höltgen, *Aspects of the Emblem: Studies in the English Emblem Tradition and the European Context* (Kassel: Reichenberger, 1986), ch. 3. On scientific title pages in particular, see Volker R. Remmert, *Widmung, Welterklärung und Wissenschaftslegitimierung: Titelbilder und ihre Funktionen in der Wissenschaftlichen Revolution* (Wolfenbüttel: Herzog August Bibliothek, 2006); and Roberto Paolo Ciardi and Lucia Tongiorgi Tomasi, "La scienza 'illustrata': Osservazioni sui frontespizi delle opere di Athanasius Kircher e di Galileo Galilei," *Annali dell'Istituto Storico Italo-Germanico in Trento* 11 (1985), p. 72.

75. On "Caesarean" births in classical historiography and mythography, see Danielle Gourevitch, "Chirurgie obstétricale dans le monde romain: Césarienne et embryotomie," in Véronique Dasen (ed.), *Naissance et petite enfance dans l'Antiquité* (Fribourg: Academic Press, 2004), pp. 240–44. I am grateful to Jessica Rosenberg for pointing out the title page's "motto."

76. Ovid, *Metamorphoses* 2.606–32.

77. One such birth plate is reproduced in Eugen Holländer, *Plastik und Medizin* (Stuttgart: Enke, 1912), p. 15. These plates were the descendants of the wooden birth trays I described in Chapter Three; see Jacqueline Marie Musacchio, *The Art and Ritual of Childbirth in Renaissance Italy* (New Haven: Yale University Press, 1999), ch. 4.

78. Fracastoro's medal, possibly by Girolamo della Torre's son Giulio, is described in George Francis Hill, *A Corpus of Italian Medals from the Renaissance before Cellini* (London: British Museum, 1930), vol. 1, no. 585, and reproduced in Fritz Saxl, "Pagan Sacrifice in the Italian Renaissance," *Journal of the Warburg and Courtauld Institutes* 2 (1939), p. 601.

79. On this tomb and its reliefs, see Leo Planiscig, *Andrea Riccio* (Vienna: Schroll and Co., 1927), pp. 371–400; Saxl, "Pagan Sacrifice," pp. 355–59; and John Pope-Hennessy, *An Introduction to Italian Sculpture*, 4th ed. (London: Phaidon, 1996), vol. 2, *Italian Renaissance Sculpture*, pp. 301–305 and 415–16. On the Paduan careers of Girolamo and Marcantonio della Torre, see Tiziana Pesenti, *Professori e promotori di medicina nello Studio di Padova dal 1405 al 1509: Repertorio bio-bibliografico* (Padua: LINT, 1984), pp. 112–14.

80. Ovid, *Metamorphoses* 2.642. This formulation also refers to the mythological status of Hygeia (Health) as the daughter of Asclepius.

81. Vesalius, preface to *Fabrica*, sig. 4r and 2v. Unless otherwise indicated, all quotations from the preface come from the online translation by Garrison and Hast. For the Hapsburgs' use of Apolline imagery, see Tanner, *Last Descendant of Aeneas*, ch. 12.

82. Vesalius, preface to *Fabrica*, sig. 3v.

83. *Ibid.*, sig. 2r.

84. *Ibid.*, sig. 3r.

85. Tanner, *Last Descendant of Aeneas*, pp. 126–34.

86. Vesalius, preface to *Fabrica*, sig. 3r.

87. *Ibid.*, sig. 3v.

88. Sawday, *Body Emblazoned*, p. 219.

89. See, in general, Sawday, *Body Emblazoned*, esp. chs. 4 and 7.

90. See also Merback, *Thief, Cross, and Wheel*, pp. 272–76, where he argues that Mantegna also straddles this divide.

91. See, for example, Robert Scribner, "Ways of Seeing in the Age of Dürer," in Dagmar Eichberger and Charles Zika (eds.), *Dürer and His Culture* (Cambridge: Cambridge University Press, 1998), esp. pp. 94–97; Suzannah Biernoff, *Sight and Embodiment in the Middle Ages* (Houndmills: Palgrave Macmillan, 2002); Hans Belting, *The Image and its Public in the Middle Ages: Form and Function of Early Paintings of the Passion*, trans. Mark Bartusis and Raymond Meyer (1981; New Rochelle, NY: Caratzas, 1990); and Neff, "Pain of *Compassio*."

92. See images in Edgerton, *Pictures and Punishment*, ch. 5.

93. O'Malley, *Andreas Vesalius*, pp. 148–49; cf. Celsus, *De re medicinae* 3.4.1.

94. Berengario, *Commentaria*, fol. 519v.

95. Vesalius, *Fabrica* 2.43, p. 304.

96. Bette Talvacchia, *Taking Positions: On the Erotic in Renaissance Culture* (Princeton, NJ: Princeton University Press, 1999), esp. ch. 8.

97. For other early examples, see Linda Hentschel, *Pornotopische Techniken des Betrachtens: Raumwahrnehmung und Geschlechterordnung in visuellen Apparaten der Moderne* (Marburg: Jonas, 2001), pp. 28–34 and, in general, ch. 1.

98. Frances A. Yates, *The Art of Memory* (London: Routledge and Kegan Paul, 1966); Karen Rosoff Encarnación, "The Proper Uses of Desire: Sex and Procreation in Reformation Anatomical Fugitive Sheets," in Anne L. McClanan and Karen Rosoff Encarnación (eds.), *The Material Culture of Sex, Precreation, and Marriage in Premodern Europe* (New York: Palgrave, 2001), pp. 221–49.

99. On the coexistence of similarity and difference in medieval and early modern theories of sex difference, see Katharine Park and Robert A. Nye, "Destiny is Anatomy" (essay-review of Thomas Laqueur, *Making Sex*), *New Republic*, 18 February 1991, pp. 53–57.

100. Vesalius, *Fabrica* 5.15, p. 539.

101. See also Musacchio, *Art and Ritual*, figures 101 and 105.

102. Monica H. Green, *Making Women's Medicine Masculine: The Rise of Male Authority in Pre-Modern Gynecology* (Oxford: Oxford University Press, 2007). Mary E. Fissell has explored the later development of some of these issues in

the English context, see Mary E. Fissell, *Vernacular Bodies: The Politics of Repro-duction in Early Modern England* (Oxford: Oxford University Press, 2004), esp. ch. 5.

103. For critiques of this myth, see Monica H. Green, "Bodies, Gender, Health, Disease: Recent Work on Medieval Women's Medicine," *Studies in Medieval and Renaissance History*, ser. 3, vol. 2 (2005), p. 17, and David Harley, "Historians as Demonologists: The Myth of the Midwife-Witch," *Social History of Medicine* 3 (1990), pp. 1–26.

104. Green, "Bodies, Gender, Health, Disease," pp. 15–16.

105. Green, *Making Women's Medicine Masculine*, conclusion.

106. Cf. Laurel Thatcher Ulrich, *A Midwife's Tale: The Life of Martha Bal-lard, Based on Her Diary, 1785–1812* (New York: Knopf, 1990).

107. See Chapter Two n. 30.

EPILOGUE

1. For good introductions to this way of understanding the body, see, for example, Gianna Pomata, *Contracting a Cure: Patients, Healers, and the Law in Early Modern Bologna*, trans. Gianna Pomata (Baltimore: Johns Hopkins Univer-sity Press, 1998), ch. 5; Gail Kern Paster, *The Body Embarrassed: Drama and the Disciplines of Shame in Early Modern England* (Ithaca, NY: Cornell University Press, 1993); and Michael C. Schoenfeldt, *Bodies and Selves in Early Modern England: Physiology and Inwardness in Spenser, Shakespeare, Herbert, and Milton* (Cambridge: Cambridge University Press, 1999), esp. ch. 1. For its ancient roots, see Shigehisa Kuriyama, *The Expressiveness of the Body and the Divergence of Greek and Chinese Medicine* (New York: Zone Books, 1999), ch. 5.

2. On the mining of the *Fabrica*'s text and images by a wide range of ver-nacular writers, see, for example, Andrea Carlino, *Paper Bodies: A Catalogue of Anatomical Fugitive Sheets, 1538–1687*, trans. Noga Arikha (London: Wellcome Institute for the History of Medicine, 1999), ch. 2 and the references therein.

3. Jonathan Sawday, *The Body Emblazoned: Dissection and the Human Body in Renaissance Culture* (London: Routledge, 1995), p. 7. Kuriyama discusses the importance of anatomical visualization in Western self-understanding in

Expressiveness of the Body, ch. 3, emphasizing the muscles rather than the viscera.

4. See John Jeffries Martin, *Myths of Renaissance Individualism* (New York: Palgrave Macmillan, 2004), esp. ch. 6, and Fernando Vidal, "Brains, Bodies, Selves, and Science: Anthropologies of Identity and the Resurrection of the Body," *Critical Inquiry* 28, no. 4 (2002), esp. pp. 937–39 and 966–69.

5. Juan Valverde de Amusco, *Historia de la composición del cuerpo humano* (Rome: Antonio Salamanca and Antonio Lafreri, 1556), which was published in Italian by the same press in 1560; later editions also appeared in Dutch, Latin, and Japanese. On the sixteenth-century trope of anatomy as self-knowledge, see Carlino, *Paper Bodies*, ch. 4 and pp. 333–34. On Valverde, see Andrea Carlino, "Tre piste per l'anatomia de Juan de Valverde: Logiche d'edizione, solidarietà nazionali e cultura artistica a Roma nel Rinascimento," *Mélanges de l'Ecole Française de Rome* 114 (2002), pp. 513–41, and Cynthia Klestinec, "Juan Valverde de (H)Amusco and Print Culture: The Editorial Apparatus in Vernacular Anatomy Texts," *Zeitsprünge* 9 (2005), pp. 101–17.

Bibliography

Manuscript sources

Bologna, Biblioteca Communale dell'Archiginnasio di Bologna (BCAB)

B4314	[*Leggenda anonima di Elena Dugioli*]	sixteenth century
Gozzadini 292	Pietro Ritta of Lucca, *Narrativa della vita e morte della Beata Elena Duglioli dall'Oglio...*, and Letter to Pope Leo X (12.x.1520); anon., *Del cuore di Helena vergine dal signore realmento tolto con molte cose circa quello sopra ogni natura accadute...*	eighteenth century

Florence, Archivio di Stato di Firenze (ASF)

Carte Strozziane

2, 16bis	*Ricordanza* of Tommaso di Francesco Giovanni	1444–58
2, 22	*Ricordanza* of Cambio di Manno Petrucci	1492–1520
3, 108, fol. 77r–88v	Letters of Filippo di Filippo Strozzi	i–ii.1525/6
5, 22	*Ricordanza* of Filippo di Matteo Strozzi	1471–83

Corporazioni Religiose Soppresse

95, 212	*Ricordanza* of Bernardo di Stoldo Rinieri	1457–1503

Deliberazioni dei Signori e Collegi, ordinaria autorità

42	1433

359

Gheri Goro Copialettere

	4	Letters of Goro Gheri	1517–19

Mediceo Avanti il Principato

35, 746	Giovanni Tornabuoni to Lorenzo de' Medici	24.i.1477
30, 274	Clarice Orsini to Lorenzo de' Medici	7.ix.1478
31, 283	Clarice to Lorenzo	8.ix.1478
31, 282	Maestro Stefano della Torre to Lorenzo	7.ix.1478
31, 284	Maestro Stefano to Lorenzo	8.ix.1478

Florence, Biblioteca Laurenziana (BLF)

Plut. 73, 51, fol. 52r–56v	*Le segrete cose delle donne*	fourteenth century
Redi 172, vol. 1, fol. 73r–82v	*Le segrete cose delle donne*	fourteenth century

Florence, Biblioteca Nazionale Centrale di Firenze (BNCF)

Fondo Nazionale

2, 1, 363, fol. 63v–75v	Brief history of the Roman emperors	late fifteenth century
2, 2, 357	*Ricordanza* of Tribaldo d'Amerigo Rossi	1481–1501
2, 4, 280, fol. 79r–90r	Brief history of the Roman emperors	1455

Fondo Palatino

557, pp. 214–35	*I segreti delle femine*	early fifteenth century
471, fol. 71v–82r	Brief history of the Roman emperors	1479

Florence, Biblioteca Riccardiana (BRF)

2350, fol. 79r–88r	*I segreti delle femine*	fifteenth century
2165, fol. 70r–76r	*Le segrete cose delle donne*	1433
2228, fol. 71r–74r	*I secreti come sta el corpo della femmina*	fifteenth century
2500, fol. 60r–64r	*Le segrete cose delle donne*	fifteenth century
2175, fol. 44r–45v	*Le segrete cose delle donne*	second half of fifteenth century

London, Wellcome Library for the History and Understanding of Medicine

5265	Letter from the medical faculty of the	c. 1464–1465
	University of Pavia to Francesco Sforza,	
	Duke of Milan	

Published sources

For the sake of consistency, authors are listed in alphabetical order of their family names, if such are known, or their first names, if not. A date in brackets after the title of a secondary work indicates that it has been republished or translated from its original language of publication; the bracketed date is the original date of publication.

Primary sources

Acta sanctorum quotquot toto orbe coluntur. Antwerp: Michael Cnobarus, 1665; repr. Brussels: Culture et Civilisation, 1965–1970.

Alberti, Leon Battista. *The Family in Renaissance Florence.* Trans. Renée Neu Watkins. Columbia: University of South Carolina Press, 1969.

———. *I libri della famiglia.* Ed. Ruggiero Romano and Alberto Tenenti, new ed. by Francesco Furlan. Turin: Einaudi, 1994.

Albertus Magnus. *Book of Minerals.* Trans. Dorothy Wyckoff. Oxford: Clarendon, 1967.

Pseudo-Albertus Magnus. *De secretis mulierum et virorum.* Leipzig: Melchior Lotter, 1505.

———. *Women's Secrets: A Translation of Pseudo-Albertus Magnus's De secretis mulierum with Commentaries.* Trans. Helen Rodnite Lemay. Albany: State University of New York Press, 1992.

Alderotti, Taddeo. *Expositio in Aphorismorum Ipocratis volumen. Expositiones in arduum Aphorismorum Ipocratis volumen, in divinum Pronosticorum Ipocratis librum.* Venice: Luca Antonio Giunta, 1527.

———. *Expositiones in arduum Aphorismorum Ipocratis volumen, in divinum Pronosticorum Ipocratis librum, in preclarum Regiminis acutorum Ipocratis opus, in subtilissimum Joannitii Isagogarum libellum.* Venice: Luca Antonio Giunta, 1527.

Ambrose. *Corpus scriptorum ecclesiasticorum latinorum*, vol. 32, *Sancti Ambrosii opera*. Ed. Karl Schenkl. Vienna: Tempsky, 1897.

Anatomies de Mondino de' Luzzi et de Guido da Vigevano. Ed. Ernest Wicker-sheimer. Paris: Droz, 1926.

Avicenna. *Liber canonis* [1507]. Repr. Hildesheim: Olms, 1964.

Benedetti, Alessandro. *Historia corporis humani, sive Anatomice*. Ed. and trans. Giovanna Ferrari. Florence: Giunti, 1998.

———. *Opera omnia*. Basel: Henricus Petrus, 1549.

Benivieni, Antonio. *De abditis nonnullis ac mirandis morborum et sanationum causis*. Florence: Filippo Giunta, 1507.

———. *De abditis nonnullis ac mirandis morborum et sanationum causis*. Ed. and trans. Giorgio Weber. Florence: Olschki, 1994.

———. *De abditis nonnullis ac mirandis morborum et sanationum causis*. Trans. Charles Singer. Springfield, IL: Thomas, 1954.

———. *De regimine sanitatis ad Laurentium Medicem*. Ed. Luigi Belloni. Turin: Società Italiana di Patologia, 1951.

Béranger of Saint-Affrique (Berengario di Sant'Africano). *Life of Saint Clare of Montefalco*. Trans. Matthew J. O'Connell. Ed. John E. Rotelle. Villanova, PA: Augustinian Press, 1999.

———. *Vita Sanctae Clarae de Cruce: Ordinis Eremitarum S. Augustini*. Ed. Alfonso Semenza. *Analecta Augustiniana* 17–18 (1939–41): 87–102, 169–76, 287–99, 393–409, 445–57, 513–17.

Berengario da Carpi, Jacopo. *Carpi commentaria cum amplissimis additionibus super Anatomia Mundini…* Bologna: Girolamo de' Benedetti, 1521.

———. *Isagogae breves perlucidae ac uberimae in anatomiam humani corporis*. Bologna: Girolamo de' Benedetti, 1522.

———. *A Short Introduction to Anatomy*. Trans. L.R. Lind. Chicago: University of Chicago Press, 1959.

Bevegnati, Giunta. *Legenda de vita et miraculis Beatae Margaritae de Cortona*. Ed. Fortunato Iozzelli. Rome: Collegii S. Bonaventurae ad Claras Aquas, 1997.

Boccadiferro, Ludovico. *Lectiones in quartum Meteororum Aristotelis librum*. Venice: Francesco de' Franceschi da Siena, 1563.

Bonacciolo, Ludovico. *Enneas muliebris.* Ferrara: L. de Rubeis, 1502

Bontempi of Perugia, Sebastiano. "Vita beatae Colombae Reatinae." In *Acta sanctorum quotquot toto orbe coluntur,* vol. 5, *Maii.* Antwerp: Michael Cnobarus, 1665; repr. Brussels: Culture et Civilisation, 1968, 320*–98*.

Bruno da Longobucco, *Cyrurgia magna.* In Guy of Chauliac, *Cyrurgia Guidonis de Cauliaco et Cyrurgia Bruni, Theodorici, Rolandi, Lanfranci, Rogerii, Bertapalie.* Venice: Bonetus Locatellus, 1498, fol. 83ra–102vb.

Burgundio of Pisa. See Galen.

Cardano, Girolamo. *Encomium Neronis.* In *Opera omnia...* Lyon: Huguetan and Ravaud, 1663, vol. 1, 179–220.

Ce sont les secrés des dames, deffendus à révéler. Ed. Al[exandre] C[olson] and Ch[arles]-Ed[ouard] C[azin]. Paris: Edouard Rouveyre, 1880.

Chronicon parmense ab anno MXXXVIII usque ad annum MCCCXXXVIII. Ed. Giuliano Bonazzi. Città di Castello: Lapi, 1902.

Corner, George (ed. and trans.). *Anatomical Texts of the Earlier Middle Ages: A Study in the Transmission of Culture.* Washington: Carnegie Institution of Washington, 1927.

Estienne, Charles. *De dissectione partium corporis humani.* Paris: Simon de Colines, 1545.

Fabrici, Girolamo, of Aquapendente. *The Embryological Treatises of Hieronymus Fabricius of Aquapendente.* Ed. and trans. Howard B. Adelmann. Ithaca, NY: Cornell University Press, 1942.

Fasiculo de medicina. Attr. (erroneously) to Johannes de Ketham. Ed. and trans. Sebastiano Manilio. Venice: Giovanni and Gregorio de' Gregori, 1494.

Fasiculo de medicina. In Tiziana Pesenti, *Fasiculo de medicina in volgare,* vol. 1. Padua: Università degli Studi, 2001.

The Fasciculo di medicina, Venice 1493. Ed. Charles Singer. Florence: Lier, 1925.

The Fasciculus medicinae of Johannes de Ketham, Alemanus: Facsimile of the First (Venetian) Edition of 1491. Ed. and trans. Charles Singer. Milan: Lier, 1924.

Fasciculus medicine. Venice: Giovanni and Gregorio de' Gregori, 1495.

Ficino, Marsilio. *Supplementum Ficinianum: Marsilii Ficini florentini philosophi platonici opuscula inedita et dispersa.* Ed. Paul Oskar Kristeller. Florence: Olschki, 1937.

Fior di virtù historiale. Florence: Jacopo di Carlo?, 1491.

The Florentine Fior di virtù of 1491. Trans. Nicholas Fersen. Washington, DC: Library of Congress, 1953.

Louis-Fernand Flutre and Cornelius Sneyders de Vogel (eds.). *Li Fet des Romains: Compilé ensemble de Saluste et de Suétoine et de Lucan; Texte du XIIIe siècle.* Paris: Droz, 1938.

Francesco of Barberino. *Reggimento e costumi di donna.* Ed. Giuseppe E. Sanson. Turin: Loescher-Chiantore, 1957.

Francis of Assisi. *Opuscula sancti patris Francisci Assisiensis.* Quaracchi: Collegium S. Bonaventurae, 1904.

Galen. *On the Anatomy of the Uterus.* Trans. Charles Mayo Goss. *Anatomical Record* 144 (1962): 77–83.

———. *Claudii Galeni Pergameni secundum Hippocratem medicorum facile principis opus De usu partium corporis human.* Trans. Niccolò of Reggio. Paris: Simon de Colines, 1528.

———. *Galieni Pergamensis medicorum omnium principis Opera.* Ed. Diomedes Bonardus. Venice: For Filippo Pinzio da Caneto, 1490.

———. *De interioribus.* Trans. Burgundio of Pisa. In Richard J. Durling and Fridolf Kudlien (eds.), *Galenus Latinus,* vol. 1, *Burgundio of Pisa's Translation of Galen's ... De interioribus* Stuttgart: Steiner, 1992.

———. *On the Usefulness of the Parts of the Body.* Trans. Margaret Tallmadge May. Ithaca, NY: Cornell University Press, 1968.

Garbo, Tommaso del. *Expositio super capitulo de generatione embrionis....* In Jacopo of Forlì, *Expositio Jacobi supra capitulum de generatione embrionis cum questionibus eiusdem,* ed. Bassianus Politus. Venice: Bonetus Locatellus, 1502, fol. 33r–45r.

Gherardi, Alessandro (ed.). *Statuti della Università et studio fiorentino dell' anno MCCCLXXXVII, seguiti da un'appendice di documenti dal MCCCXX al MCCCCLXXII.* Florence: Cellini, 1881.

Giordano of Pisa (Giordano of Rivalto). *Prediche del Beato Fra Giordano da*

Rivalto, dell'Ordine dei predicatori recitate in Firenze dal MCCCIII al MCCCVI. Ed. Domenico Moreni. Florence: Magheri, 1831.

Gordon, Bernard de. *Practica, seu Lilium medicinae.* Venice: Giovanni and Gregorio de' Gregori, 1496.

Guaineri, Antonio. *Commentariolus de egritudinibus matrices.* In *Opus praeclarum ad praxim non mediocriter necessarium cum Joannis Falconis nonnullis non inutiliter adjunctis.* Lyon: Constantinus Fradin, 1525, fol. 133v–65v.

Guglielmo of Saliceto. *Summa conservationis et curationis.* Venice: Marinus Saracenus, 1490.

Guido of Vigevano. See *Anatomies di Mondino de' Luzzi et de Guido da Vigevano.*

Guillaume de Lorris and Jean de Meun. *Le roman de la rose.* Ed. Daniel Poirion. Paris: Garnier-Flammarion, 1974.

———. *The Romance of the Rose.* Trans. Charles Dahlberg. Hanover, NH: University Press of New England, 1983.

Guy of Chauliac, *Cyrurgia Guidonis de Cauliaco et Cyrurgia Bruni, Theodorici, Rogerii, Rolandi, Bertapalie, Lanfranci.* Venice: Bonetus Locatellus, 1498.

———. *Inventarium sive chirurgia magna.* Ed. Michael R. McVaugh. Leiden: Brill, 1997.

Heseler, Baldasar. *Vesalius' First Public Anatomy at Bologna, 1540: An Eyewitness Report...* Ed. and trans. Ruben Eriksson. Uppsala: Almqvist and Wiksells, 1959.

Incisa della Rocchetta, Giovanni, and Nello Vian (eds.), with P. Carlo Gasbarri. *Il primo processo per San Filippo Neri nel Codice Vaticano Latino 3798 e in altri esemplari dell'Archivio dell'Oratorio di Roma.* Vatican City: Biblioteca Apostolica Vaticana, 1957–1963.

Jacobus de Voragine. *The Golden Legend: Readings on the Saints.* Trans. William Granger Ryan. Princeton, NJ: Princeton University Press, 1993.

———. *Sermones pulcherrimi variis scripturarum doctrinis referti de sanctis per anni totius circulum concurrentibus.* Paris: n.p., 1510.

Jacopo of Forlì. *Expositio Jacobi supra capitulum de generatione embrionis cum questionibus eiusdem.* Ed. Bassianus Politus. Venice: Bonetus Locatellus, 1502.

Jean de Meun. See Guillaume de Lorris and Jean de Meun.

Keele, Kenneth, and Carlo Pedretti. *Leonardo da Vinci: Corpus of the Anatomical Studies in the Collection of Her Majesty the Queen at Windsor Castle.* London: Johnson Reprint, 1978–1980.

Ketham, Johannes de. See *Fasiculo de medicina*, *The Fasciculus medicinae*.

Lanfranco of Milan. *Practica … que dicitur ars completa totius cyrurgie.* In Guy of Chauliac, *Cyrurgia Guidonis de Cauliaco et Cyrurgia Bruni, Theodorici, Rogerii, Rolandi, Bertapalie, Lanfranci.* Venice: Bonetus Locatellus, 1498, fol. 166va–210vb.

Laurent, M.H. (ed.). "La plus ancienne légende de la B. Marguerite de Città di Castello." *Archivum fratrum praedicatorum* 10 (1940): 109–31.

Leonardo da Vinci. See Keele, Kenneth, and Carlo Pedretti.

Il libro delle segrete cose delle donne. Ed. Giuseppe Manuzzi. Florence: Tipografia del Vocabolario, 1863 [actually, post-1874].

Liuzzi, Mondino de'. *Anothomia di Mondino de' Liuzzi da Bologna, XIV secolo.* Eds. Piero P. Giorgi and Gian Franco Pasini. Bologna: Istituto per la Storia dell'Università di Bologna, 1992.

———. *Expositio super capitulum de generatione embrionis Canonis Avicennae cum quibusdam quaestionibus.* Ed. Romana Martorelli Vico. Rome: Istituto Storico Italiano per il Medio Evo, 1993.

Malagola, Carlo (ed.). *Statuti delle Università e dei collegi dello studio Bolognese.* Bologna: Zanichelli, 1888.

Manfredi, Girolamo. *Anothomia.* Trans. A. Mildred Westland. In Charles Singer, "A Study in Early Renaissance Anatomy, with a New Text: The *Anothomia* of Hieronymo Manfredi (1490)," in Charles Singer, ed., *Studies in the History and Method of Science.* Oxford: Clarendon, 1917, 79–164.

Menestò, Enrico, with Silvestro Nessi (eds.). *Il processo di canonizzazione di Chiara da Montefalco.* Florence: Nuova Italia, 1984.

Mondino de' Liuzzi. See Liuzzi, Mondino de'.

Nadi, Gaspare. *Diario Bolognese.* Eds. Corrado Ricci and A. Bacchi della Lega. Bologna: Romagnoli dall'Acqua, 1886.

Nessi, Silvestro (ed.). *Chiara da Montefalco, badessa del monastero di S. Croce: Le sue testimonianze — i suoi "dicti"*. Montefalco: Associazione dei Quartieri di Montefalco, 1981.

Paolo of Certaldo. *Libro di buoni costumi*. Ed. Alfredo Schiaffini. Florence: Le Monnier, 1945.

Pliny. *Natural History*, Trans. H. Rackham. Cambridge, MA: Harvard University Press, 1938–62.

Polentone, Sicco Ricci. *Sancti Antonii ... vita*. In Vergilio Gamboso, "La *Sancti Antonii confessoris de Padua vita* di Sicco Ricci Polentone (ca. 1435)," *Il Santo* 11 (1971): 199–283.

Poliziano, Angelo. *Prose volgari inedite e poesie latine e greche edite e inedite*. Ed. Isidoro Del Lungo. Florence: Barbèra, 1867.

Pomponazzi, Pietro. *Dubitationes in quartum Meteorologicorum Aristotelis librum*. Venice: Francisco de' Franceschi da Siena, 1563.

———. *De naturalium effectuum causis; sive, De incantationibus*. Basel: Henricus Petrus, 1567; repr. Hildesheim: Olms, 1970.

Problemata varia anatomica: The University of Bologna MS 1165. Ed. L.R. Lind. Lawrence: University of Kansas Publications, 1968.

Raymond of Capua. *The Life of Catherine of Siena*. Trans. Conleth Kearnes. Wilmington, DE: Glazier, 1980.

Salimbene de Adam. *Cronica*. Ed. Giuseppe Scalia. Bari: Laterza, 1966.

Savonarola, Michele. *Il trattato ginecologico-pediatrico in volgare: Ad mulieres ferrarienses de regimine pregnantium et noviter natorum usque ad septennium*. Ed. Luigi Belloni. Milan: Società Italiana di Ostetricia e Ginecologia, 1952.

Scalza, Jacopo. *Leggenda latina della B. Giovanna detta Vanna d'Orvieto*. Ed. Vincenzo Marreddu. Orvieto: Sperandio Pompei, 1853.

Suetonius. *Suetonius Tranquillus [De vita duodecim caesarum]*... Eds. Filippo Beroaldo and Marco Antonio Sabellico. Venice: Giovanni Rossi, 1506.

Teodorico Borgognoni da Lucca. *Cyrurgia*. In Guy of Chauliac, *Cyrurgia Guidonis de Cauliaco et Cyrurgia Bruni, Theodorici, Rogerii, Rolandi, Bertapalie, Lanfranci*. Venice: Bonetus Locatellus, 1498, fol. 106ra–146vb.

Tommaso del Garbo. See Garbo, Tommaso del.

Torni, Bernardo. *Opuscoli filosofici e medici*. Ed. Marina Messina Montelli. Florence: Nuova Italia, 1982.

The Trotula: A Medieval Compendium of Women's Medicine. Ed. and trans. Monica H. Green. Philadelphia: University of Pennsylvania Press, 2001.

Vesalius, Andreas. *De humani corporis fabrica*. Basel: Joannes Oporinus, 1543.

———. *De humani corporis fabrica*. 2nd ed. Basel: Oporinus, 1555.

———. *De humani corporis fabrica*. Trans. Daniel H. Garrison and Malcolm Hast (in preparation). http://vesalius.northwestern.edu.

"Vita B. Margaritae virginis de Civitate Castelli." *Analecta Bollandiana* 19 (1900): 23–36.

Zerbi, Gabriele de'. *Anathomia matricis pregnantis: Et est sermo de anathomia et generatione embrionis*. Separately foliated treatise in *Liber anathomie corporis humani et singulorum membrorum illius*. Venice: Bonetus Locatellus for Octavianus Scotus, 1502.

———. *Liber anathomie corporis humani et singulorum membrorum illius*. Venice: Bonetus Locatellus for Octavianus Scotus, 1502.

Secondary sources

Adelman, Janet. *Suffocating Mothers: Fantasies of Maternal Origin in Shakespeare's Plays, Hamlet to the Tempest*. New York: Routledge, 1992.

Agrimi, Jole, and Chiara Crisciani. *Les consilia médicaux*. Trans. Caroline Viola. Turnhout: Brepols, 1994.

———. *Edocere medicos: Medicina scolastica nei secoli XIII–XV*. Naples: Guerini, 1988.

———. "Immagini e ruoli della 'vetula' tra sapere medico e antropologia religiosa (secoli XIII-XV)." In Agostino Paravicini Bagliani and André Vauchez (eds.), *Poteri carismatici e informali: Chiesa e società medioevali*. Palermo: Sellerio, 1992, 224–61.

———. "Medici e 'vetulae' dal Duecento al Quattrocento: Problemi di una ricerca." In Paolo Rossi, Lucilla Borcelli, Chiaretta Poli, and Giancarlo Carabelli, *Cultura popolare e cultura dotta nel Seicento*. Milan: Angeli, 1983, 144–59.

————. "Per una ricerca su *experimentum-experimenta*: Riflessione epistemologica e tradizione medica (secoli XIII–XV)." In Pietro Janni and Innocenzo Mazzini (eds.), *Presenza del lessico greco e latino nelle lingue contemporanee*. Macerata: Università degli Studi di Macerata, 1990, 9–49.

————. "Savoir médical et anthropologie religieuse: Les représentations et les fonctions de la *vetula* (XIIIe–XVe siècle)," *Annales: Economies, sociétés, civilisations* 48 (1993): 1281–1308.

————. "The Science and Practice of Medicine in the Thirteenth Century According to Guglielmo da Saliceto, Italian Surgeon." In Luis García-Ballester, Roger French, Jon Arrizabalaga, and Andrew Cunningham (eds.), *Practical Medicine from Salerno to the Black Death*. Cambridge: Cambridge University Press, 1994, 60–87.

Alston, Mary Niven. "The Attitude of the Church Towards Dissection before 1500." *Bulletin of the History of Medicine* 16 (1944): 221–38.

Ariès, Philippe. *The Hour of Our Death* [1977]. Trans. Helen Weaver. New York: Oxford University Press, 1981.

Artelt, Walter. "Die ältesten Nachrichten über die Sektion menschlicher Leichen im mittelalterlichen Abenland," *Abhandlungen zur Geschichte der Medizin und der Naturwissenschaften* 34 (1949): 3–25.

————. "Das Titelbild zur 'Fabrica' Vesals und seine kunstgeschichtlichen Voraussetzungen." *Centaurus* 1 (1950): 66–77.

Azzolini, Monica. "Leonardo da Vinci's Anatomical Studies in Milan: A Re-examination of Sites and Sources." In Jean A. Givens, Karen M. Reeds, and Alain Touwaide (eds.), *Visualizing Medieval Medicine and Natural History, 1200–1550*. Burlington, VT: Ashgate, 2006, 147–76.

————. "Leonardo in Context: Medical Ideas and Practices in Renaissance Milan." Ph.D. dissertation, University of Cambridge, 2001.

Baskins, Cristelle. *Cassone Painting, Humanism, and Gender in Early Modern Italy*. Cambridge: Cambridge University Press, 1998.

Bedos-Rezak, Brigitte Miriam. "Medieval Identity: A Sign and a Concept." *American Historical Review* 105 (2000): 1489–1533.

Belloni, Luigi. "Gli schemi anatomici trecenteschi (serie dei cinque sistemi e occhio) del Codice Trivulziano 836." *Rivista di storia delle scienze mediche e naturali* 41 (1950): 193–207.

Belting, Hans. *The Image and Its Public in the Middle Ages: Form and Function of Early Paintings of the Passion* [1981]. Trans. Mark Bartusis and Raymond Meyer. New Rochelle, NY: Caratzas, 1990.

———. *Likeness and Presence: A History of the Image before the Era of Art* [1990]. Trans. Edmund Jephcott. Chicago: University of Chicago Press, 1994.

Beltrami, Luca. "L'annullamento del contratto di matrimonio fra Galeazzo M. Sforza e Dorotea Gonzaga (1463)." *Archivio storico lombardo* ser. 2, 6 (1889): 126–32.

Benvenuti Papi, Anna. "Mendicant Friars and Female Pinzochere in Tuscany: From Social Marginality to Models of Sanctity" [1983]. In Daniel Bornstein and Roberto Rusconi (eds.), *Women and Religion in Medieval and Renaissance Italy*, trans. Margery J. Schneider. Chicago: University of Chicago Press, 1996, 84–103.

Bertagna, Martino. *L'Osservanza di Siena: Studi storici*. Siena: Osservanza, 1964.

Bestor, Jane Fair. "Ideas about Procreation and Their Influence on Ancient and Medieval Views of Kinship." In David I. Kertzer and Richard P. Saller (eds.), *The Family in Italy from Antiquity to the Present*. New Haven: Yale University Press, 1991, 150–67.

Biagioli, Mario. "Galileo the Emblem Maker." *Isis* 81 (1990): 230–58.

Biernoff, Suzannah. *Sight and Embodiment in the Middle Ages*. New York: Palgrave Macmillan, 2002.

Bliquez, Lawrence J., and Alexander Kazhdan. "Four Testimonia to Human Dissection in Byzantine Times." *Bulletin of the History of Medicine* 58 (1984): 554–57.

Blume, Dieter. "Ordenskonkurrenz und Bildpolitik: Franziskanische Programme nach dem theoretischen Armutsstreit." In Hans Belting and Dieter Blume (eds.), *Malerei und Stadtkultur in der Dantezeit*. Munich: Hirmer, 1989, 149–71.

Blumenfeld-Kosinski, Renate. *Not of Woman Born: Representations of Caesarean Birth in Medieval and Renaissance Culture*. Ithaca, NY: Cornell University Press, 1990.

Bollone, Pierluigi Baima. "Autopsia e conservazione del cuore: 'Un evento unico.'" In Giuseppe Zois (ed.), *S. Chiara da Montefalco: Dove ci porta il cuore*, n.p.: Ritter, 1995, 119–23.

Bordin, Bernardino. "La devozione popolare a S. Antonio di Padova: Documenti e testimonianze." In Antonino Poppi (ed.), *Liturgia, pietà et ministeri al Santo*. Verona: Pozza, 1978, vol. 2, 88–215.

Bornstein, Daniel. "The Uses of the Body: The Church and the Cult of Santa Margherita of Cortona," *Church History* 62 (1993): 163–77.

Bornstein, Daniel, and Roberto Rusconi (eds). *Mistiche e devote nell'Italia tardo-medievale*. Naples: Liguori, 1992.

———. *Women and Religion in Medieval and Renaissance Italy*. Trans. Margery J. Schneider. Chicago: University of Chicago Press, 1996.

Bourdillon, F.W. *The Early Editions of The Roman de la rose*. London: Bibliographical Society of the Chiswick Press, 1906.

Bouvier, Michel. "De l'incorruptibilité des corps saints." In Jacques Gélis and Odile Redon (eds.), *Les miracles, miroirs des corps*. Paris: Université de Paris VIII–Vincennes, 1983, 193–221.

Bronfen, Elisabeth. *Over Her Dead Body: Death, Femininity and the Aesthetic*. New York: Routledge, 1992.

Brown, Elizabeth A.R. "Death and the Human Body in the Later Middle Ages: The Legislation of Boniface VIII on the Division of the Corpse." *Viator* 12 (1981): 221–70.

Brown, Judith C. *Immodest Acts: The Life of a Lesbian Nun in Renaissance Italy*. New York: Oxford University Press, 1986.

Brown, Peter. *The Cult of the Saints: Its Rise and Function in Latin Christianity*. Chicago: University of Chicago Press, 1981.

Browning, Robert. "A Further Testimony to Human Dissection in the Byzantine World." *Bulletin of the History of Medicine* 59 (1985): 518–20.

Burckhardt, Jacob. *The Civilization of the Renaissance in Italy* [1860]. Trans. S.G.C. Middlemore. New York: Harper, 1958.

Busacchi, Vincenzo. "Necroscopie trecentesche a scopo anatomo-patologico in Perugia." *Rivista di storia della medicina* 9 (1965): 160–64.

Bylebyl, Jerome J. "Interpreting the *Fasciculo* Anatomy Scene." *Journal of the History of Medicine and Allied Sciences* 45 (1990): 285–316.

Bynum, Caroline Walker. "'…and Woman His Humanity': Female Imagery in the Religious Writing of the Later Middle Ages." *Fragmentation and Redemption: Essays on Gender and the Human Body in Medieval Religion*. New York: Zone Books, 1991, 151–80.

———. "The Female Body and Religious Practice in the Later Middle Ages." *Fragmentation and Redemption*, 169–219.

———. *Holy Feast and Holy Fast: The Religious Significance of Food to Medieval Women*. Berkeley: University of California Press, 1987.

———. "Jesus as Mother and Abbot as Mother: Some Themes in Twelfth-Century Cistercian Writing." *Jesus as Mother: Studies in the Spirituality of the High Middle Ages*. Berkeley: University of California Press, 1982, 110–69.

———. "Material Continuity, Personal Survival and the Resurrection of the Body: A Scholastic Discussion in Its Medieval and Modern Contexts." *Fragmentation and Redemption*, 239–97.

———. *The Resurrection of the Body in Western Christianity, 200–1336*. New York: Columbia University Press, 1995.

———. "Seeing and Seeing Beyond: The Mass of St. Gregory in the Fifteenth Century." In Jeffrey F. Hamburger and Anne-Marie Bouché (eds.). *The Mind's Eye: Art and Theological Argument in the Middle Ages*. Princeton, NJ: Princeton Univeristy Press, 2005, 208–40.

———. "Violent Imagery in Late Medieval Piety." *Bulletin of the German Historical Institute* 30 (2002): 3–36.

———. "Why All the Fuss about the Body? A Medievalist's Perspective." *Critical Inquiry* 22 (1995): 1–33.

Caciola, Nancy. *Discerning Spirits: Divine and Demonic Possession in the Middle Ages*. Ithaca, NY: Cornell University Press, 2003.

———. "Mystics, Demoniacs, and the Physiology of Spirit Possession in Medieval Europe." *Comparative Studies in Society and History* 42 (2000): 268–306.

Cadden, Joan. *The Meanings of Sex Difference in the Middle Ages: Medicine, Science, and Culture*. Cambridge: Cambridge University Press, 1994.

Callmann, Ellen. "The Growing Threat to Marital Bliss as Seen in Fifteenth-Century Florentine Paintings." *Studies in Iconography* 5 (1979): 73–92.

Calvi, Giulia. *Il contratto morale: Madri e figli nella Toscana moderna*. Rome: Laterza, 1994.

Cannon, Joanna. "Simone Martini, the Dominicans, and the Early Sienese Polyptich." *Journal of the Warburg and Courtauld Institutes* 45 (1982): 69–93.

Cannon, Joanna, and André Vauchez. *Margherita of Cortona and the Lorenzetti: Sienese Art and the Cult of a Holy Woman in Medieval Tuscany*. University Park, PA: Pennsylvania State University Press, 1998.

Carlino, Andrea. *Books of the Body: Anatomical Ritual and Renaissance Learning* [1994]. Trans. John Tedeschi and Anne C. Tedeschi. Chicago: University of Chicago Press, 1999.

———. "Marsia, Sant'Antonio ed altri indizi: Il corpo punito e la dissezione tra Quattro e Cinquecento." In Jean Céard, Marie-Madeleine Fontaine, and Jean-Claude Margolin (eds.), *Le corps à la Renaissance*. Paris: Aux Amateurs de Livres, 1990, 129–38.

———. *Paper Bodies: A Catalogue of Anatomical Fugitive Sheets, 1538–1687*. Trans. Noga Arikha. London: Wellcome Institute for the History of Medicine, 1999.

———. "Tre piste per l'anatomia de Juan de Valverde: Logiche d'edizione, solidarietà nazionali e cultura artistica a Roma nel Rinascimento." *Mélanges de l'Ecole Française de Rome* 114 (2002): 513–41.

Carmichael, Ann G. "The Health Status of Florentines in the Fifteenth Century." In Marcel Tetel, Ronald G. Witt, and Rona Goffen (eds.), *Life and Death in Fifteenth-Century Florence*. Durham, NC: Duke University Press, 1989, 28–45.

Carroll, Margaret D. "The Erotics of Absolutism: Rubens and the Mystification of Sexual Violence." *Representations* 25 (1989): 3–30.

Casagrande, Giovanna. "Movimenti religiosi umbri e Chiara da Montefalco." In Claudio Leonardi and Enrico Menestò (eds.), *S. Chiara da Montefalco e il suo tempo*. Florence: Nuova Italia, 1985, 53–70.

———. "Terziarie domenicane a Perugia." In Casagrande and Menestò (eds.), *Una santa, una città*, 109–60.

Casagrande, Giovanna, and Enrico Menestò (eds.). *Una santa, una città*. Spoleto: Centro Italiano di Studi sull'Alto Medioevo, 1991.

Cassuto, Umberto. *Gli ebrei a Firenze nell'età del Rinascimento*. Florence: Galletti e Cocci, 1918.

Castelli, Elizabeth A. *Visions and Voyeurism: Holy Women and the Politics of Sight in Early Christianity*. Ed. Christopher Ocker. Berkeley: Center for Herme- neutical Studies, 1995.

Cavazza, Silvano. "Double Death: Resurrection and Baptism in a Seventeenth- Century Rite" [1982]. Trans. Mary M. Gallucci. In Edward Muir and Guido Ruggiero (eds.), *History from Crime*. Baltimore: Johns Hopkins University Press, 1994, 1–31.

Caviness, Madeline H. *Visualizing Women in the Middle Ages: Sight, Spectacle, and Scopic Economy*. Philadelphia: University of Pennsylvania Press, 2001.

Cazort, Mimi, Monique Kornell, and K.B. Roberts. *The Ingenious Machine of Nature: Four Centuries of Art and Anatomy*. Ottawa: National Gallery of Canada, 1996.

Chiappelli, Alberto. "Di un singolare procedimento medico-legale tenuto in Pistoia nell'anno 1375 per supposizione d'infante." *Rivista di storia critica delle scienze mediche e naturali* 10 (1919): 129–35.

Chojnacki, Stanley. "Daughters and Oligarchs: Gender and the Early Renais- sance State." In Judith C. Brown and Robert C. Davis (eds.), *Gender and Society in Renaissance Italy*. London: Longman, 1998, ch. 3.

Choulant, Ludwig. *Geschichte und Bibliographie der anatomischen Abbildung nach ihrer beziehung auf anatomische Wissenschaft und bildende Kunst*. Leipzig: Weigel, 1952.

Ciardi, Roberto Paolo, and Lucia Tongiorgi Tomasi. "La 'scienza illustrata': Osservazioni sui frontespizi delle opere di Athanasius Kircher e di Galileo Galilei." *Annali dell'Istituto Storico Italo-Germanico in Trento* 11 (1985): 69–78.

Clark, Stuart. *Thinking with Demons: The Idea of Witchcraft in Early Modern Europe*. Oxford: Clarendon, 1997.

Cline, Ruth H. "Heart and Eyes." *Romance Philology* 25 (1971–72): 262–97.

Coakley, John. "Friars as Confidants of Holy Women in Medieval Dominican Hagiography." In Renate Blumenfeld-Kosinski and Timea Szell (eds.), *Images of Sainthood in Medieval Europe*. Ithaca, NY: Cornell University Press, 1991, 222–46.

———. "Friars, Sanctity, and Gender: Mendicant Encounters with Saints, 1250–1325." In Clare A. Lees (ed.), *Medieval Masculinities: Regarding Men in the Middle Ages*. Minneapolis: University of Minnesota Press, 1994, 91–110.

Connolly, Thomas. *Mourning into Joy: Music, Raphael, and Saint Cecilia*. New Haven: Yale University Press, 1994.

Cook, Harold. "Medicine." In Katharine Park and Lorraine Daston (eds.), *The Cambridge History of Science*, vol. 3, *Early Modern Science*. Cambridge: Cambridge University Press, 2006, 307–34.

Corbett, Margery, and Ronald Lightbown, *The Comely Frontispiece: The Emblematic Title-page in England, 1550–1660*. London: Routledge and Kegan Paul, 1979.

Corradi, Alfonso. *Annali delle epidemie occorse in Italia dalle prime memorie fino al 1850*. Bologna: Gamberini e Parmeggiani, 1865–94.

———. "Degli esperimenti tossicologici in *animae nobili* nel Cinquecento." *Rendiconti del Reale Istituto Lombardo di Scienze e Lettere*, ser. 2, 19 (1886): 361–63.

Corsi, Dinora. "'Les secrés des dames': Tradition, traductions." *Médiévales* 14 (1988): 47–57.

Corsini, Andrea. *Malattia e morte di Lorenzo de' Medici, duca d'Urbino: Studio critico di medicina storica*. Florence: Istituto Micrografico Italiano, 1913.

Crabb, Ann. *The Strozzi of Florence: Widowhood and Family Solidarity in the Renaissance*. Ann Arbor: University of Michigan Press, 2000.

Cunningham, Andrew. *The Anatomical Renaissance: The Resurrection of the Anatomical Projects of the Ancients*. Aldershot: Scolar Press, 1997.

Dall'Osso, Eugenio. *L'organizzazione medico-legale a Bologna e a Venezia nei secoli XII–XIV*. Cesena: Addolorata, 1956.

Daston, Lorraine, and Katharine Park. *Wonders and the Order of Nature, 1150–1750*. New York: Zone Books, 1998.

Davidson, Arnold. "Miracles of Bodily Transformation, or, How St. Francis

Received the Stigmata." In Caroline A. Jones and Peter Galison (eds.), *Picturing Science, Producing Art*. New York: Routledge, 1998, 101–24.

Demand, Nancy. *Birth, Death, and Motherhood in Classical Greece*. Baltimore: Johns Hopkins University Press, 1994.

Derbes, Anne. *Picturing the Passion in Late Medieval Italy: Narrative Painting, Franciscan Ideologies, and the Levant*. Cambridge: Cambridge University Press, 1996.

Dunstan, G.R. (ed.). *The Human Embryo: Aristotle and the Arabic and European Traditions*. Exeter: Exeter University Press, 1990.

Durling, Richard J. "A Chronological Census of Renaissance Editions and Translations of Galen." *Journal of the Warburg and Courtauld Institutes* 24 (1961): 230–305.

Eamon, William. *Science and the Secrets of Nature: Books of Secrets in Medieval and Early Modern Culture*. Princeton, NJ: Princeton University Press, 1994.

Edgerton, Samuel Y., Jr. *Pictures and Punishment: Art and Criminal Prosecution During the Florentine Renaissance*. Ithaca, NY: Cornell University Press, 1985.

Elliott, Dyan. "The Physiology of Rapture and Female Spirituality." In Peter Biller and A.J. Minnis (eds.), *Medieval Theology and the Natural Body*. Rochester, NY: York Medieval Press, 1997, 141–73.

Elsheikh, Mahmoud Salem. *Medicina e farmacologia nei manoscritti della Biblioteca Riccardiana di Firenze*. Rome: Vecchiarelli, 1990.

Encarnación, Karen Rosoff. "The Proper Uses of Desire: Sex and Procreation in Reformation Anatomical Fugitive Sheets." In Anne L. McClanan and Karen Rosoff Encarnación (eds.), *The Material Culture of Sex, Precreation, and Marriage in Premodern Europe*. New York: Palgrave, 2001, 221–49.

Fabbri, Lorenzo. *Alleanza matrimoniale e patriziato nella Firenze del '400: Studio sulla famiglia Strozzi*. Florence: Olschki, 1991.

Faietti, Marzia, and Daniela Scaglietti Kelescian. *Amico Aspertini*. Modena: Artioli, 1995.

Ferckel, Christoph. "Die *Secreta mulierum* und ihr Verfasser." *Sudhoffs Archiv für Geschichte der Medizin und der Naturwissenschaften* 38 (1954): 267–74.

376

Ferrari, Giovanna. *L'esperienza del passato: Alessandro Benedetti filologo e medico umanista*. Florence: Olschki, 1996.

———. "Public Anatomy Lessons and the Carnival: The Anatomy Theatre of Bologna." *Past and Present* 117 (1987): 50–106.

Filippini, Nadia Maria. *La nascita straordinaria: Tra madre e figlio, la rivoluzione del taglio cesareo (sec. XVIII–XIX)*. Milan: Angeli, 1995.

Fissell, Mary E. *Vernacular Bodies: The Politics of Reproduction in Early Modern England*. Oxford: Oxford University Press, 2004.

Flemming, Rebecca. *Medicine and the Making of Roman Women: Gender, Nature, and Authority from Celsus to Galen*. Oxford: Oxford University Press, 2000.

Flügge, Sibylla. *Hebammen und heilkundige Frauen: Recht und Rechtswirklichkeit im 15. und 16. Jahrhundert*. Frankfurt: Stroemfeld, 1998.

Flutre, Louis-Fernand. *Li Fait des Romains dans les littératures française et italienne du XIIIe au XVe siècle*. Paris: Hachette, 1932.

———. "La naissance de César." *Aesculape* 24 (1934): 244–50.

Fornaciari, Gino. "Renaissance Mummies in Italy." *Medicina nei secoli* 11 (1999): 85–105.

Foucault, Michel. *The Birth of the Clinic: An Archaeology of Medical Perception* [1973]. Trans. A.M. Sheridan Smith. New York: Vintage, 1975.

———. *Discipline and Punish: The Birth of the Prison* [1975]. Trans. Alan Sheridan. New York: Pantheon, 1977.

Freedberg, David. *The Power of Images: Studies in the History and Theory of Response*. Chicago: University of Chicago Press, 1989.

French, R.K. "Berengario da Carpi and the Use of Commentary in Anatomical Teaching." In Andrew Wear, R.K. French, and Iain M. Lonie (eds.), *The Medical Renaissance of the Sixteenth Century*. Cambridge: Cambridge University Press, 1985, 42–74.

———. *Canonical Medicine: Gentile da Foligno and Scholasticism*. Leiden: Brill, 2001.

———. "*De juvamentis membrorum* and the Reception of Galenic Physiological Anatomy." *Isis* 70 (1979): 96–109.

———. *Dissection and Vivisection in the European Renaissance*. Aldershot: Ashgate, 1999.

————. *Medicine before Science: The Rational and Learned Doctor from the Middle Ages to the Enlightenment.* Cambridge: Cambridge University Press, 2003.

Freuler, Gaudenz. "Andrea di Bartolo, Fra Tommaso d'Antonio Caffarini, and Sienese Dominicans in Venice." *Art Bulletin* 69 (1987): 570–86.

Frugoni, Chiara. "The Cities and the 'New' Saints." In Anthony Molho, Kurt Raaflaub, and Julia Emlen (eds.), *City-States in Classical Antiquity and Medieval Italy.* Ann Arbor: University of Michigan Press, 1991, 71–91.

————. "Female Mystics, Visions, and Iconography" [1983]. In Daniel Bornstein and Roberto Rusconi (eds.), *Women and Religion in Medieval and Renaissance Italy*, trans. Margery J. Schneider. Chicago: University of Chicago Press, 1996, 130–64.

————. *Francesco e l'invenzione delle stimmate: Una storia per parole e immagini fino a Bonaventura e Giotto.* Turin: Einaudi, 1993.

Gamboso, Vergilio. "La *Sancti Antonii confessoris de Padua vita* di Sicco Ricci Polentone (c. 1435)." *Il Santo* 11 (1971): 199–283.

Gasparini, Giuseppina, de Sandre. "La confraternità di S. Giovanni Evangelista della Morte in Padova e una 'riforma' ispirata dal vescovo Pietro Barozzi (1502)." In *Miscellanea Gilles Gérard Meersseman.* Padua: Antenore, 1970, vol. 1, 765–815.

Gennaro, Clara. "Clare, Agnes, and Their Earliest Followers: From the Poor Ladies of San Damiano to the Poor Clares" [1980]. In Daniel Bornstein and Roberto Rusconi (eds.), *Women and Religion in Medieval and Renaissance Italy.* Trans. Margery J. Schneider. Chicago: University of Chicago Press, 1996, 39–55.

Georges, Patrice, "Mourir c'est pourrir un peu…: Techniques contre la corruption des cadavres à la fin du Moyen Age." *Micrologus* 7 (1999): 359–82.

Gilman, Sander L. *Sexuality: An Illustrated History; Representing the Sexual in Medicine and Culture from the Middle Ages to the Age of AIDS.* New York: Wiley, 1989.

Goffen, Rona. *Spirituality in Conflict: Saint Francis and Giotto's Bardi Chapel.* University Park: Pennsylvania State University Press, 1988.

————. *Titian's Women.* New Haven: Yale University Press, 1997.

Goldberg, Jonathan. "Fatherly Authority: The Politics of Stuart Family Images." In Margaret W. Ferguson, Maureen Quilligan, and Nancy J. Vickers (eds.), *Rewriting the Renaissance: The Discourses of Sexual Difference in Early Modern Europe*. Chicago: University of Chicago Press, 1986, 3–32.

Gourevitch, Danielle. "Chirurgie obstétricale dans le monde romain: Césarienne et embryotomie." In Véronique Dasen (ed.), *Naissance et petite enfance dans l'Antiquité*. Fribourg: Academic Press, 2004, 239–64.

Graf, Arturo. *Roma nella memoria e nelle immaginazioni del medio evo*. Turin: Loescher, 1923; repr. Sala Bolognese: Arnaldo Forni, 1987.

Grafton, Anthony, and Nancy G. Siraisi. "Between the Election and My Hopes: Girolamo Cardano and Medical Astrology." In William R. Newman and Anthony Grafton (eds.), *Secrets of Nature: Astrology and Alchemy in Early Modern Europe*. Cambridge, MA: MIT Press, 2001, 69–131.

Grant, Edward. *God and Reason in the Middle Ages*. Cambridge: Cambridge University Press, 2001.

Green, Monica H. "Bodies, Gender, Health, Disease: Recent Work on Medieval Women's Medicine." *Studies in Medieval and Renaissance History*, ser. 3, vol. 2 (2005): 1–46.

———. "Constantinus Africanus and the Conflict between Religion and Science." In Dunstan, *The Human Embryo*, 47–69.

———. "The Development of the *Trotula*." *Revue d'histoire des textes* 26 (1996): 119–203. Repr. in Green, *Women's Healthcare in the Medieval West: Texts and Contexts*, ch. 5.

———. "Documenting Medieval Women's Medical Practice." In Luis García-Ballester, Roger French, Jon Arrizabalaga, and Andrew Cunningham (eds.), *Practical Medicine from Salerno to the Black Death*. Cambridge: Cambridge University Press, 1994, 322–52. Repr. in Green, *Women's Healthcare in the Medieval West*, ch. 2.

———. "From 'Diseases of Women' to 'Secrets of Women': The Transformation of Gynecological Literature in the Later Middle Ages." *Journal of Medieval and Early Modern Studies* 30 (2000): 5–39.

———. "A Handlist of the Latin and Vernacular Manuscripts of the So-Called

Trotula Texts, Part II: The Vernacular Texts and Latin Re-Writings." *Scriptorium* 51 (1997): 80–103.

———. *Making Women's Medicine Masculine: The Rise of Male Authority in Pre-Modern Gynecology*. Oxford: Oxford University Press, forthcoming 2007.

———. "Medieval Gynecological Literature: A Handlist." *Women's Healthcare in the Medieval West: Texts and Contexts*, Appendix.

———. "The Possibilities of Literacy and the Limits of Reading: Women and the Gendering of Medical Literacy." *Women's Healthcare in the Medieval West: Texts and Contexts*, ch. 7.

———. Review of Renate Blumenfeld-Kosinski, *Not of Woman Born: Representations of Caesarean Birth in Medieval and Renaissance Culture* (1990). *Speculum* 67 (1992): 380–81.

———. "Secrets of Women." In Margaret Schaus (ed.), *Women and Gender in Medieval Europe: An Encyclopedia*. New York: Routledge, 2006.

———. "'Traittié tout de mençonges': The *Secrés des dames*, 'Trotula,' and Attitudes toward Women's Medicine in Fourteenth- and Early-Fifteenth-Century France." In Marilynn Desmond (ed.), *Christine de Pizan and the Categories of Difference*. Minneapolis: University of Minnesota Press, 1998, 146–78. Repr. in Green, *Women's Healthcare in the Medieval West: Texts and Contexts*, ch. 6.

———. *Women's Healthcare in the Medieval West: Texts and Contexts*. Burlington, VT: Ashgate, 2000.

———. "Women's Medical Practice and Health Care in Medieval Europe." In Judith M. Bennett (ed.), *Sisters and Workers in the Middle Ages*. Chicago: University of Chicago Press, 1989, 39–78. Repr. in Green, *Women's Healthcare in the Medieval West: Texts and Contexts*, ch. 1.

Guibert, Joseph de. "La componction du coeur." *Revue d'ascétique et de mystique* 15 (1934): 225–40.

Haas, Louis. *The Renaissance Man and His Children: Childbirth and Early Childhood in Florence 1300–1600*. Houndmills: Macmillan, 1998.

Hamburger, Jeffrey F. *Nuns as Artists: The Visual Culture of a Medieval Convent*. Berkeley: University of California Press, 1997.

————. "Seeing and Believing: The Suspicion of Sight and the Authentication of Vision in Late Medieval Art." In Alessandro Nova and Klaus Krüger (eds.), *Imagination und Wirklichkeit: Zum Verhältnis von mentalen und realen Bildern in der Kunst der Frühen Neuzeit*. Mainz: Von Zabern, 2000, 47–69.

Hanson, Ann Ellis. "A Long-Lived 'Quick Birther' (*okytokion*)." In Véronique Dasen (ed.), *Naissance et petite enfance dans l'Antiquité*. Fribourg: Academic Press, 2004, 265–80.

Hanson, Ann Ellis, and Monica H. Green. "Soranus of Ephesus: *Methodicorum princeps*." In Wolfgang Haase and Hildegard Temporini (eds.), *Aufstieg und Niedergang der Römischen Welt*, pt. 2, *Principat*, vol. 37.2. Berlin: Walter de Gruyter, 1994, 968–1075.

Harley, David. "Historians as Demonologists: The Myth of the Midwife-Witch." *Social History of Medicine* 3 (1990): 1–26.

Heckscher, William S. *Rembrandt's Anatomy of Dr. Nicolaas Tulp: An Iconological Study*. New York: New York University Press, 1958.

Hentschel, Linda. *Pornotopische Techniken des Betrachtens: Raumwahrnehmung und Geschlechterordnung in visuellen Apparaten der Moderne*. Marburg: Jonas, 2001.

Herrlinger, Robert. *History of Medical Illustration from Antiquity to A.D. 1600* [1967]. London: Pitman Medical, 1970.

Hill, George Francis. *A Corpus of Italian Medals from the Renaissance before Cellini*. London: British Museum, 1930.

Höltgen, Karl Josef. *Aspects of the Emblem: Studies in the English Emblem Tradition and the European Context*. Kassel: Reichenberger, 1986.

Holdefleiss, Erich. *Der Augenscheinsbeweis im mittelalterlichen deutschen Strafverfahren*. Stuttgart: Kohlhammer, 1933.

Holländer, Eugen. *Plastik und Medizin*. Stuttgart: Enke, 1912.

Hollywood, Amy. "Inside Out: Beatrice of Nazareth and Her Hagiographer." In Catherine M. Mooney (ed.), *Gendered Voices: Medieval Saints and Their Interpreters*. Philadelphia: University of Pennsylvania Press, 1999, 78–98.

————. *The Soul as Virgin Wife: Mechthild of Magdeburg, Marguerite Porete, and Meister Eckhart*. Notre Dame: University of Notre Dame Press, 1995.

Hughes, Diane Owen. "Representing the Family: Portraits and Purposes in Early Modern Italy." *Journal of Interdisciplinary History* 17 (1986): 7–38.

Ingegno, Alfonso. "La storia come teatro della violenza: L'*Encomium Neronis.*" *Saggio sulla filosofia di Cardano.* Florence: Nuova Italia, 1980, 184–208.

Innocenti, Cristiana. "Introduzione: Magia religione e superstizione nel Rinascimento." In Pietro Pomponazzi, *Gli incantesimi,* ed. Cristiana Innocenti. Florence: Nuova Italia, 1997, 9–41.

Innocenti, Giancarlo. *L'immagine significante: Studio sull'emblematica cinquecentesca.* Padua: Liviana, 1981.

Jacquart, Danielle. "La morphologie du corps féminin selon les médecins de la fin du Moyen Age." *Micrologus* 1 (1993): 81–98.

Jacquart, Danielle, and Claude Thomasset. *Sexuality and Medicine in the Middle Ages* [1985]. Trans. Matthew Adamson. Princeton, NJ: Princeton University Press, 1988.

Jager, Eric. *The Book of the Heart.* Chicago: University of Chicago Press, 2000.

——. "The Book of the Heart: Reading and Writing the Medieval Subject." *Speculum* 71 (1996): 1–26.

Janson, H.W. *The Sculpture of Donatello.* Princeton, NJ: Princeton University Press, 1963.

——. "Titian's *Laocoon Caricature* and the Vesalian-Galenist Controversy." *Art Bulletin* 28 (1946): 49–53.

Jeauneau, Edouard. *Quatre thèmes érigéniens.* Montreal: Institut d'Etudes Médiévales Albert-le-Grand, 1978.

Johnson, Geraldine A. "Approaching the Altar: Donatello's Sculpture in the Santo." *Renaissance Quarterly* 52 (1999): 626–66.

Jordanova, Ludmilla. *Sexual Visions: Images of Gender in Science and Medicine between the Eighteenth and Twentieth Centuries.* Madison: University of Wisconsin Press, 1989.

Kay, Sarah. "Women's Body of Knowledge: Epistemology and Misogyny in the *Romance of the Rose.*" In Sarah Kay and Miri Rubin (eds.), *Framing Medieval Bodies.* Manchester: Manchester University Press, 1994, 211–35.

Keele, Kenneth D. *Leonardo da Vinci's Elements of the Science of Man*. New York: Academic Press, 1983.

Keele, Kenneth D., and Carlo Pedretti. *Leonardo da Vinci: Corpus of the Anatomical Studies in the Collection of Her Majesty the Queen at Windsor Castle*. London: Johnson Reprint, 1978–1980.

Kemp, Martin. "A Drawing for the *Fabrica* and Some Thoughts upon the Vesalius Muscle-Men." *Medical History* 14 (1970): 277–88.

———. "'The Mark of Truth': Looking and Learning in Some Anatomical Illustrations from the Renaissance and Eighteenth Century." In W.F. Bynum and Roy Porter (eds.), *Medicine and the Five Senses*. Cambridge: Cambridge University Press, 1993, 85–121.

Kent, Francis William. "La famiglia patrizia fiorentina nel Quattrocento: Nuovi orientamenti nella storiografia recente." In *Palazzo Strozzi, metà millennio, 1489–1989*. Rome: Istituto dell Enciclopedia Italiana, 1991, 70–91.

———. *Household and Lineage in Renaissance Florence: The Family Life of the Capponi, Ginori, and Rucellai*. Princeton, NJ: Princeton University Press, 1977.

King, Helen. *Hippocrates' Woman: Reading the Female Body in Ancient Greece*. London: Routledge, 1998.

Kintzinger, Marion. *Chronos und Historia: Studien zur Titelblattikonigraphie historiographischer Werke vom 16. bis zum 18. Jahrhundert*. Wiesbaden: Harrassowitz, 1995.

Klapisch-Zuber, Christiane. "Blood Parents and Milk Parents: Wet Nursing in Florence, 1300–1530." *Women, Family, and Ritual in Renaissance Italy*, 132–64.

———. "Les clefs florentines de Barbe-Bleue: L'apprentissage de la lecture." *La maison et le nom*, 309–30.

———. "Le dernier enfant: Fécondité et vieillissement chez les Florentines (XIVe–XVe siècles)." In Jean-Pierre Bardet, François Lebrun, and René Le Mée (eds.), *Mesurer et comprendre: Mélanges offerts à Jacques Dupâquier*. Paris: Presses Universitaires de France, 1993, 277–90.

———. "Les femmes et la mort à la fin du Moyen Age." In Stéphane Toussaint (ed.), *Ilaria del Carretto e il suo monumento: La donna nell'arte, la cultura et la società del '400*. Lucca: S. Marco, 1995, 207–21.

———. *La maison et le nom: Stratégies et rituels dans l'Italie de la Renaissance.* Paris: Editions de l'Ecole des Hautes Etudes en Sciences Sociales, 1990.

———. "La mère cruelle: Maternité, veuvage et dot." *La maison et le nom,* ch. 12.

———. *Women, Family, and Ritual in Renaissance Italy.* Trans. Lydia Cochrane. Chicago: University of Chicago Press, 1985.

Kleinberg, Aviad M. *Prophets in Their Own Country: Living Saints and the Making of Sainthood in the Later Middle Ages.* Chicago: University of Chicago Press, 1992.

———. "Proving Sanctity: Selection and Authentication of Saints in the Later Middle Ages." *Viator* 20 (1989): 183–205.

Klestinec, Cynthia. "A History of Anatomy Theaters in Sixteenth-Century Padua." *Journal of the History of Medicine and Allied Sciences* 59 (2004): 375–412.

———. "Juan Valverde de (H)Amusco and Print Culture: The Editorial Apparatus in Vernacular Anatomy Texts." *Zeitsprünge* 9 (2005): 101–17.

Kudlien, Fridolf. "The Seven Cells of the Uterus: The Doctrine and Its Roots." *Bulletin of the History of Medicine* 39 (1965): 415–23.

Kuehn, Thomas. "Some Ambiguities of Female Inheritance Ideology in the Renaissance." *Continuity and Change* 2 (1987): 11–36.

Kuriyama, Shigehisa. *The Expressiveness of the Body and the Divergence of Greek and Chinese Medicine.* New York: Zone Books, 1999.

Kusukawa, Sachiko. "Illustrating Nature." In Marina Frasca-Spada and Nick Jardine (eds.), *Books and the Sciences in History.* Cambridge: Cambridge University Press, 2000, 90–113.

Lambert, Samuel W. "The Initial Letters of the Anatomical Treatise, *De humani corporis fabrica*, of Vesalius." In Lambert, Willy Wiegand, and William M. Ivins Jr., *Three Vesalian Essays to Accompany the Icones anatomicae of 1934.* New York: Macmillan, 1952, 3–24.

Landau, David, and Peter Parshall. *The Renaissance Print, 1470–1550.* New Haven: Yale University Press, 1994.

Langlois, Ernest. *Origines et sources du Roman de la Rose.* Paris: Thorin, 1891.

———. "La traduction de Boèce par Jean de Meun." *Romania* 42 (1913): 331–69.

Laqueur, Thomas. *Making Sex: Body and Gender from the Greeks to Freud.* Cambridge, MA: Harvard University Press, 1990.

Laurenza, Domenico. *De figura umana: Fisiognomica, anatomia e arte in Leonardo.* Florence: Olschki, 2001.

———. *Leonardo nella Roma di Leone X (c. 1513–16): Gli studi anatomici, la vita, l'arte.* Vinci: Biblioteca Leonardiana, 2004.

———. *La ricerca dell'armonia: Rappresentazioni anatomiche nel Rinascimento.* Florence: Olschki, 2003.

Lemay, Helen Rodnite, "Anthonius Guainerius and Medieval Gynecology." In Julius Kirshner and Suzanne F. Wemple (eds.), *Women of the Medieval World: Essays in Honor of John H. Mundy.* Oxford: Blackwell, 1985, 317–36.

———. "William of Saliceto on Human Sexuality." *Viator* 12 (1981): 165–81.

———. "Women and the Literature of Obstetrics and Gynecology." In Joel T. Rosenthal (ed.), *Medieval Women and the Sources of Medieval History.* Athens: University of Georgia Press, 1990, 189–209.

Leonardi, Claudio. "Chiara e Berengario: L'agiografia sulla santa di Montefalco." In Claudio Leonardi and Enrico Menestò (eds.), *S. Chiara da Montefalco e il suo tempo.* Florence: Nuova Italia, 1985, 369–86.

———. "Committenze agiografiche nel Trecento." In Vincent Moleta (ed.), *Patronage and Public in the Trecento.* Florence: Olschki, 1986, 37–58.

Leonardi, Claudio, and Enrico Menestò (eds.). *S. Chiara da Montefalco e il suo tempo.* Florence: Nuova Italia, 1985.

Lightbown, Ronald. *Sandro Botticelli.* London: Elek, 1978.

Lind, L.R. *Studies in Pre-Vesalian Anatomy: Biography, Translations, Documents.* Philadelphia: American Philosophical Society, 1975.

Linhardt, Robert. *Die Mystik des hl. Bernhard von Clairvaux.* Munich: Verlag Natur und Kultur, 1923.

Lochrie, Karma. *Covert Operations: The Medieval Uses of Secrecy.* Philadelphia: University of Pennsylvania Press, 1999.

Long, Pamela O. *Openness, Secrecy, Authorship: Technical Arts and the Culture of*

Knowledge from Antiquity to the Renaissance. Baltimore: Johns Hopkins University Press, 2001.

Lubkin, Gregory. *A Renaissance Court: Milan under Galeazzo Maria Sforza.* Berkeley: University of California Press, 1994.

Maccagni, Carlo. "Frammento di un codice di medicina del secolo XIV (manoscritto N. 735, già Codice Roncioni N. 99) della Biblioteca Universitaria di Pisa." *Physis* 11 (1969): 311–78.

MacKinney, Loren C., and Boyd H. Hill, Jr. "A New Fünfbilderserie Manuscript — Vatican Palat. Lat. 1110." *Sudhoffs Archiv für Geschichte der Medizin und der Naturwissenschaften* 48 (1964): 323–30.

Maclean, Ian. *Logic, Signs, and Nature in the Renaissance: The Case of Learned Medicine.* Cambridge: Cambridge University Press, 2002.

————. *The Renaissance Notion of Woman: A Study in the Fortunes of Scholasticism and Medical Science in European Intellectual Life.* Cambridge: Cambridge University Press, 1980.

Mandach, Conrad de. *Saint Antoine de Padoue et l'art italien.* Paris: Laurens, 1899.

Martin, John Jeffries. *Myths of Renaissance Individualism.* New York: Palgrave Macmillan, 2004.

Maurer, Armand. "Between Reason and Faith: Siger of Brabant and Pomponazzi on the Magic Arts." *Mediaeval Studies* 18 (1956): 1–18.

Mayor, A. Hyatt. *Artists and Anatomists.* New York: Metropolitan Museum of Art, 1984.

McVaugh, Michael. "Surgical Education in the Middle Ages." *Dynamis* 20 (2000): 283–304.

————. "Two Montpellier Recipe Collections." *Manuscripta* 20 (1976): 175–80.

Menestò, Enrico. "The Apostolic Canonization Proceedings of Clare of Montefalco, 1318–1319." In Daniel Bornstein and Roberto Rusconi (eds.), *Women and Religion in Medieval and Renaissance Italy.* Trans. Margery J. Schneider. Chicago: University of Chicago Press, 1996, 107–26.

————. "La 'legenda' della beata Colomba e il suo biografo." In Casagrande and Menestò (eds.), *Una santa, una città,* 161–76.

————. "La 'legenda' di Margherita da Città di Castello." In Roberto Rusconi (ed.), *Il movimento religioso femminile in Umbria nei secoli XIII-XIV*. Florence: Nuova Italia, 1984, 217–37.

————. "Il processo apostolico per la canonizzazione di Chiara da Montefalco (1318–1319)." In Claudio Leonardi and Enrico Menestò (eds.), *S. Chiara da Montefalco e il suo tempo*. Florence: Nuova Italia, 1985, 269–301.

Menestò, Enrico, and Roberto Rusconi. *Umbria sacra e civile*. Turin: Nuova Eri Edizioni Rai, 1989.

Merback, Mitchell B. *The Thief, the Cross, and the Wheel: Pain and the Spectacle of Punishment in Medieval and Renaissance Europe*. Chicago: University of Chicago Press, 1999.

Merchant, Carolyn. *The Death of Nature: Women, Ecology, and the Scientific Revolution* [1980]. 2nd ed. New York: HarperSanFrancisco, 1990.

Mooney, Catherine M. "*Imitatio Christi* or *Imitatio Mariae*? Clare of Assisi and Her Interpreters." In Catherine M. Mooney (ed.), *Gendered Voices: Medieval Saints and Their Interpreters*. Philadelphia: University of Pennsylvania Press, 1999, 52–77.

————. "Women's Visions, Men's Words: The Portrayal of Holy Women and Men in Fourteenth-Century Italian Hagiography." Ph.D. dissertation, Yale University, 1991.

Münster, Ladislao. "Alcuni episodi sconosciuti o poco noti sulla vita e sull'attività di Bartolomeo da Varignana." *Castalia: Rivista di storia della medicina* 10 (1954): 207–15.

Muraro, Michelangelo. "Tiziano e le anatomie del Vesalio." In *Tiziano e Venezia: Convegno internazionale di studi, Venezia, 1976*. Vicenza: Neri Pozza, 1980, 307–16.

Murray, Jacqueline. "On the Origins and Role of 'Wise Women' in Causes for Annulment on the Grounds of Male Impotence." *Journal of Medieval History* 16 (1990): 235–49.

Musacchio, Jacqueline Marie. *The Art and Ritual of Childbirth in Renaissance Italy*. New Haven: Yale University Press, 1999.

————. "Imaginative Conceptions in Renaissance Italy." In Geraldine A. Johnson and Sara F. Matthews Grieco (eds.), *Picturing Women in Renaissance and Baroque Italy*. Cambridge: Cambridge University Press, 1997, 42–60.

———. "The Rape of the Sabine Women on Quattrocento Marriage Panels." In Trevor Dean and K.J.P. Lowe (eds.), *Marriage in Renaissance Italy, 1300–1650*. Cambridge: Cambridge University Press, 1998, 66–82.

Neff, Amy. "The Pain of *Compassio*: Mary's Labor at the Foot of the Cross." *Art Bulletin* 80 (1998): 254–73.

Nessi, Silvestro. *Chiara da Montefalco*. Città di Castello: Edimond, 1999.

———. "Primi appunti sull'iconografia clariana dei secoli XIV e XV." In Silvestro Nessi (ed.), *La spiritualità di S. Chiara da Montefalco*. Montefalco: Monastero S. Chiara, 1986, 313–38.

Neumeister, Sebastian. "Das Bild der Geliebten im Herzen." In Ingrid Kasten, Werner Paravicini, and René Pérennec (eds.), *Kultureller Austausch und Literaturgeschichte im Mittelalter/Transferts culturels et histoire littéraire au moyen age*. Sigmaringen: Jan Thorbecke, 1998, 315–30.

Niccoli, Ottavia. "'Menstruum quasi monstruum:' Monstrous Births and Menstrual Taboo in the Sixteenth Century." Trans. Mary M. Gallucci, in Edward Muir and Guido Ruggiero (eds.), *Sex and Gender in Historical Perspective*. Baltimore: Johns Hopkins University Press, 1990, 1–25.

Nutton, Vivian. "André Vésale et l'anatomie parisienne." *Cahiers de l'Association Internationale des Etudes Françaises* 55 (2003): 240–41.

———. "Continuity or Rediscovery? The City Physician in Classical Antiquity and Medieval Italy." In Andrew W. Russell (ed.), *The Town and State Physician in Europe from the Middle Ages to the Enlightenment*. Wolfenbüttel: Herzog August Bibliothek, 1981, 9–46.

———. "The Diffusion of Ancient Medicine in the Renaissance." *Medicina nei secoli* 14 (2002): 461–78.

———. "Introduction." In Andreas Vesalius, *De humani corporis fabrica*, trans. Daniel Garrison and Malcolm Hast. http://vesalius.northwestern.edu.

Nutton, Vivian, and Christine Nutton. "The Archer of Meudon: A Curious Absence of Continuity in the History of Medicine." *Journal of the History of Medicine and Allied Sciences* 58 (2003): 401–27.

O'Boyle, Cornelius. "Gesturing in the Early Universities." *Dynamis* 20 (2000): 249–81.

Offner, Richard. *A Critical and Historical Corpus of Florentine Painting*. New York: College of Fine Arts, New York University, 1947, pt. 3, vol. 5.

O'Malley, Charles D. *Andreas Vesalius of Brussels, 1514–1564*. Berkeley: University of California Press, 1964.

O'Neill, Ynez Violé. "Innocent III and the Evolution of Anatomy." *Medical History* 20 (1976): 429–33.

———. "The Fünfbilderserie – A Bridge to the Unknown." *Bulletin of the History of Medicine* 51 (1977): 538–49.

———. "The Fünfbilderserie Reconsidered." *Bulletin of the History of Medicine* 43 (1969): 236–45.

Ongaro, Giuseppe. "La medicina nello studio di Padova e nel Veneto." In Girolamo Arnaldi (ed.), *Storia della cultura veneta*. Vicenza: Pozza, 1981, vol. 3, pt. 3, 75–134.

Origo, Iris. *The Merchant of Prato, Francesco di Marco Datini*. London: Cape, 1957.

Ortalli, Edgardo. "La perizia medica a Bologna nei secoli XIII e XIV." *Atti e memorie della Deputazione di storia patria per le province di Romagna*, n.s. 17–19 (1969): 223–59.

Paglia, Vincenzo. *La morte confortata: Rita della paura e mentalità religiosa a Roma nell'età moderna*. Rome: Edizioni di Storia e Letteratura, 1982.

Pansier, Pierre. "Un manuel d'accouchements du XVe siècle." *Janus* 14 (1909): 217–20.

Paravicini Bagliani, Agostino. "La papauté du XIIIe siècle et la renaissance de l'anatomie." *Medicina e scienze della natura alla corte dei papi nel Duecento*. Spoleto: Centro Italiano di Studi sull'Alto Medioevo, 1991, 269–79.

———. *The Pope's Body* [1994]. Trans. David S. Peterson. Chicago: University of Chicago Press, 2000.

———. *I testamenti dei cardinali del Duecento*. Rome: Società Romana di Storia Patria, 1980.

Pardo-Tomás, José. "L'anatomia rinascimentale: Un soggetto storiografico rinnovato." In Maurizio Rippa Bonati and José Pardo-Tomás (eds.), *Il teatro dei corpi: Le pitture colorate d'anatomia di Girolamo Fabrici d'Acquapendente*. Milan: Mediamed, 2004, 31–44.

Park, Katharine. "The Criminal and the Saintly Body: Autopsy and Dissection in Renaissance Italy." *Renaissance Quarterly* 47 (1994): 1–33.

———. *Doctors and Medicine in Early Renaissance Florence*. Princeton, NJ: Princeton University Press, 1985.

———. "Holy Autopsy: Saintly Bodies and Medical Expertise, 1300–1600." In Julia Hairston and Walter Stephens (eds.), *The Body in Early Modern Italy*. Baltimore, MD: Johns Hopkins University Press, forthcoming 2007.

———. "Impressed Images: Reproducing Wonders." In Caroline A. Jones and Peter Galison (eds.), *Picturing Science, Producing Art*. New York: Routledge, 1998, 255–71.

———. "Itineraries of the 'One-Sex Body': A History of an Idea." Forthcoming.

———. "The Life of the Corpse: Division and Dissection in Late Medieval Europe." *Journal of the History of Medicine and Allied Sciences* 50 (1995): 111–32.

———. "Masaccio's Skeleton: Art and Anatomy in Early Renaissance Italy." In Rona Goffen (ed.), *Masaccio's Trinity*. Cambridge: Cambridge University Press, 1998, 119–40.

———. "Medicine and Magic: The Healing Arts." In Judith C. Brown and Robert C. Davis (eds.), *Gender and Society in Renaissance Italy*. London: Longman, 1998, 129–48.

———. "The Rediscovery of the Clitoris: French Medicine and the *Tribade*, 1570–1620." In David Hillman and Carla Mazzio (eds.), *The Body in Parts: Fantasies of Corporeality in Early Modern Europe*. New York: Routledge, 1997, 171–93.

———. "Relics of a Fertile Heart: The 'Autopsy' of Clare of Montefalco." In Anne L. McClanan and Karen Rosoff Encarnación (eds.), *The Material Culture of Sex, Procreation, and Marriage in Premodern Europe*. New York: Palgrave, 2002, 115–34.

———. "Was There a Renaissance Body?" In Allen J. Grieco, Michael Rocke, and Fiorella Gioffredi Superbi (eds.), *The Italian Renaissance in the Twentieth Century*. Florence: Olschki, 2002, 321–35.

Park, Katharine, and Lorraine Daston. "Introduction." In Park and Daston (eds.),

The Cambridge History of Science, vol. 3, *Early Modern Science*. Cambridge: Cambridge University Press, 2006, 1–17.

Park, Katharine, and Robert A. Nye. "Destiny is Anatomy" (essay-review of Thomas Laqueur, *Making Sex*). *New Republic* (18 February 1991): 53–57.

Parodi, Ernesto Giacomo. "Le storie di Cesare nella letteratura italiana dei primi secoli." *Studi di filologia romanza* 4 (1889): 237–501.

Paster, Gail Kern. *The Body Embarrassed: Drama and the Disciplines of Shame in Early Modern England*. Ithaca, NY: Cornell University Press, 1993.

Paul, Jacques, and Mariano D'Alatri. *Salimbene da Parma: Testimone e cronista*. Rome: Istituto Storico dei Cappuccini, 1992.

Paxton, Frederick S. *Christianizing Death: The Creation of a Ritual Process in Early Medieval Europe*. Ithaca, NY: Cornell University Press, 1990.

Pazzini, Adalberto. "La medicina alla corte dei Gonzaga a Mantova." In *Mantova e i Gonzaga nella civiltà del Rinascimento*. Mantua: Mondadori, 1977.

Pereira, Michela. "Un trattato medievale sul corpo delle donne: Il *De secretis mulierum*." *Memoria: Rivista di storia delle donne* 3 (1982): 108–13.

Pesenti, Tiziana. "Editoria medica tra Quattro e Cinquecento." In Ezio Riondato, *Trattati scientifici nel Veneto fra il XV e XVI secolo*. Vicenza: Neri Pozza, 1985, 1–28.

———. *Fasiculo de medicina in volgare: Venezia, Giovanni e Gregorio de Gregori, 1494*. Padua: Università degli Studi, 2001.

———. *Professori e promotori di medicina nello studio di Padova dal 1405 al 1509: Repertorio bio-bibliografico*. Padua: LINT, 1984.

Piana, Celestino. "I processi di canonizzazione sulla vita di S. Bernardino da Siena." *Archivum Franciscanum Historicum* 44 (1951): 87–160, 383–435.

Pieraccini, Gaetano. *La stirpe de' Medici di Cafaggiolo: Saggio di ricerche sulla trasmissione ereditaria dei caratteri biologici*. Florence: Vallechi, 1924–25.

Pine, Martin L. *Pietro Pomponazzi: Radical Philosopher of the Renaissance*. Padua: Antenore, 1986.

Pitkin, Hanna Fenichel. *Fortune Is a Woman: Gender and Politics in the Thought of Niccolò Machiavelli*. Berkeley: University of California Press, 1984.

Planiscig, Leo. *Andrea Riccio*. Vienna: Schroll and Co., 1927.

Polo de Beaulieu, Marie-Anne. "La légende du coeur inscrit dans la littérature religieuse et didactique." In Le "Cuer" au Moyen Age: Réalité et Senefiance (1991): 299–312.

Pomata, Gianna. "A Christian Utopia of the Renaissance: Elena Duglioli's Spiritual and Physical Motherhood (ca. 1510–1520)." In Kaspar von Greyerz, Hans Medick, and Patrice Veit (eds.), Von der dargestellten Person zum erinnerten Ich: Europäische Selbstzeugnisse als historische Quellen (1500–1850). Cologne: Böhlau, 2002, 323–53.

———. Contracting a Cure: Patients, Healers, and the Law in Early Modern Bologna. Trans. Gianna Pomata. Baltimore: Johns Hopkins University Press, 1998.

———. "Legami di sangue, legami di seme: Consanguineità e agnazione nel diritto romano." Quaderni storici 86 (1994): 299–334.

———. "Menstruating Men: Similarity and Difference of the Sexes in Early Modern Medicine." In Valeria Finucci and Kevin Brownlee (eds.), Generation and Regeneration: Tropes of Reproduction in Literature and History from Antiquity to Early Modern Europe. Durham, NC: Duke University Press, 2001, 109–52.

———. "La 'meravigliosa armonia': Il rapporto fra seni ed utero dall'anatomia vascolare all'endocrinologia." In Giovanna Fiume (ed.), Madri: Storia di un ruolo sociale. Venice: Marsilio, 1995, 45–81.

Pope-Hennessy, John. An Introduction to Italian Sculpture. 4th ed. London: Phaidon, 1996.

Pouchelle, Marie-Christine. The Body and Surgery in the Middle Ages [1983]. Trans. Rosemary Morris. New Brunswick, NJ: Rutgers University Press, 1990.

Povolo, Claudio. "Aspetti e problemi dell'amministrazione della giustizia penale nella Repubblica di Venezia, secoli XVI–XVII." In Gaetano Cozzi (ed.), Stato, società e giustizia nella Repubblica Veneta (sec. XV–XVIII). Rome: Jouvence, 1980, vol. 1, 153–258.

Premuda, Loris. Storia della iconografia anatomica. Milan: Martelli, 1957.

Premuda, Loris, and Giuseppe Ongaro. "I primordi della dissezione anatomica in Padova." Acta medicae historiae patavina 12 (1965–66): 117–42.

Prosperi, Adriano. "Il sangue e l'anima: Ricerche sulle compagnie di giustizia in Italia." *Quaderni storici* 51 (1982): 959–99.

Puccinotti, Francesco. *Storia della medicina.* Livorno: Wagner, 1850–1859.

Puff, Helmut. *Sodomy in Reformation Germany and Switzerland, 1400–1600.* Chicago: University of Chicago Press, 2003.

Putti, Vittorio. *Berengario da Carpi: Saggio biografico e bibliografico, seguito dalla traduzione del "De fractura calvae sive cranei."* Bologna: Cappelli, 1937.

Rahner, Hugo. "Die Gottesgeburt: Die Lehre der Kirchenväter von der Geburt Christi im Herzen des Gläubigen." *Zeitschrift für katholische Theologie* 59 (1935): 333–418.

Randolph, Adrian. "Regarding Women in Sacred Space." In Geraldine A. Johnson and Sara F. Matthews Grieco (eds.), *Picturing Women in Renaissance and Baroque Italy.* Cambridge: Cambridge University Press, 1997, 17–41.

Reguardati, Fausto de'. *Benedetto de' Reguardati da Norcia: "Medicus tota Italia celeberrimus."* Trieste: Lint, 1977.

Reisert, Robert. *Der siebenkammerige Uterus: Studien zur mittelalterlichen Wirkungsgeschichte und Entfaltung eines embryologischen Gebärmuttermodells.* Hannover: Wellm, 1986.

Remmert, Volker R. "In the Sign of Galileo: Pictorial Representation in the Seventeenth-Century Copernican Debate." *Endeavour* 27 (2003): 26–31.

———. *Widmung, Welterklärung und Wissenschaftslegitimierung: Titelbilder und ihre Funktionen in der Wissenschaftlichen Revolution.* Wolfenbüttel: Herzog August Bibliothek, 2006.

Ricci, Giovanni. *Il principe e la morte: Corpo, cuore, effigie nel Rinascimento.* Bologna: Mulino, 1998.

Rippa Bonati, Maurizio. "*Manuum munus:* Per una iconografia del 'toccar con mano.'" In A. Olivieri and M. Rinaldi (eds.), *All'incrocio dei saperi: La mano.* Padua: CLUEB, 2004, 325–36.

Roberts, K.B., and J.D.W. Tomlinson, *The Fabric of the Body: European Traditions of Anatomical Illustration.* Oxford: Clarendon, 1992.

Romby, Giuseppina Carla. *Descrizioni e rappresentazioni della città di Firenze nel XV secolo.* Florence: Libreria Editrice Fiorentina, 1976.

Rosand, David, and Michelangelo Muraro. *Titian and the Venetian Woodcut.* Washington, DC: International Exhibitions Foundation, 1976.

Rudloff, Ernst von. *Über das Konservieren von Leichen im Mittelalter: Ein Beitrag zur Geschichte der Anatomie und des Bestattungswesens.* Freiburg: Karl Henn, 1921.

Ruggiero, Guido. *Violence in Early Renaissance Venice.* New Brunswick, NJ: Rutgers University Press, 1980.

Rusconi, Roberto. "Colomba da Rieti: La signoria dei Baglioni e la 'seconda Caterina.'" In Enrico Menestò and Roberto Rusconi, *Umbria sacra e civile.* Turin: Nuova Eri Edizioni Rai, 1989, 211–26.

————. "Pietà, povertà e potere: Donne e religione nell'Umbria tardome-dievale." In Daniel Bornstein and Roberto Rusconi (eds.), *Mistiche et devote nell'Italia tardomedievale.* Naples: Liguori, 1992, 11–24.

Russell, Daniel S. *Emblematic Structures in Renaissance French Culture.* Toronto: University of Toronto Press, 1995.

Russell, Jeffrey Burton. *Inventing the Flat Earth: Columbus and Modern Historians.* New York: Praeger, 1991.

Sallmann, Jean-Michel. *Naples et ses saints à l'âge baroque: 1540–1750.* Paris: Presses Universitaires de France, 1994.

Salmón, Fernando, and Montserrat Cabré. "Fascinating Women: The Evil Eye in Medical Scholasticism." In Roger French *et al.* (eds.), *Medicine from the Black Death to the French Disease.* Aldershot: Ashgate, 1998, 53–84.

Samoggia, Luigi. "Lodovico Bonacciolo medico ostetrico di Lucrezia Borgia in Ferrara." *Atti della Accademia dei Fisiocratici di Siena, Sezione medico-fisica* n.s. 13, fasc. 1 (1974): 513–31.

Sappol, Michael. *A Traffic of Dead Bodies: Anatomy and Embodied Social Identity in Nineteenth-Century America.* Princeton, NJ: Princeton University Press, 2002.

Saunders, J.B. de C.M., and Charles D. O'Malley. *The Illustrations from the Works of Andreas Vesalius of Brussels.* Cleveland: World Publishing, 1950.

Savage-Smith, Emilie. "Attitudes toward Dissection in Medieval Islam." *Journal of the History of Medicine and Allied Sciences* 50 (1995): 67–110.

Sawday, Jonathan. *The Body Emblazoned: Dissection and the Human Body in Renaissance Culture*. London: Routledge, 1995.

———. "The Fate of Marsyas: Dissecting the Renaissance Body." In Lucy Gent and Nigel Llewellyn (eds.), *Renaissance Bodies: The Human Figure in English Culture, c. 1540–1660*. London: Reaktion, 1990, 112–35.

Saxl, Fritz. "Pagan Sacrifice in the Italian Renaissance." *Journal of the Warburg and Courtauld Institutes* 2 (1939): 346–67.

Schäfer, Daniel. *Geburt aus dem Tod: Der Kaiserschnitt an Verstorbenen in der abendländischen Kultur*. Hürtgenwald: Guido Pressler, 1999.

Schäfer, Dietrich. "Mittelalterlicher Brauch bei der Überführung von Leichen." *Sitzungsberichte der Preussischen Akademie der Wissenschaften zu Berlin* 26 (1920): 478–98.

Schleissner, Margaret R. "*Secreta mulierum*." In Kurt Ruh (gen. ed.), *Die deutsche Literatur des Mittelalters: Verfasserlexikon*, 2nd ed., vol. 8. Berlin: Walter de Gruyter, 1992, 986–93.

———. "Sexuality and Reproduction in the Late Medieval 'Problemata Aristotelis.'" In Josef Domes, Werner Gerabek, Bernhard D. Haage, Christoph Weisser, and Volker Zimmermann (eds.), *Licht der Natur: Medizin in Fachliteratur und Dichtung*. Göppingen: Kümmerle, 1994, 383–98.

Schoenfeldt, Michael C. *Bodies and Selves in Early Modern England: Physiology and Inwardness in Spenser, Shakespeare, Herbert, and Milton*. Cambridge: Cambridge University Press, 1999.

Schullian, Dorothy M. "An Anatomical Demonstration by Giovanni Lorenzo of Sassoferrato, 19 November 1519." In *Miscellanea di scritti di bibliografia ed erudizione in memoria di Luigi Ferrari*. Florence: Olschki, 1952, 487–94.

Schultz, Bernard. *Art and Anatomy in Renaissance Italy*. Ann Arbor, MI: UMI Research Press, 1985.

Scribner, Robert. "Ways of Seeing in the Age of Dürer." In Dagmar Eichberger and Charles Zika (eds.), *Dürer and His Culture*. Cambridge: Cambridge University Press, 1998, 93–117.

Sensi, Mario. "Anchoresses and Penitents in Thirteenth- and Fourteenth-Century Umbria" [1982]. In Daniel Bornstein and Roberto Rusconi (eds.), *Women and Religion in Medieval and Renaissance Italy*, trans. Margery J. Schneider. Chicago: University of Chicago Press, 1996, 56–83.

———. "La monacazione delle recluse nella valle Spoletina." In Claudio Leonardi and Enrico Menestò, *S. Chiara da Montefalco e il suo tempo*. Florence: Nuova Italia, 1985, 71–94.

Serra-Zanetti, Alberto. *L'arte della stampa in Bologna nel primo ventennio del Cinquecento*. Bologna: A spese del Comune, 1959.

Shatzmiller, Joseph. "The Jurisprudence of the Dead Body: Medical Practition [*sic*] at the Service of Civic and Legal Authorities." *Micrologus* 7 (1999): 223–30.

Sigerist, Henry. "A Doctor's Family in the Fifteenth Century." *Bulletin of the Institute of the History of Medicine* 3 (1935): 159–62.

Simons, Patricia. "Anatomical Secrets: *Pudenda* and the *Pudica* Gesture." In Gisela Engel (ed.), *Das Geheimnis am Beginn der europäischen Moderne*. (*Zeitsprünge: Forschungen zur frühen Neuzeit*, vol. 2, fasc. 1–4.) Frankfurt: Klostermann, 2002, 302–27.

Singer, Charles. "A Study in Early Renaissance Anatomy, with a New Text: The *Anothomia* of Hieronymo Manfredi (1490)." In Charles Singer (ed.), *Studies in the History and Method of Science*. Oxford: Clarendon, 1917, 79–164.

Siraisi, Nancy G. *The Clock and the Mirror: Girolamo Cardano and Renaissance Medicine*. Princeton, NJ: Princeton University Press, 1997.

———. "La comunicazione del sapere anatomico ai confini tra diritto e agiografia: Due casi del secolo XVI." In Massimo Galuzzi, Gianni Micheli, and Maria Teresa Monti (eds.), *Le forme della comunicazione scientifica*. Milan: FrancoAngeli, 1998, 419–38.

———. "Girolamo Cardano and the Art of Medical Narrative." *Journal of the History of Ideas* 52 (1991): 581–602.

———. "How to Write a Latin Book on Surgery: Organizing Principles and Authorial Devices in Guglielmo da Saliceto and Dino del Garbo." In Luis García-Ballester, Roger French, Jon Arrizabalaga, and Andrew Cunningham (eds.), *Practical Medicine from Salerno to the Black Death*. Cambridge: Cambridge University Press, 1994, 88–109.

————. *Medieval and Early Renaissance Medicine: An Introduction to Knowledge and Practice*. Chicago: University of Chicago Press, 1990.

————. "'Remarkable' Diseases, 'Remarkable' Cures, and Personal Experience in Renaissance Medical Texts." *Medicine and the Italian Universities, 1250–1600*. Leiden: Brill, 2001, 226–52.

————. "Taddeo Alderotti and Bartolomeo da Varignana on the Nature of Medical Learning." *Isis* 68 (1977): 27–39.

————. *Taddeo Alderotti and His Pupils: Two Generations of Italian Medical Learning*. Princeton, NJ: Princeton University Press, 1981.

————. "Vesalius and Human Diversity in *De humani corporis fabrica*." *Journal of the Warburg and Courtauld Institutes* 57 (1994): 60–88.

Smith, Pamela H. *The Body of the Artisan: Art and Experience in the Scientific Revolution*. Chicago: University of Chicago Press, 2004.

Stefanutti, Ugo. "Benivieni, Antonio," in *Dizionario biografico degli italiani*, vol. 8. Rome: Istituto della Enciclopedia Italiana, 1966, 543–45.

————. *Documentazioni cronologiche per la storia della medicina, chirurgia e farmacia in Venezia dal 1258 al 1332*. Venice: Ongania, 1961.

Steinberg, Leo. *The Sexuality of Christ in Renaissance Art and in Modern Oblivion*. Rev. ed. Chicago: University of Chicago Press, 1996.

Stolberg, Michael. "A Woman Down to Her Bones: The Anatomy of Sexual Difference in the Sixteenth and Early Seventeenth Centuries." *Isis* 94 (2003): 274–99.

Strocchia, Sharon T. *Death and Ritual in Renaissance Florence*. Baltimore: Johns Hopkins University Press, 1992.

Sudhoff, Karl. "Abbildungen zur Anatomie des Maître Henri de Mondeville (ca. 1260 bis ca. 1320)." *Ein Beitrag zur Geschichte der Anatomie im Mittelalter speziell der anatomischen Graphik: Nach Handschriften des 9. bis 15. Jahrhunderts*. Leipzig: Barth, 1908, 82–89.

Tabanelli, Mario. *La chirurgia italiana nell'Alto Medioevo*. Florence: Olschki, 1965.

Talbot, Charles H. *Medicine in Medieval England*. London: Oldbourne, 1967.

Talvacchia, Bette. *Taking Positions: On the Erotic in Renaissance Culture*. Princeton, NJ: Princeton University Press, 1999.

Tanner, Marie. *The Last Descendant of Aeneas: The Hapsburgs and the Mythic Image of the Emperor*. New Haven: Yale University Press, 1993.

Tardito, E. "Tagli cesarei di altri tempi." *Minerva medica* 59/supplement to n. 94 (24 November 1968): 5079–5106.

Tenenti, Alberto. *Il senso della morte e l'amore della vita nel Rinascimento (Francia et Italia)*. Turin: Einaudi, 1957.

Thomasset, Claude. "Le corps féminin ou le regard empêché." *Micrologus* 1 (1993): 99–114.

Thorndike, Lynn. "Further Consideration of the *Experimenta*, *Speculum Astronomiae*, and *De secretis mulierum* Ascribed to Albertus Magnus." *Speculum* 30 (1955): 413–43.

———. *Science and Thought in the Fifteenth Century: Studies in the History of Medicine and Surgery, Natural and Mathematical Science, Philosophy and Politics*. New York: Hafner, 1967.

Tozzi, Ileana. "Tra mistica e politica: L'esperienza femminile nel Terz'Ordine della Penitenza di San Domenico." *Rassegna storica online* 1 n.s. 4 (2003).

Ulrich, Laurel Thatcher. *A Midwife's Tale: The Life of Martha Ballard, Based on Her Diary, 1785–1812*. New York: Knopf, 1990.

Valori, Alessandro. "L'onore femminile attraverso l'epistolario di Margherita e Francesco Datini da Prato." *Giornale storico della letteratura italiana* 175 (1998): 53–83.

Valverde de Amusco, Juan. *Anatomia del corpo umano*. Rome: Antonio Salamanca and Antonio Lafreri, 1560.

———. *Historia de la composición del cuerpo humano*. Rome: Antonio Salamanca and Antonio Lafreri, 1556.

Vauchez, André. "La nascita del sospetto." Trans. Monica Turi. In Gabriella Zarri (ed.), *Finzione e santità tra medioevo e età moderna*. Turin: Rosenberg and Sellier, 1991, 39–51.

———. *Sainthood in the Later Middle Ages* [1988]. Trans. Jean Birrell. Cambridge: Cambridge University Press, 1997.

———. "Les stigmates de Saint François et leurs détracteurs dans les derniers siècles du Moyen Age." *Mélanges d'archéologie et d'histoire* 80 (1968): 595–625.

Vedder, Ursula. "Frauentod-Kriegertod im Spiegel der attischen Grabkunst des 4. Jhr. v. Chr." *Mitteilungen des Deutschen Archäologischen Instituts* 103 (1988): 161–91.

Verde, Armando F. *Lo studio fiorentino, 1473–1503: Ricerche e documenti*. Florence: Istituto Nazionale di Studi sul Rinascimento, 1973–77.

Vidal, Fernando. "Brains, Bodies, Selves, and Science: Anthropologies of Identity and the Resurrection of the Body." *Critical Inquiry* 28, no. 4 (2002): 930–74.

Von Simpson, Otto. "*Compassio* and *Co-Redemptio* in Roger van der Weyden's *Descent from the Cross*." *Art Bulletin* 35 (1953): 9–16.

Von Staden, Heinrich. "The Discovery of the Body: Human Dissection and its Cultural Contexts in Ancient Greece." *Yale Journal of Biology and Medicine* 65 (1992): 223–41.

Weindler, Fritz. *Geschichte der gynäkologisch-anatomischen Abbildung*. Dresden: Zahn and Jaensch, 1908.

White, Andrew Dickson. *A History of the Warfare of Science with Theology in Christendom*. New York: Appleton, 1897.

White, John. "Donatello." In Giovanni Lorenzoni (ed.), *Le sculture del Santo di Padova*. Vicenza: Pozza, 1984, 51–94.

Wickersheimer, Ernest. "L'Anatomie' de Guido de Vigevano, médecin de la reine Jeanne de Bourgogne (1345)." *Archiv für Geschichte der Medizin* 7 (1913): 2–25.

———. *Anatomies de Mondino dei Liuzzi et de Guido de Vigevano*. Paris: Droz, 1926.

Wolf-Heidegger, Gerhard, and Anna Maria Cetto. *Die anatomische Sektion in bildlicher Darstellung*. Basel: Karger, 1967.

Wood, Charles T. "The Doctors' Dilemma: Sin, Salvation, and the Menstrual Cycle in Medieval Thought." *Speculum* 56 (1981): 710–27.

Wood, Jeryldene. "Perceptions of Holiness in Thirteenth-Century Italian Painting: Clare of Assisi." *Art History* 14 (1991): 301–28.

Woods-Marsden, Joanna. "*Ritratto al naturale:* Questions of Realism and Idealism in Early Renaissance Portraits." *Art Journal* 46 (1987): 209–16.

Wright, Stephen K. *The Vengeance of Our Lord: Medieval Dramatizations of the Destruction of Jerusalem*. Toronto: Pontifical Institute of Mediaeval Studies, 1989.

Yates, Frances A. *The Art of Memory*. London: Routledge and Kegan Paul, 1966.

Zambelli, Paola. "L'immaginazione e il suo potere (da al-Kindi, al-Farabi e Avicenna al Medioevo latino e al Rinascimento)." In Albert Zimmermann and Ingrid Craemer-Ruegenberg (eds.), *Orientalische Kultur und europäisches Mittelalter*. Berlin: Walter de Gruyter, 1985, 188–206.

Zarri, Gabriella. "L'altra Cecilia: Elena Duglioli dall'Olio (1472–1520)" [1984]. *Le sante vive*, 165–96.

———. "Colomba da Rieti e i movimenti religiosi femminili del suo tempo." In Casagrande and Menestò (eds.), *Una santa, una città*, 89–108.

———. *L'Estasi di Santa Cecilia di Raffaello da Urbino nella Pinacoteca Nazionale di Bologna*. Bologna: ALFA, 1983, 21–42.

——— (ed.). *Finzione e santità tra medioevo e età moderna*. Turin: Rosenberg and Sellier, 1991.

———. *Le sante vive: Cultura e religiosità femminile nella prima età moderna*. Turin: Rosenberg and Sellier, 1990.

———. "'Vera' santità, 'simulata' santità: Ipotesi e riscontri." In Gabriella Zarri (ed.), *Finzione e santità tra medioevo e età moderna*. Turin: Rosenberg and Sellier, 1991, 9–36.

Zglinicki, Friedrich von. *Geburt: Eine Kulturgeschichte in Bildern*. Braunschweig: Westermann, 1983.

Ziegler, Joseph. "Practitioners and Saints: Medical Men in Canonization Processes in the Thirteenth to Fifteenth Centuries." *Social History of Medicine* 12 (1999): 191–225.

Zois, Giuseppe (ed.). *S. Chiara da Montefalco: Dove ci porta il cuore*. N.p.: Ritter, 1995.

Zorzi, Domenico. "Sull'amministrazione della giustizia penale nell'età delle riforme: Il reato di omicidio nella Padova di fine Settecento." In Luigi Berlinguer and Floriana Colao (eds.), *Crimine, giustizia e società veneta in età moderna*. Milan: Giuffrè, 1989, 237–308.

Zorzi, Marino. "Stampa, illustrazione libraria e le origini dell'incisione figurativa a Venezia." In Mauro Lucco (ed.), *La pittura nel Veneto: Il Quattrocento*. Milan: Electa, 1990, vol. 2, 686–702.

Photo credits

Figures I.1, I.6, 2.1, 2.5, 4.1–4.4, 4.6, 5.1–5.3, 5.12, 5.13, 5.20. Courtesy of the National Library of Medicine, Bethesda, MD.

Figures I.2–I.5, 2.2, 4.8, 4.9, 5.16. Reproduction made from originals in the Boston Medical Library in the Francis A. Countway Library of Medicine, Harvard University, Cambridge, MA.

Figure I.7. Royal Collection Picture Library, Windsor Castle, Windsor. Alinari/Art Resource, NY.

Figure 1.1, 1.5, 4.7, 5.7. Scala/Art Resource, NY.

Figure 1.8, 3.3, 5.10. Alinari/Art Resource, NY.

Figure 1.9. Foto Toso.

Figure 2.3, 5.4, 5.5. The Wellcome Trust Medical Photographic Library, London.

Figure 2.4. Réunion des Musées Nationaux/Art Resource, NY.

Figure 2.6, 3.7. Houghton Library, Harvard University, Cambridge, MA.

Figure 3.1, 5.11, 5.18, 5.19. Erich Lessing/Art Resource, NY.

Figure 3.2, 5.14. Bibliothèque Nationale de France, Paris.

Figure 3.6. Attributed to Francesco del Cossa, Italian (Ferrarese), about 1436–1478, *The Meeting of Solomon and the Queen of Sheba*, third Quarter of the fifteenth century, tempera and oil on panel, 61 cm x 61 cm (24 x 24 in.), Museum of Fine Arts, Boston, Gift of Mrs. W. Scott Fitz, 17.198. Photograph © 2006, Museum of Fine Arts, Boston, MA.

Figure 3.8. British Library, London.

Figure 4.5. Uffizi, Florence. Scala/Art Resource, NY.

Figure 5.6. The Library of Congress, Lessing J. Rosenwald Collection, Washington, DC.

Figure 5.17. © Trustees of the British Museum, London.

Figure 5.21. Accademia Nazionale dei Lincei, per gentile concessione del Ministero per i Beni e le Attività Culturali, Rome.

Index